International Conference on Engineering, Applied Sciences and Technology (ICEAST 2019)

IOP Conference Series: Materials Science and Engineering
Volume 639

Luang Prabang, Laos
2 – 5 July 2019

ISBN: 978-1-5108-9841-7
ISSN: 1757-8981

Printed from e-media with permission by:

Curran Associates, Inc.
57 Morehouse Lane
Red Hook, NY 12571

Printed with permission by Curran Associates, Inc. (2020)

For permission requests, please contact the Institute of Physics at the address below.

Institute of Physics
Dirac House, Temple Back
Bristol BS1 6BE UK

Phone: 44 1 17 929 7481
Fax: 44 1 17 920 0979

techtracking@iop.org

Additional copies of this publication are available from:

Curran Associates, Inc.
57 Morehouse Lane
Red Hook, NY 12571 USA
Phone: 845-758-0400
Fax: 845-758-2633
Email: curran@proceedings.com
Web: www.proceedings.com

Table of contents

Volume 639

International Conference on Engineering, Applied Sciences and Technology 2019

2–5 July 2019, Luang Prabang, Laos

Accepted papers received: 29 August 2019
Published online: 25 October 2019

Preface

Papers

Industrial Engineering, Materials and Manufacturing

Mechanical, Mechatronics and Civil Engineering

Food, Chemical and Agricultural Engineering

IOP Conf. Series: Materials Science and Engineering **639** (2019) 011001 doi:10.1088/1757-899X/639/1/011001

Preface

On behalf of the Technical Program Committee (TPC), it is my great pleasure to welcome you all to participate in the 5th International Conference on Engineering, Applied Sciences and Technology (ICEAST 2019). This year we have received 270 papers from various countries. After the rigorous reviews process, only 152 papers have been accepted with the acceptance rate of 56.3%. ICEAST 2019 will provide a technical program consisting of 3 keynote sessions, 2 poster presentation sessions, and 23 oral presentation sessions. All presented papers in track 1-3 will be published in the IOP conference series and all presented papers in track 4-5 will be published in the IEEE Xplore Digital Library.

I would like to sincerely appreciate great contribution and success of all organizing committee members and our staff as well as numerous reviewers for their volunteering reviews and their great contribution in shaping the technical program. I would like to express my deepest gratitude to all authors who kindly submitted their paper to ICEAST 2019.

I hope that all of you will enjoy in ICEAST 2019 as well as the beautiful scenery of Luang Prabang, Lao PDR.

Asst. Prof. Dr. Sathaporn Promwong
Technical Program Chair
King Mongkut's Institute of Technology Ladkrabang

IOP Conf. Series: Materials Science and Engineering **639** (2019) 011002 doi:10.1088/1757-899X/639/1/011002

Organizing Committee

Honorary Chairs

Suchatvee Suwansawat (KMITL, Thailand)
Somsy Gnophanxay (NUOL, Lao PDR)
Si Anouvong Savathvong (SU, Lao PDR)

General Chairs

Komsan Maleesee (KMITL, Thailand)
Khamphoui Southisombat (NUOL, Lao PDR)

Organizing-Chairs

Chuwong Phongcharoenpanich (KMITL, Thailand)
Somphone Kanthavong (NUOL, Lao PDR)

International Advisory Committee (IAC)

Yoshikazu Miyanaka (Hokkaido University, Japan)
Takao Onoye (Osaka University, Japan)
Kazuhiko Hamamoto (Tokai University, Japan)
Kyung Sup Kwak (Inha University, South Korea)
Dusit Niyato (Nanyang Technological University, Singapore)
Tadashi Ariga (Tokai University, Japan)
Tomotaka Homae (National Institute of Technology, Toyama College, Japan)
Nobutaka Ito (Chiang Mai University, Thailand)
Satoru Seto (National Institute of Technology, Ishikawa College, Japan)
Kuniaki Yajima (National Institute of Technology, Sendai College, Japan)
Phonekeo Chanthamaly (NUOL, Lao PDR)
Sengpasong Phrakonekham (NUOL, Lao PDR)
Eryk Dutkiewicz (University of Technology Sydney, Australia)
Myo Myint Maw (Mandalay Technological University, Myanmar)

ICEAST Steering Committee

Anantawat Kunakorn (KMITL, Thailand)
Vanchai Riewruja (KMITL, Thailand)
Monai Krairiksh (KMITL, Thailand)
Chuwong Phongcharoenpanich (KMITL, Thailand)
Uma Seeboonruang (KMITL, Thailand)
Pornchai Supnithi (KMITL, Thailand)

Sathaporn Promwong (KMITL, Thailand)
Supan Tungjitkusolmun (KMITL, Thailand)
Chaowalit Hamontree (KMITL, Thailand)
Souliphanh Kannitha (NUOL, Lao PDR)
Sounthone Phommasone (NUOL, Lao PDR)
Khamphuey (NUOL, Lao PDR
Myint Thein (Yangon Technology University, Myanmar)
Sint Soe (Mandalay Technology University, Myanmar)

Technical Committee

Pakorn Wattanachaturaporn (KMITL, Thailand)
Sompob Polmai (KMITL, Thailand)
Surapan Airphaiboon (KMITL, Thailand)
Arthit Petchsasithon (KMITL, Thailand)
Jedsada Chaishom (KMITL, Thailand)
Chaowalit Hamontree (KMITL, Thailand)
Nattakarn Puttarak (KMITL, Thailand)
Supan Gulpanich (KMITL, Thailand)
Jitraporn Wongsa-Ngam (KMITL, Thailand)
Apinan Namkanisorn (KMITL, Thailand)
Vasu Udompetaikul (KMITL, Thailand)
Jarruwat Charoensuk (KMITL, Thailand)
Sawai Pongswatd (KMITL, Thailand)
Ekkapong Cheevitsopon (KMITL, Thailand)
Panmanas Sirisomboon (KMITL, Thailand)
Ranon Jientrakul (KMITL, Thailand)
Sorawat Chivapreecha (KMITL, Thailand)
Pearawut Yutthagowith (KMITL, Thailand)
Vanvisa Chutchavon (KMITL, Thailand)
Montree Siriprachyanun (KMUTNB, Thailand)
Nantakrit Yodpijit (KMUTNB, Thailand)
Sukanya Praesomboon (KMUTNB, Thailand)
Banhan Lila (BUU, Thailand)
Supphachan Rajsiri (MU, Thailand)
Anupun Terdwongworakul (KU, Thailand)
Anirut Chaijaruwanich (CMU, Thailand)
Wassanai Wattanutchariya (CMU, Thailand)
Wasawat Nakkiew (CMU, Thailand)
Chusit Pradabpet (ARU, Thailand)
Vimane Phoummavong (NUOL, Lao PDR)

IOP Conf. Series: Materials Science and Engineering **639** (2019) 011002 doi:10.1088/1757-899X/639/1/011002

Phoumy Indarack (NUOL, Lao PDR)
Korakan Pasomsouk (NUOL, Lao PDR)
Khampha Sihanakhone (NUOL, Lao PDR)
Yinxay Visane (NUOL, Lao PDR)
Khampasith Thammathevo (NUOL, Lao PDR)
Khammoun Sitthilat (NUOL, Lao PDR)
Chanhkhachone Sonemanivong (NUOL, Lao PDR)
Win Saw (YTU, Myanmar)
Soe Linn Aung (Magway, Myanmar)

Technical Program Chairs

Sathaporn Promwong (KMITL, Thailand)

Track 1 Industrial Engineering, Materials and Manufacturing

Kannachai Kanlayasiri (KMITL, Thailand)
Tossapol Kiatcharoenpol (KMITL, Thailand)
Sounthisack Phommachanh (NUOL, Lao PDR)
Timnoy Salithxay (SU, Lao PDR)

Track 2 Mechanical, Mechatronics, and Civil Engineering

Worapong Tangsrirat (KMITL, Thailand)
Monsak Pimsarn (KMITL, Thailand)
Khamhoo Xaphouvong (NUOL, Lao PDR)
Kongsy Phimmavong (SU, Lao PDR)

Track 3 Food, Chemical and Agricultural Engineering

Vasu Udompetaikul (KMITL, Thailand)
Jedsada Chaishome (KMITL, Thailand)
Keonakhone Khounvilay (NUOL, Lao PDR)
Vilayphone Somsamone (SU, Lao PDR)
Xoxiong Briatia (SU, Lao PDR)

Special Session Chairs

Chayapol Kamyod (MFU, Thailand)
Vimontha Khiewvongphachanh (NUOL, Lao PDR)
Donekeo Lakanchanh (NUOL, Lao PDR)
Sengphet Keokangdong (NUOL, Lao PDR)
Phetsadachanh Saypaseuth (SU, Lao PDR)
Xoxiong Briatia (SU, Lao PDR)

IOP Conf. Series: Materials Science and Engineering **639** (2019) 011002 doi:10.1088/1757-899X/639/1/011002

Publication Chair

Sumek Wisayataksin (KMITL, Thailand)
Kriangkrai Sooksood (KMITL, Thailand)
Khanthanou Luangxaysana (NUOL, Lao PDR)

Information System Chair/ Performance Publicity/Public Relation Chairs

Rathachai Chawuthai (KMITL, Thailand)
Sisongkham Phimmasean (NUOL, Lao PDR)
Manolom Chanthavong (NUOL, Lao PDR)
Thipphakone Thavonesack (SU, Lao PDR)
Phothong Xathamma (SU, Lao PDR)

Financial Chair

Wiboon Promphanich (KMITL, Thailand)
Daoheung Siboriboun (NUOL, Lao PDR)
Sun Misisouk (SU, Lao PDR)

Registration Chair

Wiboon Promphanich (KMITL, Thailand)
Thavisone Moonlasane (NUOL, Lao PDR)
Phutsavanh Thongphanh (NUOL, Lao PDR)

Sponsor & Exhibition Chair

Sumek Chanthala (NUOL, Lao PDR)

General Administration Chairs

Palitha Douangchack (SU, Lao PDR)
Sonesack Thammavong (SU, Lao PDR)
Chanthavong Mua (SU, Lao PDR)

Local Arrangements Chairs

Wiboon Promphanich (KMITL, Thailand)
Khamphao Sisaat (NUOL, Lao PDR)
Nikone Thammachareun (SU, Lao PDR)

General Secretary

Theerayod Wiangtong (KMITL, Thailand)
Somsanouk Pathoumvanh (NUOL, Lao PDR)

IOP Conf. Series: Materials Science and Engineering **639** (2019) 011002 doi:10.1088/1757-899X/639/1/011002

List of Reviewers

Aditep Chaisang
Akapot Tantrapiwat
Alexis Guilloteau
Boonchai Boonchu
Boontee Kruatrachue
Bundit Ruckveratham
Chak Chantalakhana
Chanin Bunlaksananusorn
Chaowalit Hamontree
Chayapol Kamyod
Chompoonoot Hirunyaprik
Chompoonuch Jinjakam
Chukiet Sodsri
Chumpol Yuangyai
Chutham Sawigan
Chutimet Srinilta
Jakrapong Pongpeng
Jarotwan Koiwanit
Jarruwat Charoensuk
Jirapat Sangthong
Jittraporn Wongsa-Ngam
Jonathan David Sands
Jukkrit Tagapanij
Kajornsak Kittimathaveenan
Kannachai Kanlayasiri
Kei Eguchi
Khemapat Tontiwattanakul
Khounhack Lorvannger
Kietikul Jearanaitanakij
Kittipitch Meesawat
Lonesy Thammavong
Mettaya Kitiwan
Monchai Chamchoy
Monsak Pimsarn
Montree Kumngern
Montri Wiboonrat
Nachanant Chitanont
Nattapan Suwansukho
Natthawut Ruangtrakoon
Nirand Pisutha-Arnond
Nobphadon Suksangpanya
Nopporn Suttiwong
Norasage Pattanadech
Pakorn Watanachaturaporn
Pasu Poonpakdee
Patchariya Petchpong
Peerapol Yuvapoositanon
Phacharaphon Tunthawiroon
Pholchai Chotiprayanakul
Phonlasit Thinnakorn Na Ayuthaya
Pichaya Supanakoon
Pipat Prommee
Pitak Thumwarin
Pitikhate Sooraksa
Prajuab Pawarangkoon
Prawit Chumchu
Rachen Kanahna
Ranon Jientrakul
Rathachai Chawuthai
Sakchai Thipchaksurat
Samphan Prompichai
Sanit Teawchim
Sarun Duangsuwan
Sathaporn Promwong
Siraphop Tooprakai
Sittiporn Pimsakul
Smith Eiamsa-Ard
Sompob Polmai
Sorasak Danworaphong
Souphavady Praseuth
Suchada Sitjongsataporn
Sumek Wisayataksin
Supat Kittiratsacha
Surapan Airphaiboon
Surin Kittitornkun
Thanadol Tiengthong
Theera Leeudomwong
Theerayod Wiangtong
Tossapol Kiatcharoenpol
Unnat Pinsopon
Veerachai Malyavej
Vongkeo Kingsakda
Wiboon Promphanich
Wipoo Sriseubsai
XIV
Worapong Tangsrirat
Yuttapong Rangsanse

IOP Conf. Series: Materials Science and Engineering **639** (2019) 011003 doi:10.1088/1757-899X/639/1/011003

Peer review statement

All papers published in this volume of *IOP Conference Series: Materials Science and Engineering* have been peer reviewed through processes administered by the proceedings Editors. Reviews were conducted by expert referees to the professional and scientific standards expected of a proceedings journal published by IOP Publishing.

IOP Conf. Series: Materials Science and Engineering **639** (2019) 012001 doi:10.1088/1757-899X/639/1/012001

Supplier selection for hospital medical equipment using fuzzy multicriteria decision making approach

Pattareeya Khumpang[1] and Sirawadee Arunyanart[1,*]

[1]Supply Chain and Logistics System Research Unit, Department of Industrial Engineering, Faculty of Engineering, Khon Kean University, Khon Kaen 40002, Thailand

Email: sirawadee@kku.ac.th

Abstract. Medical equipment is important in the healthcare service system and plays a major role in global markets. Meanwhile, selection of suitable medical equipment supplier is also essential because medical equipment relates to the patient's life. Moreover, there are several qualitative and quantitative evaluation factors that may be conflicting in nature while assessing optimum supplier among several alternatives. This study has objectives to discover the criteria for selection of appropriate medical equipment suppliers and to illustrate methods for selection of a suitable medical equipment supplier. The ROC (Rank Order Centroid) method was applied to determine the weight of each criterion. Fuzzy TOPSIS (Fuzzy Technique for Order Preference by Similarity to Ideal Solution) was used to select the optimal supplier. A case study was chosen concerning a hospital in the northeast of Thailand. Findings indicated the main criteria for consideration of supplier selection as quality, price, reliability, agility, compliance, service, benefits/bargaining, and transport/delivery and the results showed that out of five candidates, supplier A was the optimal choice.

1. Introduction

Today, technological advances in medical devices and equipment play a huge role in health service systems which are constantly adapting and changing to keep pace with new scientific discoveries. Good, reliable equipment can increase human life quality. On the other hand, if the medical devices are not dependable or safe, this may cause problems in health economy and society.

In 2016, 131 medical supplies and equipment companies were registered in Thailand which made the country a major importer and exporter of medical equipment in the ASEAN region [1]. The medical equipment market in Thailand is continuously growing. In 2015, Thailand exported medical materials, supplies and equipment worth more than ten billion baht. Figure 1 shows the budget for the Thailand Ministry of Public Health 2007-2018 which has been increasing every year [2]. In a similar vein, the global medical devices and equipment market is becoming increasingly important. Value of the worldwide market is expected to reach US$409.5 billion by 2023, while projected CAGR (Compound Annual Growth Rate) is 4.5% from 2018 to 2023 [3]. Therefore, medical equipment is extremely important in healthcare.

Although medical devices and equipment markets both in Thailand and worldwide show strong growth, research concerning the selection of medical equipment suppliers is limited. This study aims to investigate the important factors that affect selection of medical equipment suppliers and illustrate the methods for selection of an appropriate medical equipment supplier.

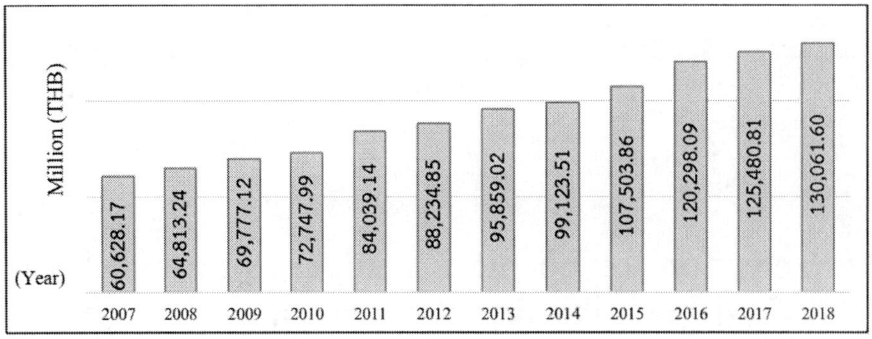

Figure 1. Budget of ministry of public health (Thailand) in 2007 – 2018

2. Literature review

Limited academic research exists concerning selection of medical equipment suppliers. K Diaconu et al. [4] stated that 40%-70% of the medical equipment in low- and middle-income countries is unfit for purpose or unused. Misuse of devices leads to incomplete costing and insufficient maintenance services. Therefore, procurement officials should prioritize devices by experience and needs-based methods to reach optimal purchase decisions. P. Ghadimi et al. [5] selected suppliers in the medical device manufacturing industry using a Fuzzy Inference System (FIS). Decision criteria consisted of environmental, economic and social dimensions to achieve sustainability. I. Ivlev [6] combined health technology assessment (HTA) and multiple-criteria decision analysis (MCDA) in the assessment of medical devices. The analytic hierarchy process (AHP) and Delphi method were applied to select a magnetic resonance imaging (MRI) system in the Czech Republic. C. Hajdau, and A. M. Spiridonica [7] used AHP to select PET-CT equipment. They concluded that AHP can be used to avoid inconsistency in decision-making preferences when the choice involves conflicting qualitative and quantitative criteria.

3. Methodology

The ROC (rank order centroid) method was used to weight criteria and fuzzy technique for order preference by similarity to the ideal solution (Fuzzy TOPSIS method) for multicriteria decision- making.

- **ROC.** ROC is a methodical approach which estimates weight that minimizes the maximum error of each weight by identifying the centroid of all possible weights while maintaining the rank order of objective importance [8]. Weight can be calculated using the following formula:

$$W_j = \frac{1}{n}\sum_{j=j}^{n}\frac{1}{r_j}\ ; j = 1,2,3,\ldots,n \tag{1}$$

Where W_j is a weight of criteria rank j, n is number of criteria and r_j is a rank position of W_j then we can assume that $W_1+ W_2+ W_3+\ldots+ W_n = 1$

- **Fuzzy TOPSIS.** Fuzzy set theory is integrated with TOPSIS to minimize the effects of hesitancy and variation in the decision makers' preference. It is a method that determines order preference by using linguistic variables. Fuzzy TOPSIS can be applied as follows [9,10].

***Step 1*:** Aggregate fuzzy ratings of the alternatives and criteria in the metrics (X_{ij});

$$\tilde{x}_{ij}^{k} = (\tilde{a}_{ij}^{k}, \tilde{b}_{ij}^{k}, \tilde{c}_{ij}^{k}) \tag{2}$$

$$a_{ij} = \min_{k}\{a_{ij}^{k}\},\ b_{ij} = \frac{1}{k}\sum_{k=1}^{K} b_{ij}^{k},\ c_{ij} = \max_{k}\{c_{ij}^{k}\} \tag{3}$$

Where m is alternatives, n is criteria from decision makers, i and j = 1, 2, 3, ..., n, r is number of decision makers from 1 to k.

Step 2: Assign and compute fuzzy rating for each criterion to alternatives matrix $(\tilde{D})$ by using fuzzy preference scale as shown in Table 1.

Table 1. Fuzzy preference scale.

Linguistic value	Triangular Fuzzy number
(1, 1, 3)	Very Poor (VP)
(1, 3, 5)	Poor (P)
(3, 5, 7)	Fair (F)
(5, 7, 9)	Good (G)
(7, 9, 9)	Very Good (VG)

Step 3: Normalize fuzzy decision matrix R by using the following formula. Where $i = 1, 2, 3, \ldots, m$ and $j = 1, 2, 3, \ldots, n$ and $\tilde{R} = [\tilde{r}_{ij}]$

$$\tilde{r}_{ij} = \left(\frac{a_{ij}}{c_j^*}, \frac{b_{ij}}{c_j^*}, \frac{c_{ij}}{c_j^*}\right);\ c_j^* = \max_i\{c_{ij}\} \tag{4}$$

$$\tilde{r}_{ij} = \left(\frac{a_{ij}}{a_j^*}, \frac{b_{ij}}{b_j^*}, \frac{a_{ij}}{c_j^*}\right);\ a_j^- = \min_i\{a_{ij}\} \tag{5}$$

Step 4: Compute the weighted normalized fuzzy decision matrix V and determine the weight of each criterion to be in the range of 0-1 by following formula.

$$\tilde{v}_{ij} = W_j \tilde{r}_{ij} \tag{6}$$

Where W_j is weight of criteria C_j. and $\tilde{v}_{ij}$ is normalized data.

Step 5: Define the fuzzy positive-ideal solution (A^*) and negative-ideal solution (A^-) of each criterion following formula:

$$A^* = (\tilde{v}_1^*, \tilde{v}_2^*, \ldots, \tilde{v}_n^*);\ \ \tilde{v}_j^* = \max_i\ \{v_{ij3}\} \tag{7}$$

$$A^- = (\tilde{v}_1^-, \tilde{v}_2^-, \ldots, \tilde{v}_n^-);\ \ \tilde{v}_j^- = \min_i\ \{v_{ij1}\} \tag{8}$$

Step 6: Define the distances of fuzzy positive-ideal solution (FPIS) A^* and fuzzy negative-ideal solution (FNIS) A^- of each criterion by using the following formula:

$$d(\bar{x}, \bar{y}) = \sqrt{\frac{1}{3}[(a_1 - a_2)^2 + (b_1 - b_2)^2 + (c_1 - c_2)^2]} \tag{9}$$

$$d_i^* = \sum_{j=1}^{n} d(\tilde{v}_{ij}, \tilde{v}_{ij}^*) \tag{10}$$

$$d_i^- = \sum_{j=1}^{n} d(\tilde{v}_{ij}, \tilde{v}_{ij}^-) \tag{11}$$

Step 7: Compute the closeness coefficient (CC_i) for each alternative using the following formula:

$$CC_i = \frac{d_i^-}{d_i^+ + d_i^-} \tag{12}$$

Step 8: Rank all alternatives according to the closeness coefficient (CC_i) in decreasing order. The alternative with the highest closeness to FPIS and farthest from the FNIS is the best.

4. Result and Discussion

Criteria were investigated from over 25 literature on supplier selection to classified into 23 main groups. Then, experts were asked to complete a questionnaire concerning the importance of these 23 main criteria for selection of medical equipment suppliers. Results in Table 2 show eight main criteria that impact highly on medical equipment supplier selection as quality, price, reliability, agility, compliance, service, benefits/ bargaining, and transport/delivery. Each main criterion has its own sub-criteria as shown in Table 2.

ROC was used to identify the weight of the main criteria and sub-criteria. Following equation (1), weights of the main criteria and sub-criteria are shown in Table 2.

Table 2. Criteria and weight for medical equipment supplier selection.

Main criteria	Weight	Sub criteria	Normalized weight
Quality	0.34	C1: Product quality/performance	0.13
		C2: Rejection rate	0.02
		C3: Quality certificate	0.05
		C4: Suitable product characteristics	0.13
Price	0.22	C5: Product price	0.22
Reliability	0.15	C6: Performance history/History	0.06
		C7: Reliability	0.09
Agility	0.11	C8: React quickly	0.03
		C9: Handling and availability support by technical experts	0.08
Compliance	0.08	C10: Compliance standard	0.02
		C11: Proper record on complaints and follow up	0.02
		C12: Provide sample before first ordering	0.04
Service	0.05	C13: Guarantee & Warranty (GW)	0.02
		C14: Service rate	0.01
		C15: Training	0.01
Benefits/Bargaining	0.03	C16: Quantity discount is offered by each supplier	0.03
Transport/Delivery	0.02	C17: On time delivery	0.02
		C18: Products delivered in good condition	0.01

Fuzzy TOPSIS was applied to select the optimal supplier. Each supplier was scored by the decision makers using fuzzy numbers mentioned in methodology section. By using equation (2), the aggregated fuzzy rating of each alternative supplier in relation to 18 sub-criteria obtained from decision makers is shown in Table 3. The normalize by fuzzy rating multiply with the normalized weight of each criterion according to equation (7) is shown in Table 4.

Table 3. The aggregated fuzzy rating.

Supplier	C1	C2	C3	C4	C5	C6	C7	C8	C9
A	5,8,9	1,1,3	5,8,9	5,8,9	5,7,9	7,9,9	7,9,9	7,9,9	5,8,9
B	5,7,9	1,2,5	5,7,9	5,7,9	5,7,9	3,6,9	3,5,7	5,8,9	5,7,9
C	5,8,9	1,2,5	5,8,9	3,7,9	3,5,7	5,7,9	5,7,9	5,7,9	3,6,9
D	5,7,9	1,1,3	5,8,9	3,6,9	1,4,5	3,6,7	3,6,7	3,5,7	1,4,5
E	5,7,9	1,1,3	5,7,9	3,6,9	3,7,7	7,9,9	5,8,9	5,7,9	5,7,9
Supplier	C10	C11	C12	C13	C14	C15	C16	C17	C18
A	7,9,9	5,8,9	7,9,9	7,9,9	5,8,9	7,9,9	5,8,9	5,8,9	5,8,9
B	5,7,9	3,6,7	5,7,9	5,7,9	5,7,9	5,8,9	5,7,9	5,7,9	5,7,9
C	3,6,7	5,7,9	5,7,9	5,7,9	3,6,7	3,6,7	5,7,9	3,6,7	5,7,9
D	3,6,7	3,5,7	3,6,7	3,6,7	3,6,9	3,6,9	3,5,7	1,5,5	3,6,7
E	5,8,9	5,7,9	5,8,9	5,8,9	5,7,9	5,8,9	5,8,9	5,7,9	5,7,9

Table 4. The normalized fuzzy.

Supplier	C1	C2	C3	C4	C5	C6
A	0.07,0.12,0.13	0.01,0.02,0.02	0.03,0.04,0.05	0.07,0.12,0.13	0.02,0.03,0.04	0.10,0.13,0.06
B	0.07,0.10,0.13	0.00,0.01,0.02	0.03,0.04,0.05	0.07,0.10,0.13	0.02,0.03,0.04	0.04,0.09,0.06
C	0.07,0.12,0.13	0.00,0.01,0.02	0.03,0.04,0.05	0.04,0.10,0.13	0.03,0.04,0.07	0.07,0.10,0.06
D	0.07,0.10,0.13	0.01,0.02,0.02	0.03,0.04,0.05	0.04,0.09,0.13	0.04,0.06,0.22	0.04,0.04,0.04
E	0.07,0.10,0.13	0.01,0.02,0.02	0.03,0.04,0.05	0.04,0.09,0.13	0.03,0.03,0.07	0.10,0.06,0.06
Supplier	C7	C8	C9	C10	C11	C12
A	0.07,0.09,0.09	0.02,0.04,0.04	0.06,0.18,0.2	0.01,0.02,0.02	0.01,0.02,0.02	0.03,0.04,0.04
B	0.03,0.05,0.07	0.02,0.03,0.04	0.11,0.16,0.20	0.01,0.01,0.02	0.01,0.01,0.01	0.02,0.03,0.04
C	0.05,0.07,0.09	0.02,0.03,0.04	0.07,0.14,0.20	0.01,0.01,0.01	0.01,0.01,0.02	0.02,0.03,0.04
D	0.03,0.06,0.07	0.01,0.02,0.03	0.02,0.09,0.11	0.01,0.01,0.01	0.01,0.01,0.01	0.01,0.03,0.03
E	0.05,0.08,0.09	0.02,0.03,0.04	0.11,0.16,0.20	0.01,0.02,0.02	0.01,0.01,0.02	0.02,0.04,0.04
Supplier	C13	C14	C15	C16	C17	C18
A	0.02,0.02,0.02	0.01,0.01,0.02	0.01,0.01,0.01	0.02,0.03,0.03	0.01,0.01,0.01	0.01,0.01,0.01
B	0.01,0.02,0.02	0.01,0.01,0.02	0.01,0.01,0.01	0.02,0.02,0.03	0.01,0.01,0.01	0.01,0.01,0.01
C	0.01,0.02,0.02	0.01,0.01,0.01	0.00,0.01,0.01	0.02,0.02,0.03	0.00,0.01,0.01	0.00,0.01,0.01
D	0.01,0.01,0.02	0.01,0.01,0.02	0.00,0.01,0.01	0.01,0.02,0.02	0.00,0.01,0.01	0.00,0.01,0.01
E	0.01,0.02,0.02	0.01,0.01,0.02	0.01,0.01,0.01	0.02,0.03,0.03	0.01,0.01,0.01	0.01,0.01,0.02

Table 5 shows the distances of FPIS and FNIS calculated by using equation (5), and CC_i value of each alternative by using equation (8). The best alternative is the one with the highest CC_i value. From result in the table, it can be seen that supplier A is the optimal supplier.

Table 5. Distances of the ratings of each alternative from A^+ and A^- and supplier ranking.

Supplier	d^+	d^-	CC_i	Rank
A	1.866	3.022	0.618	1
B	2.024	2.927	0.591	4
C	1.968	2.902	0.596	2
D	2.095	2.642	0.558	5
E	1.979	2.891	0.594	3

The current method for selecting medical equipment supplier of a case study hospital is to use subjective judgment of decision makers which supplier A was selected. The proposed method provides a systematic evaluation process to select medical equipment supplier based on important criteria which proves that the result is acceptable as supplier A gets the highest closeness coefficient value.

5. Conclusion

Eight main criteria are important in selection of the medical equipment supplier. These are quality, price, reliability, agility, compliance, service, benefits/bargaining, and transport/delivery. Quality is the most important criterion with weight of 34%. However, product price has highest weight when considering all sub-criteria. This means that using efficient equipment is necessary for medical treatment because it relates to the patient's life while the price is still important due to the procurement budget. Fuzzy logic was applied in the selection of supplier due to the number of qualitative criteria which each required the

IOP Conf. Series: Materials Science and Engineering **639** (2019) 012001 doi:10.1088/1757-899X/639/1/012001

opinion of the decision maker. The fuzzy TOPSIS method ranked the suppliers which helped in selecting the optimal alternative.

This study identifies criteria for supplier selection of medical equipment. Findings will be beneficial for medical equipment suppliers to evaluate and develop their products and service. Moreover, other hospitals and organizations can use the criteria and the decision-making method as guidelines to apply for selecting the optimal medical equipment supplier.

References

[1] National science and technology development agency August 2016 Thailand in https://waa.inter.nstda.or.th/prs/pub/Final-Report-Medical-Device.pdf

[2] Information http://dataservices.mof.go.th/Dataservices/GovernmentExpenditureEconomyMinistry

[3] Medical device market report: Trends, forecast and competitive analysis Information on https://www.lucintel.com/medical-device-market-2018.aspx

[4] Diaconu K, Chen Y F, Cummins C, Moyao G J, Manaseki-Holland S and Lilford R 2017 Methods for medical device and equipment procurement and prioritization within low-and middle-income countries: findings of a systematic literature review *Globalization and health*, **13(1)** 59

[5] Ghadimi P and Heavey C 2014 Sustainable supplier selection in medical device industry: toward sustainable manufacturing *Procedia CIRP* **15** 165-170

[6] Ivlev I, Vacek J and Kneppo P 2015 Multi-criteria decision analysis for supporting the selection of medical devices under uncertainty *Eur. J. Oper. Res.* **247(1)** 216-228

[7] Hajdau C and Spiridonica A M 2015 AHP - Based weighting of criteria for medical equipment selection *E-Health and Bioengineering Conference (EHB)* 1-5

[8] Roszkowska E 2013 Rank ordering criteria weighting methods–a comparative overview *Optimum Studia Ekonomiczne* NR **5** 65

[9] Nădăban S, Dzitac S, and Dzitac I 2016 Fuzzy topsis: A general view. *Procedia Comput Sci* **91** 823-831

[10] Aziz A, Aikhuele D O and Souleman F S 2015 A fuzzy topsis model to rank automotive suppliers *Procedia Manufacturing* **2** 159-164

IOP Conf. Series: Materials Science and Engineering **639** (2019) 012002 doi:10.1088/1757-899X/639/1/012002

A survey of criteria for a selection of logistics service providers: a case of Thailand's automotive industry

Khanitta Norkaew[1] and Panitas Sureeyatanapas[1]*

[1] Supply Chain and Logistics System Research Unit, Department of Industrial Engineering, Faculty of Engineering, Khon Kean University, Khon Kaen 40002, Thailand

E-mail: panisu@kku.ac.th

Abstract. This study aims to investigate the criteria that the automotive manufacturing companies currently consider when selecting a logistics service provider; and to prioritise the importance level of those criteria based on service buyers' perspectives. The criteria were synthesised from the relevant literature review, and they were then used to develop the questionnaire. In this current study, the questionnaire was divided into two parts: the manufacturers' current criteria for considering the service provider: and, the levels of importance for the criteria. It was then sent to five large-scale manufacturers in the automotive industry in Thailand. Following up with questionnaire responses, the researchers also conducted an interview with each participating manufacturer to elicit additional insightful information. According to the results, there were seven criteria that all five manufacturers collectively used to consider the logistics service: 1) Price, 2) Relationship, 3) Management system, 4) Flexibility, 5) Delivery, 6) Physical facilities, equipment and technology, and 7) Service areas. In this regard, the most important criteria were Price and Delivery, respectively. For logistics service providers, the findings of the study might be of use to design strategic planning, to improve their organization for a future service providing, and to maximize the competitive advantage. Additionally, the government could also implement these study findings to formulate the policy to support the logistics industry in Thailand to be able to grow sustainably and compete in the global market.

1. Introduction

During the past decade, the logistics industry in Thailand has been expanding rapidly. This may be partly influenced by the emerging of the ASEAN Economic Community (AEC) that stimulates the growing movements of goods among border countries. The Era of Digital Economy and online commerce also lead to a more intense competition regarding transportation and distribution, and this increases the demand for logistics services. Logistics now becomes a key factor for the successfulness of a plurality of industries [1].

With these reasons, a great number of companies start to consider outsourcing logistic service providers (LSPs) for their transportation, warehousing, and inventory management, etc. [2,3]. According to [4], there are multiple reasons for outsourcing a logistics service provider, namely reducing logistics cost, enhancing business performance since expertise and competent persons operated logistics functions, enhancing flexibility in operations, and accessing to emerging technology. Moreover, firms need to focus only on their core business process without worrying about transportation and warehouse management.

According to the review of relevant literature, it is seen that a selection of LSPs is a complicated decision-making problem as it contains a large number of criteria, in which more than ten criteria have always been considered. Additionally, the importance weights of the criteria are definitely unequal based on service buyers' perspectives, and these tend to be different among various industries or business sectors. This issue makes it difficult for LSPs to decide which points they should improve themselves in order to meet customer needs in a particular industry.

This study aims to investigate the criteria that the automotive manufacturing companies currently consider when selecting a logistics service provider; and to prioritise the importance level of those criteria based on service buyers' perspectives. This study focuses on the Thai automotive industry since Thailand has been well known as a hub of automotive manufacturing within the region [5]. Note that this paper reports merely an initial part of a full questionnaire-based survey, as its pilot study.

2. Methodology

The study started by gathering the criteria for selecting LSPs from the literature review. Research studies that mainly proposed methods or discussed issues about the selection of LSPs were reviewed. The reviewed papers were all taken from www.scopus.com through the keywords: logistics service provider selection, logistics service third party selection, and logistics 3PL selection. In this regard, the selected papers in this study were the ones with high citations or published during the past few years. There were selected 15 studies, accordingly [1-4,6-16]. After examining, a large number of criteria were collected and were synthesized into 13 criteria, as shown in Table 1.

To develop the questionnaire, it consisted of two parts. The first part asked the respondents about which criteria their companies have used when selecting LSPs. The respondents were then asked, in the second part, to rate the importance of each criterion towards 1-5 Likert Scale (5 = extremely important to 1 = never considered at all). After this questionnaire was constructed, five automotive industry managers from different manufacturing companies were contacted for the interviews. The five participating companies include both automakers and part suppliers. The managers' reasons to support their questionnaire answers were also elicited during the interviews.

3. Result and Discussion

The questionnaire response of five automobile and parts manufacturers was shown in Table 2. All five manufacturers used seven criteria for considering the logistics service provider: 1) Price, 2) Relationship, 3) Management system, 4) Flexibility, 5) Delivery, 6) Physical facilities, equipment and technology, and 7) Service areas. The results also yielded that two criteria, which were Human resource policies and Environmental management, were least taken into account.

Concerning the levels of importance, it was found that Price and Delivery were the two most important criteria, respectively, based on the average rating scores. The interviews further demonstrated that a majority of interviewed participants considerably taken a cost reduction into account to attain the maximum profits. Employing the relatively inexpensive logistics providers helps manufacturers to save their total costs. Regarding the delivery, the participants reckoned that the goods and raw materials extensively needed quick and punctual delivery. Most informants in this study were from the leading automakers and parts manufacturers in Thailand; thus, a goods delivery to the manufacturing lines must be prompt. The goods quantity must be fully accurate, and the delivery to manufacturer mills must be well time so as not to cause problems or delay the production process.

Another group of criteria that also received high average scores consisted of Service types, Management system, and Service areas. However, the Service types criterion had generated a high standard deviation (1.2) because one interviewee from a company had different comments from the rest that there was no demand for various services. Based on the experiences in employing logistics service providers, this interviewee found that services of most logistics services were not distinctly different and they had already met the company's demands, such as the delivery and warehouse management services together with holding goods.

IOP Conf. Series: Materials Science and Engineering **639** (2019) 012002 doi:10.1088/1757-899X/639/1/012002

Table 1. The criteria.

Criteria	Definitions
Price	The cost of employing a logistics service provider, such as price, service price, and cost of service.
Relationship	The relationship, trust, compatibility, assistance between customers and logistics service providers.
Service types	The scope of service, various kinds of offered service, such as transportation and warehouse.
Management system	The quality management, KPIs, quality certification, continuous improvement policy, technical controlling, and resource management policy.
Information & communication system	Information, equipment and technology system for contacting customers (such as Electronic Data Interchange: EDI, tracking system), and information safety.
Flexibility	Availability for customer needs, such as the capability to handle special requirement, instant needs, adaptation for future needs, and service cancellation.
Delivery	Time to transport, on-time shipments and deliveries, and on-time delivery performance.
Reputation and professionalism	Knowledge and experiences, reputation, experiences in the same industry, and local or global market ranking.
Physical facilities, equipment, and technology	The availability of materials, equipment and technology for logistics services.
Service areas	Comprehensive service areas, service locations.
Human resource policies	Human resource policies, employee satisfaction record.
Finance	Market share, asset ownership, and financial stability.
Environmental management	Environmental awareness, environmental certification, and environmental management policy.

Table 2. The survey results.

Criteria	Number of companies using each criterion	Importance level	
		Mean	S.D.
Price	5	5	0.0
Relationship	5	3.8	1.1
Service types	4	4	1.2
Management system	5	4	0.7
Information & communication system	4	3.8	0.8
Flexibility	5	3.8	1.6
Delivery	5	4.6	0.5
Reputation and professionalism	4	3.6	1.1
Physical facilities, equipment, and technology	5	3.8	0.8
Service areas	5	4	1.0
Human resource policies	3	2.8	0.8
Finance	4	3.8	0.8
Environmental management	3	3	1.2

The criterion, which had the highest standard deviation, was Flexibility. This was because some manufacturers always had a definite manufacturing plan. They, therefore, attached no importance to those delivery providers who were able to respond quickly to urgently changed sale orders while the other manufacturers still regarded this criterion as the crucial issue.

The least two important criteria were Human resource policies and Environmental management. This was resonated with the first part of questionnaire results that yielded no favours to these two criteria at the present. However, the Environment management criterion had reached the high standard deviation of 1.2 because one of five participating interviewees ranked this criterion at 5. This interviewee further added that the current environmental laws are playing more roles in Thailand and gradually becoming effective in the warehouses and transportation. Logistics service providers needed to well maintain truck conditions to be constantly ready and efficient. Trucks must not have belched out black smoke. Manufacturing automobile parts needed the most decent goods holding; for example, the warehouses must have been operated in closed premises which were free of dust, insects, birds, and rats.

4. Conclusion

This study aims to survey the criteria that five automotive manufacturers currently used to choose the logistics services and rank the importance level of the criteria. The study employs the questionnaire as a research tool to collect the quantitative data and follows up with the interviews to elicit the in-depth information. According to the data collection, there were seven criteria that all five manufacturers used for selecting the service providers: 1) Price, 2) Relationship, 3) Management system, 4) Flexibility, 5) Delivery, 6) Physical facilities, equipment and technology, and 7) Service areas. The two most important criteria were Price and Delivery, respectively; these two criteria were considered relatively crucial in which logistics service providers needed to consider so as to compete with other providers. A good-quality and punctual delivery with a low cost would enable logistics service providers to increase their competitiveness.

Among all criteria, the least important ones were Human resource policies and Environmental management. However, although these two criteria were not the prime considerations that the respondents mainly used to choose the logistics services at present, these criteria deserved no ignorance, particularly for the environmental management, because service users were increasingly aware of it. In the future, the environmental law enforcement seemed to be more rigorous. If logistics service providers have an earlier preparation than others, this will construct the uniqueness to increase the sales and receive more market shares.

The study findings might be of use for logistics service providers to formulate the strategic plan in order to improve the service performances for the future business, and to maximize their competitive advantage. Moreover, if the government wants to level up the competitive edge in the logistics industry of Thailand, it should formulate the policy to enhance two aspects for the logistics providers that manufacturers consider them important, which are cost and delivery. In this regard, the government possibly launches a campaign to entice logistics service providers to use the alternative energy for the cost reduction, or to invest in an alternative transportation system, namely more convenient train and ship delivery. Perceivably, the total delivery cost is lower than a truck delivery. If the government wants to support the Thai logistics system to compete with the global market, it needs to motivate the manufactures and logistics service providers in Thailand to have an awareness of the environment. For example, the government promotes the evaluation of carbon dioxide emission or carbon footprint as a means to prepare Thai entrepreneurs for the global competition. Ultimately, this will thus be regarded as sustainable economy development and helps to alleviate the environmental problems in Thailand.

IOP Conf. Series: Materials Science and Engineering **639** (2019) 012002 doi:10.1088/1757-899X/639/1/012002

Refferences

[1] Aguezzoul A, Rabenasolo B and Jolly-Desodt A M 2006 Multicriteria decision aid tool for third-party logistics providers' selection *Proceeding of the International Conference Service Systems and Service Management (ICSSSM),* Oct 2006, Troyes, France

[2] Narkhede B E, Raut R, Gardas B, Luong H T and Jha M 2017 Selection and evaluation of third party logistics service provider (3PLSP) by using an interpretive ranking process (IRP) *Benchmarking: An International Journal* **24(6)** pp 1597-1648

[3] Sharma S K and Kumar V 2015 Optimal selection of third-party logistics service providers using quality function deployment and Taguchi loss function *Benchmarking: An International Journal* **22(7)** pp 1281-1300

[4] Khan H, Hussainy S K, Khan A, Khan K, Farooq U and Mir M M 2017 Drivers of outsourcing and selection criteria of 3rd party logistics service providers *Operations and Supply Chain Management* **10(4)** pp 240-249

[5] Information on https://thaipublica.org/2018/11/krungsri-research-future-automotive-industry03/

[6] Aguezzoul A 2014 Third-party logistics selection problem: a literature review on criteria and methods *Omega* **49** pp 69-78

[7] Bhatti R S, Kumar P and Kumar D 2010 Analytical modeling of third party service provider selection in lead logistics provider environments *Journal of Modelling in Management* **5(3)** pp 275-286

[8] Celik E, Erdogan M and Gumus A T 2016 An extended fuzzy TOPSIS–GRA method based on different separation measures for green logistics service provider selection *International Journal of Environmental Science and Technology* **13** pp 1377–1392

[9] Datta S, Samantra C, Mahapatra S S, Mandal G and Majumdar G 2013 Appraisement and selection of third party logistics service providers in fuzzy environment *Benchmarking: An International Journal* **20(4)** pp 537-548

[10] Gol H and Catay B 2007 Third-party logistics provider selection: insights from a Turkish automotive company *Supply Chain Management: An International Journal* **12(6)** pp 379-384

[11] Govindan K, Khodaverdi R and Vafadarnikjoo A 2016 A grey DEMATEL approach to develop third-party logistics provider selection criteria *Industrial Management & Data Systems* **116(4)** pp 690-722

[12] Jharkharia S and Shankar R 2007 Selection of logistics service provider: an analytic network process (ANP) approach *Omega* **35(3)** 274-289

[13] Liao H, Wu D, Huang Y, Ren P, Xu Z and Verma M 2018 Green logistic provider selection with a hesitant fuzzy linguistic thermodynamic method integrating cumulative prospect theory and PROMETHEE *Sustainability* **10(1291)** pp 1-16

[14] Liu H T and Wang W K 2009 An integrated fuzzy approach for provider evaluation and selection in third-party logistics *Expert Systems with Applications* **36(3)** pp 4387-4398

[15] Raut R, Kharat M, Kamble S and Kumar C S 2018 Sustainable evaluation and selection of potential third-party logistics (3PL) providers: an integrated MCDM approach *Benchmarking: An International Journal* **25(1)** pp 76-97

[16] Thornton L M, Autry C W, Gligor D M and Brik A B 2013 Dose socially responsible supplier selection pay off for customer firm? a cross-cultural comparison *Journal of Supply Chain Management* **49(3)** pp 66-89

A Mathematical Programming Approach for Product Selection with Multiple Criteria

U Janjarassuk[1] and S Puengrusme[2]

[1] Faculty of Engineering, King Mongkut's Institute of Technology Ladkrabang, Bangkok 10520, Thailand
[2] Stamford International University, Bangkok 10110, Thailand

E-mail: udom.ja@kmitl.ac.th

Abstract. Product selection is very important for customer who looks for functionalities of the product. To get the best product with the available budget, multiple criteria should be considered simultaneously while making decision. In this paper, we propose a multi-objective decision model for selecting the most suitable product for the customers. The model is tested by using a real case of power unit selection in studio recording, and the results are presented.

1. Introduction

The decision making of choosing a right product for the customer can be difficult due to multiple factors involved in the decision process. For examples, the variety of product, the decision making criteria, the available budget, the product requirement, and etc. Some of these factors are often conflict to each other. For example, a feature-rich product is usually more expensive.

In general, the decision making process is commonly divided into seven steps. It starts with identifying the decision, gathering the relevant information, and identifying the alternatives. The next steps are evaluating the evidence and then choosing among alternatives. Finally, the corresponding action is taken, and the result of the decision is reviewed.

Decision making tools can be used to simplify the decision making process and to help decision makers for making better decision. Decision making tools are widely used in many areas. Among them, the decision tree is probably the most commonly used method in decision analysis. It can be used to help decision makers identify the most valuable choice to the goal. The decision tree uses a flowchart-like structure to track for the possible decisions and their consequences. Each option is "tested" to evaluate its expected value (or utility) and the option with the highest expected value is selected. The analytic hierarchy process (AHP) and the analytic network process (ANP) are widely used for multi-criteria decision making [1] in different areas such as strategic planning [2], material or equipment selection [3,4], aircraft type selection [5], nuclear power plant selection [6], and etc. They provide a comprehensive framework for structuring the decision problem in a hierarchical structure (in AHP) or in a complex interrelationships among decision levels and attributes (in ANP) by using pairwise comparisons that represent the dominance of one element over another. A numerical weight is derived for each element to allow calculation of the final numerical priority for each of the decision alternatives. Multi-objective programming is another widely used tool in multiple criteria decision making for identifying optimal solutions to decision problems which involve more than one objective. Applications of multi-objective programming include product design [7], product line design and supplier selection [8], material selection [9], and etc.

IOP Conf. Series: Materials Science and Engineering **639** (2019) 012003 doi:10.1088/1757-899X/639/1/012003

2. Problem statement

In this paper, we consider a product selection problem where customers are faced with making decision to choose the right product among many alternatives. Multiple criteria are considered in this problem including price and product requirements. In particular, we consider a power unit selection problem in studio recording for this study. There are fifteen products or alternatives to choose from depending on customer's application or needs. The objective (or goal) is to maximize customer's satisfaction or preferences based on his/her available budget subject to customer's product requirements (product features). Table 1 shows the features of each product and the short description of each feature is provided in Table 2.

Table 1: Features of the products

Product/Feature	Price	Amp	Outlets	Surge Protection	Noise Filtering	Over-voltage Protection	Voltage Regulation	Power Boost	Power Sequencing	UPS	Isolation	Indicator	USB Charger	Software Control
M-10X E	3,780	10	11	Standard	RFI/EMI	-	-	-	-	-	-	Yes	-	-
PL-8C E	11,000	10	11	SMP	LiFT	EVS	-	-	-	-	Yes	Yes	-	-
PL-PLUS C E	14,000	10	11	SMP	LiFT	EVS	-	-	-	-	Yes	Yes	-	-
PL-PRO DMC E	22,500	10	11	SMP	LiFT	EVS	-	-	-	-	Yes	Yes	Yes	-
PS-8RE III	14,000	10	9	SMP	LiFT	EVS	-	-	Yes	-	Yes	Yes	-	-
CN-3600 SE	22,500	16	9	SMP	LiFT	EVS	-	-	Yes	-	Yes	Yes	-	Yes
P-2300 IT E	108,000	10	14	SMP	LiFT+BP	EVS	-	-	-	-	Yes	Yes	Yes	-
P-1400 AR E	44,100	6	11	SMP	LiFT	EVS	True RMS	-	-	-	Yes	Yes	Yes	-
P-6900 AR E	126,000	30	11	SMP	LiFT	EVS	True RMS	-	-	-	Yes	Yes	Yes	-
F1500-UPS E	54,000	7	10	SMP	LiFT	EVS	AVR	-	-	Yes	Yes	Yes	-	Yes
IT-REF 16EI	146,200	16	12	SMP	LiFT+BP	EVS	-	Yes	-	-	Yes	-	-	-
ELITE-16 PFE I	43,000	16	12	SMP	LiFT	EVS	-	Yes	-	-	Yes	Yes	-	-
SPR-16E I	94,600	16	12	SMP	LiFT	EVS	AVR	-	-	-	Yes	Yes	-	-
ELITE-10E I	28,337	10	8	SMP	LiFT	EVS	-	-	-	-	-	Yes	-	-
AC-210A E	6,300	10	2	SMP	LiFT	EVS	-	-	-	-	-	Yes	-	-

Table 2: Description of the product features

No.	Features	Description
1	Price	The price of the product
2	Amp	The current rating of the product
3	Outlets	The number of outlets
4	Surge protection	The type of surge protection
5	Noise filtering	The type(s) of noise filtering technology
6	Over-voltage protection	The type of over-voltage protection
7	Voltage regulation	The type of voltage regulation technology
8	Power boost	Whether the product provides power boost feature
9	Power sequencing	Whether the product provides power sequencing feature
10	UPS	Whether the product provides energy storage (UPS)
11	Isolation	Whether the product provides isolation circuits for different equipment
12	Indicator	Whether the product provides system monitoring indicator
13	USB charger	Whether the product provides built-in USB charger
14	Software control	Whether the product can be controlled by software

3. Mathematical model

To solve the product selection problem, we define set, index, decision variable and parameters as follows.

3.1. Set

P = Set of product

C = Set of criteria

F = Set of features

3.2. *Index*

i = Index of product

j = Index of criteria

k = Index of features

3.3. *Variable*

$$x_i = \begin{cases} 0, & \text{if product } i \text{ is not selected} \\ 1, & \text{if product } i \text{ is selected} \end{cases}$$

3.4. *Parameter*

b = available budget

w_j = weight associated with criteria j

c_{ij} = score of product i associated with criteria j

p_i = price of product i

r_k = requirement of feature k

s_{ik} = specification of product i associated with feature k

3.5. *Mathematical model*

The following mathematical model is proposed.

$$\max \sum_{i \in P} \sum_{j \in C} w_j c_{ij} x_i \tag{1}$$

$$\sum_{i \in P} p_i x_i \leq b \tag{2}$$

$$\sum_{i \in P} x_i \leq 1 \tag{3}$$

$$s_{ik} x_i \geq r_k x_i, \quad \forall i \in P, k \in F \tag{4}$$

$$x_i \in \{0,1\}, \quad \forall i \in P \tag{5}$$

Equation (1) is the objective function where the customer satisfaction is maximized for the selected product based on the weight of each criterion. Equation (2) ensures that the total price is within the available budget. Equation (3) enforces that at most one product can be selected. Equation (4) ensures that the selected product has met the customer's requirements. Note that the decision variable x_i on both sides of the inequality is mandatory to ensure that the inequality is feasible for the unqualified product ($x_i = 0$). Equation (5) defines that the decision variable x_i is binary for each product i.

4. Experiments and results

In this section, we present experiment for testing the proposed model and the results. Five customers are asked to answer a given set of questions to obtain his/her product requirements and preferences of the product. Four criteria, i.e., price, equipment protection, noise, and voltage regulation are considered in this experiment. The features of the products are collected from the product manuals. The score of each product which associated with each criterion is provided by the experts of the field.

Table 3 shows the results gathered from the customer surveys as well as the solutions from the proposed model. The recommendations from the expert according to each customer's requirement are also provided for comparison.

Table 3: Results from customer surveys and product recommendations

Question	Result from customers				
	1	2	3	4	5
1. What is your budget?	<= 50,000	> 100,000	Not specified	Not specified	<= 20,000
2. What is your equipment current rating?	< 10	< 30	Don't know	< 30	< 6
3. What is your maximum number of equipment	<= 9	<= 11	<= 2	<= 14	<= 2
4. What level of equipment protection is required? (i.e. surge and over-voltage protections)	medium	very high	medium	high	high
5. What level of sound quality (protection from noise) do you prefer?	very high	very high	very high	very high	high
6. How often do you require for power boost?	very high	medium	medium	medium	medium
7. Do you need power sequencing?	No	No	No	Yes	No
8. Do you need energy storage? (UPS)	Not required	Not required	Required	Required	Not required
9. How important is power isolation to your equipment?	very low	very high	medium	very high	medium
10. Do you care about the status of the power system?	No	Yes	Don't know	Yes	Don't know
11. Do you need built-in USB charger?	No	No	No	Doesn't matter	Doesn't matter
12. Do you need software control for the power unit?	No	No	No	No	No
Recommendation from proposed model	ELITE-16 PFE I	P-6900 AR E	F1500-UPS E	-	AC_210AE
Recommendation from expert	PL-PRO DMC E - -	P-6900 AR E - -	F1500-UPS E - -	F1500-UPS E/ P-6900 ARE / PS-8RE III - -	PL-8C E PL-PLUS C E -

From Table 3, we can see that the suggested products from the model agree with the suggestions from the expert in Case 2 and Case 3. In Case 1 however, the model selects a more expensive product which has better features compared to the one suggested by the expert. While the more expensive alternative is still within the customer's budget, this product has a power boot feature which is rated very high in the customer survey. In contrast, the expert suggests a more economical alternative base on his experience that the 'PL-PRODMC E' is sufficient enough for most users in this level. In Case 4, the model cannot suggest any alternative since there is no product that could meet the customer's requirements. On the other hand, the expert suggests a combination of three products to the customer which in turn provide solution to the customer. In Case 5, the model suggests a less expensive alternative to the customer compared to the suggestions from the expert (see prices in Table 1). The explanation is that there are no significant differences of the features based on the criteria provided by the customer. In contrast, the suggestions from the expert are based on his personal preferences.

5. Conclusion

In this paper, we proposed a mathematical programming model to solve the product selection problem where multiple criteria are considered in the objective function. The model is tested by using a real case in studio recording for selecting a suitable power unit. We compared the results from the model with the recommendations from the expert. The results are well applicable but with limitation for recommendation with combination of the products.

References

[1] Zopounidis C and Pardalos P M 2010 *Handbook of Multicriteria Analysis* (Springer Science & Business Media)

[2] Görener A 2012 Comparing AHP and ANP: An Application of Strategic Decisions Making in a Manufacturing Company **3** 15

IOP Conf. Series: Materials Science and Engineering **639** (2019) 012003 doi:10.1088/1757-899X/639/1/012003

[3] Rao R V 2007 Material Selection for a Given Engineering Application *Decision Making in the Manufacturing Environment: Using Graph Theory and Fuzzy Multiple Attribute Decision Making Methods* Springer Series in Advanced Manufacturing (London: Springer London) pp 53–69

[4] Dağdeviren M 2008 Decision making in equipment selection: an integrated approach with AHP and PROMETHEE *J Intell Manuf* **19** 397–406

[5] Dožić S, Lutovac T and Kalić M 2018 Fuzzy AHP approach to passenger aircraft type selection *Journal of Air Transport Management* **68** 165–75

[6] Locatelli G and Mancini M 2012 A framework for the selection of the right nuclear power plant *International Journal of Production Research* **50** 4753–66

[7] Shidpour H, Shahrokhi M and Bernard A 2013 A multi-objective programming approach, integrated into the TOPSIS method, in order to optimize product design; in three-dimensional concurrent engineering *Computers & Industrial Engineering* **64** 875–85

[8] Deng S, Aydin R, Kwong C K and Huang Y 2014 Integrated product line design and supplier selection: A multi-objective optimization paradigm *Computers & Industrial Engineering* **70** 150–8

[9] Zhou C-C, Yin G-F and Hu X-B 2009 Multi-objective optimization of material selection for sustainable products: Artificial neural networks and genetic algorithm approach *Materials & Design* **30** 1209–15

Determinants of Port Performance – Case Study of Five Major Container Ports in Myanmar

Wai Phyo Paing, Jaruwit Prabnasak

International College, King Mongkut's Institute of Technology Ladkrabang, Bangkok, Thailand.

waiphyopaing009@gmail.com

Abstract. Container ports are playing a pivotal role in facilitating global logistics and supply chains. Therefore, their performance and overall competitiveness must be improved in order to retain shipping lines and shippers. This study aims to analyze the relationship between container throughput (Twenty-Foot Equivalent Unit (TEU) per year) and other important Key Performance Indicators (KPIs). This analysis can help the Myanma Port Authority (MPA) and container port operators in the country to identify which KPIs are essential for monitoring and improving their performance. The Data used in this study was collected from five major Myanmar container ports under the authority of the MPA for the period 2010-2015. This study employed a quantitative analysis method by applying a regression modelling technique. As a result, it is concluded that container throughput is influenced only by some KPIs namely total berth length and the total numbers of ship calls per year.

Keywords: Myanmar, container ports, container throughput, key performance indicators

1. Introduction

Myanmar possesses many ports, not only river ports but also seaports. These ports are undergoing significant development with processing and manufacturing projects being implemented at many of them. There are, in total, five major container ports in Myanmar, Thilawa is a sea port located 25 kilometers from Yangon city, adjacent to the Thilawa Special Economic Zone while the others are river ports (namely Asia World Port Terminal, Myanmar Industrial Port, Myanmar Terminal Port, Alone International Port Terminal) situated in the Yangon City Area. They handle 90% of the export and import containers in Myanmar [1]. There was an increase in container traffic between 2007 and 2016 [2] and since the enactment of a new Foreign Direct Investment Law, many foreign companies have been investing in ports and related infrastructure.

Myanmar potentially has numerous transshipment opportunities because of its strategic position in close proximity to India and Bangladesh and also near to China, Laos and Thailand. 75% of Myanmar's ports, known as "landlord ports" [3], are being leased to the private sector by a new government trying to promote private sector investment in coastline ports and thereby increase their traffic capacity. As ports are certainly important for a country in terms of facilitation the movement of goods [4], performance indicators are needed in order to measure the performance and grasp the operation of ports in a country [5]. There are many ways of measuring container port efficiency or productivity [6]. With regards to single output criteria, the throughput of container terminals in TEU is a popular indicator display in maritime business rankings [7]. Container throughput is a standard measure for the productivity of a seaport [8]. Therefore, in order to increase the productivity of the ports, MPA, an organization operation under the authority of the Ministry of Transport and Communications, has selected container throughput (TEU per year) as its main Key Performance Measure (KPI) for all the container ports of Myanmar. As we can see in Fig.1, total TEU/year from 2005 to 2015 shows an upward trend because during these years respective governments tried to change public service ports into landlord ports as they knew it would increase the performance of all ports. Furthermore, the improvement of the country's logistics performance allowed for further increase to the container throughputs during this 10-year period.

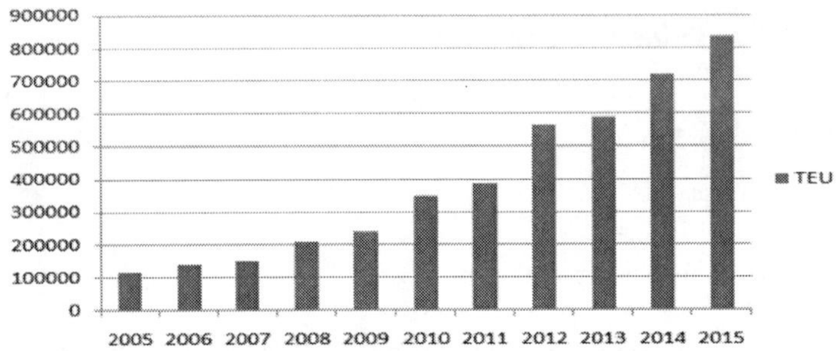

Figure 1. Total container throughput from 2005 to 2015.

As both container traffic and GDP increased between 2009 and 2014, Myanmar's ports served the country by distributing goods domestically and internationally, which served as a gateway for trade and stimulated national development and growth. However, compared to other maritime nations in Southeast Asia, Myanmar still has poorer than average port services and facilities. So, knowing which KPIs influence the container throughput at the five major container ports of Myanmar is important for the MPA and stakeholders of port industry. Regression analysis is a very valuable which can be used to model such things as the relationship between dependent variable and independent variable [9].

Therefore, the objective of the study undertaken from this paper is to help the MPA and its stakeholders to investigate whether there is any relationship between total container throughput (TEU per year) and other Key Performance Indicators (KPIs) so as to determine which predictors, among all our selected predictors can be used to determine total container throughput per year.

2. Research Methodology

The study this time, we used quantitative analysis with the aim of understanding and predicting the determinants of total throughput per year for the five major container ports in Myanmar between 2011 - 2015 by applying a Multi Linear Regression Model.

There are generally two purposes for using regression analysis. The first is to understand the relationship between variables such as container throughput and number of ship call per year. The second purpose is to predict the value of one variable based on the value of the others.

Looking at the literature data, we found that a total of 10 KPIs have been commonly used in past studies up to now. They are average turnaround time, average waiting time, container throughput, average tonnage per ship, berth occupancy rate, dwell time, vessel size, truck turnaround time, ton per crane (hook) per hours and ton per gantry crane per hours [5, 10, 11, 12, 13, 14, 15, 16, 17, 18]. For this study, in Myanmar, it was only possible to acquire data for one of these commonly used KPIs over a five-year period (container throughput) for our dependent variables. Consequently, we used the KPIs previously collected by the MPA as independent variables (numbers of berths, berth length, terminal area, total number of quay container crane, total number of transfer crane, number of reach stacker, number of frock lift, number of ship calls per year). Other paragraphs are indented.

Total throughput can be assigned as below:

$\mathbf{Y = f\,(X_1, X_2, X_3, X_4, X_5, X_6, X_7, X_8)}$

According to above equation, we can convert into MLR model to explain about the factors effecting on throughput as that:

$Y = \beta_0 + \beta_1 X_1 + \beta_2 X_2 + \beta_3 X_3 + \beta_4 X_4 + \beta_5 X_5 + \beta_6 X_6 + \beta_7 X_7 + \beta_8 X_8 + \varepsilon$

Where,

ε = random error

β_0 = intercept (value of Y when all X=0)

$\beta_1, \beta_2, \beta_3, \beta_4, \beta_5, \beta_6, \beta_7, \beta_8$ = regression coefficients

Table 1. Variables used for study.

Symbol	Variables
Y	Total container throughput (TEU per year)
X_1	Total number of berths
X_2	Total berth length (m)
X_3	Total terminal area (m^2)
X_4	Total number of quay container cranes
X_5	Total number of transfer cranes
X_6	Total number of reach stackers
X_7	Total number of forklifts
X_8	Total number of ship calls per year

In order to run a regression model by using R studio, we have to make the following assumption: the current MLR model form is reasonable, the residuals have constant variance, the residuals are normally distributed, the residuals are uncorrelated, and the predictors are not highly correlated with each other and no outlier.

3. Results and Discussion

3.1. Model 1

3.1.1. Significance test for model 1.

Table 2. Significance test results for model 1.

Coefficients:	Estimate	$Pr(>\|t\|)$
(Intercept)	-1.109e+04	0.705521
Total number of berths	-2.312e+04	0.080730 ·
Total berth length	2.522e+02	0.000371 ***
Total terminal area	-3.908e-02	0.257517
Total number of quay cranes	-1.205e+03	0.726055
Total number of transfer cranes	-8.842e+02	0.664097
Total number of reach stackers	3.748e+03	0.051641 ·
Total number of forklifts	-1.227e+04	0.063655 ·
Total number of ship calls per year	9.702e+02	2.49e-05 ***

Significant codes:0 '***'0.001 '**'0.01 '*'0.05 '.'0.1 ' '1

Residual standard error: 15380 on 13 degrees of freedom
Multiple R-squared: 0.9943, Adjusted R-squared: 0.9907
F-statistic: 282 on 8 and 13 DF, p-value: 2.66e-13

According to the above results, we observed that P-value=$2.66e^{-13}$<α = 0.05. Therefore, we could reject the null hypothesis H_0. This means that a relationship exists between Xi and Y for at least one variable. In addition, we can see that R^2=0.9943. This means that an approximate 99.43% variation in total throughput can be explained by our model. We proceeded to test which variables were most effective in forecasting total throughput.

3.1.2. Variable Significance test for each independent variable. According to the variable significance tests, we observed that total berth length and total number of ship calls per year correlated with total throughput (P-value< α = 0.05). On the other hands, total no. of berths, total terminal area, total number of quay cranes, total no. of reach stackers, total no. of forklifts and total no. of transfer cranes have no relation with total throughput (P-value> α = 0.05). Therefore, we concluded that there are two variables that can be used to predict Y. We removed the insignificant independent variables and created a new model.

3.2. Model 2

3.2.1. Significance test for model 2.

Table 3. Significance test results for model 2.

Coefficients:	Estimate	Pr(>|t|)
(Intercept)	-66387.99	2.48e-06***
Total berth length	139.57	2.41e-05***
Total number of ship calls per year	946.21	1.40e-13***
Significant codes0 '***'0.001 '**'0.01 '*'0.05 '.'0.1''1		
Residual Standard Error: 18330 on 19 degrees of freedom		
Multiple R-Squared: 0.9881, Adjusted R-Squared: 0.9869		
F-statistic: 789.4 on 2 and 19 DF, p-value: <2.2e-16		

According to the above results, we determined our new P-value=$2.2e^{-16}$<α = 0.05. Therefore, we can reject the null hypothesis H_0. This means that a relationship exists between Xi and Y for at least one variable. We proceeded to test which variables were most effective in forecasting total throughput.

3.2.2. Variable Significance test for each independent variable. According to the Variable Significance tests, we observed that total berth length and total number of ship calls per year correlated with total throughput (P-value<α = 0.05). Therefore, we could conclude that our new model is an improvement on the old one. We also concluded that there are two variables that can be used to predict Y and there was no need to remove any further variables from the model.

3.2.3. Multicollinearity test.

Table 4. Multicollinearity test results for model 2.

vif(test2)	
Total berth length	Total number of ship call per year
2.943587	2.943587

According to the above results we observed that VIF value of 2 dependence variables remain smaller than 4 (VIF > 4 suggest collinearity). Thus, we could conclude that there's no collinearity in our model.

3.2.4. Heteroskedasticity test.

Table 5. Heteroskedasticity test results for model 2.

residualPlots(test2, quadratic = FALSE)		
	Test stat	Pr(>|t|)
Total berth length	1.916	0.071
Total number of ship calls per year	-0.602	0.555
Tukey test	0.247	0.805
ncvTest(test2)		
Non-constant Variance Score Test		
Variance formula: ~ fitted values		
Chisquare = 7.345889 Df = 1 p = 0.00672163		

According to the above results, we observed that P-value=0.00672163 < $\alpha = 0.05$. Therefore, we cannot reject null hypothesis H_0. This means that there is no heteroskedasticity problem. Thus, we used log transformations for our independent variables to help to solve the heteroskedasticity problem.

3.3. Model 3

3.3.1. Significance test for model 3.

Table 6. Significance test results for model 3.

Coefficients:	Estimate	Pr(>|t|)
(Intercept)	-1390554	1.41e-09***
Log (Total berth length)	210546	1.24e-08***
Log (Total number of ships call per year)	53911	6.78e-08***
Significant codes0 '***'0.001 '**'0.01 '*'0.05 '.'0.1''1		
Residual Standard Error: 40720 on 19 degrees of freedom		
Multiple R-Squared: 0.9413, Adjusted R-Squared: 0.9351		
F-statistic: 152.3 on 2 and 19 DF, p-value: <2.009e-12		

According to the above results, we observed that P-value=2.009e-12< $\alpha = 0.05$. Therefore, we could reject the null hypothesis H_0. This means that a relationship exists between Xi and Y for at least one variable. So, we proceeded to test which variables were most effective in forecasting total throughput.

3.3.2. Variable significance test for each independent variable. According to the Variable Significance tests above, we observed that total berth length and total number of ship calls per year have a relationship with total throughput (P-value< $\alpha = 0.05$). Therefore, we concluded that there are two variables that can be used to predict Y and there was no need to remove any further variables from the model.

IOP Conf. Series: Materials Science and Engineering **639** (2019) 012004 doi:10.1088/1757-899X/639/1/012004

3.3.3. Multicollinearity test result for model 3.

Table 7. Multicollinearity test results for model 3.

vif(test2)	
Log (Total berth length)	Log (Total number of ship calls per year)
1.282395	1.282395

According to the above results, we observed that the VIF value of two dependent variables remains smaller than 4 (VIF > 4 suggests collinearity). Thus, we could conclude that there is no collinearity in our model.

3.3.4. Heteroskedasticity test.

Table 8. Heteroskedasticity test results for model 3.

Residual Plots (test 3, quadratic =FALSE)		
	Test stat	Pr(>|t|)
Log (Total berth length)	1.676	0.111
Log (Total number of ship calls per year)	3.498	0.003
Tukey test	3.652	0.000
nCVTest (test 3) Non-constant variance score test Variance formula: ~ fitted values		
Chisquare = 0.05416069	Df = 1	P = 0.8159753

According to the above results, we observed that P-value=0.8159753>α = 0.05 and. Therefore, we fail to reject null hypothesis H_0. It means that there's no heteroskedasticity occurred in our model 3.

3.3.5. Normal distribution test.

Table 9. Normal distribution test results for model 3.

Shapiro-Wilk normality test	
data: test 3$residuals	
W=0.91945	p-value=0.07405

Using the Shapiro-Wilk normality test, we observed that P-value= 0.07405>α=0.05. Therefore, we failed to reject H_0. This means that the residuals were normally distributed.

3.3.6. Residual test.

Table 10. Residual test results for model 3.

durbinwatsonTest (test3)			
Lag	Autocorrelation	D-W Statistic	p-value
1	-0.1008584	2.19831	0.56
Alternative hypothesis: rho !=0			

According to the Durbin Watson test, we observed that the p-value = 0.56>α=0.05. Therefore, we failed to reject H_0. This means that the residuals were not serially correlated.

3.3.7. Outlier test.

Table 11. Outlier test results for model 3.

Rstudent	Unadjusted p-value	Bonferonni p
103.641764	0.00089208	0.033007

According to the outlier test, we observed that P-value=0.00089208<α =0.05. Therefore, we could reject H_0. This means that the observation is no outlier problem. So, even if we found some outlier points, since the P-value is significant, any outliers that were found had no effect on our model.

4. Conclusion

In short, we can conclude that there are two independent variables that effect on total throughput such as total berth length and total ship call per year respectively. According to result tests, there are no collinearity and no heteroskedasticity found. The residuals are found to be normally distributed, not serially correlated and no outliers. This implies that these two variables can be used to determine total container throughput of the five major container ports in Myanmar.

In this study, a major limitation is the amount of data that can be obtained. Therefore, an important suggestion for future work is to gather more data and take them into the analysis. This could increase the reliability of the analysis results.

References

[1] Elly W, Enkhtsetseg G, Kichan, Nam 2015 Port governance structure: the port of Yangon. *KMI International Journal of Maritime Affairs and Fisheries*. Volume **7**, Issue 1, pp.001-015

[2] CEIC 2017, retreieved from: https://www.ceicdata.com/en/myanmar/transportation/mm-container-port-traffic-teu-20-foot-equivalent-units

[3] VDB 2016 Progress Report 2016 Myanmar, retreieved from: http://www.vdb-loi.com/wp-content/uploads/2017/01/2016-Progress-Report.pdf

[4] Tarantola 2015 *Puerotos como nodos integrados a la cadena logística, Informe especial: Intermodalismo y Logística International*. FCL Año **XI**, no. 5, 2015

[5] Atika Aqmarina, Nuzul Achjar 2017 Determinants of port performance – case study of 4 main ports in Indonesia (2005-2015). *Economis and Finance in Indonesia,* Vol. **63**, No. 2

[6] Trujillo, Lourdes, Nombela, Gustavo 1999 Privatization and Regulation of the Seaport Industry Privatization and regulation of the seaport industry. World Bank Policy Research Working Paper No. 2181. Available at SSRN: https://ssrn.com/abstract=623975

[7] Vera F 2012 Efficiency measurement of container ports: a New opportunity for hinterland integration. *Research in Logistics & Production.* Vol. **2**, No.2, pp.163-173

[8] U.S Department of Transporation 2016 *Bureau of Transportation Statistics*, retreieved from: https://definedterm.com/container_throughput

[9] Render, Stair, Hanna 2010 Quantitative analysis for management, New Jersey: Pearson Education

[10] Osman M, Eli V, Olta N 2016 Performance measurement : the case of Durres port. *European Journal of Bussiness, Economics and Accountancy* Vol.**4**, No.6, , ISSN 2065-6018

[11] Okeudo, G.N. 2015 Measurement of efficiency level in nigerian seaport after reform policy implementation. case study of Onne and Rivers seaport, Nigeria, *IORS Journal of Bussiness and Management*(IOSR-JBM), Vol. **12**, Issue 5

[12] Nazery K, 2012 Measuring the performance of Malaysian container ports. 1

[13] Kek C 1993 Port performance indicators, transportaion, water and urban development department. the World Bank, *Trasport* No. PS-6

IOP Conf. Series: Materials Science and Engineering **639** (2019) 012004 doi:10.1088/1757-899X/639/1/012004

[14] Jose L. Tongzon 1995 Determinants of port performance and efficiency. *ScienceDirect*, Vol. **29**, Issue 3
[15] UNCTAD 1976 Port performance indicator. Td/B/C.4/131/Supp.1/Eev.1
[16] Soner E 2008 Performeasure of conteiner terminal operation. *Dokuz Eylül Üniversitesi Sosyal Bilimler Enstitüsü Dergisi Cilt* 10, Sayı: 1
[17] UNTCAD 2015 Port performance. *Port Management Series*, Vol **4**, July 2015
[18] Min-Ho Ha et al., 2017 Revisition port performance measurement: A hybrit multi-stakeholder framework for the Modelling of Port Performance indicators. *Transportaion Research Part E: Logistics and Transportation Review*, Vol. **103**

Acknowledgement
We would like to show our gratitude to the Dr. Myo Nyein Aye, Deputy General Manager, Myanma Port Authority, for not only sharing data but also for providing his valuable guidance for our study.

IOP Conf. Series: Materials Science and Engineering **639** (2019) 012005 doi:10.1088/1757-899X/639/1/012005

Performance Evaluation of Hard Drive Sorting Systems Using Discrete-Event Simulation

Surached Wongnibunkit, Juta Pichitlamken and Suwiwat Suebsakul

Industrial Engineering, Faculty of Engineering Kasetsart University
50 Ngamwongwan Road, Chatuchak, Bangkok 10900, THAILAND

E-mail: surached.j@gmail.com

Abstract. We propose an approach for enhancing the throughput level and decreasing the throughput time of the hard disk drive sorting department that are congested and inefficient. We model the hard disk drive sorting department with discrete-event simulation on Arena. The proposed automated loop conveyor system can reduce the waiting time, the work-in-process and the time in system by 74.78%, 85.09% and 96.85%, respectively when comparing with the current sorting system.

1. Introduction

The technology of magnetic plate storage as we know in the hard disk drive (HDD) is now replaced with the solid state drives that is energy-saving, fast and convenient to use than HDDs. This new technology depresses the price of HDDs so some HDD manufacturers adapt by integrating the production bases. For example, Western Digital closed its plant in Malaysia and moved its operations to Thailand [1]. Thailand is now the world's largest HDD production base.

In addition to consolidating its production bases, the production must be more efficient and cost effective. For the company in this case study, the HDDs are sorted by humans via barcodes. The throughputs are 80,000-150,000 pieces/day which is not fast enough and too costly for the management, and an automatic system on the loop conveyor is being considered.

In this paper, we develop simulation models of the current manual sorting system and the automated conveyor system on ARENA (Rockwell Software) as a design prototype.

2. Simulation in the Automated Conveyor Sorting System Design Context

The DES technique is used to design the automated sorting process of the Brazilian Air Force which has problems in terms of throughput level and throughput time in the sorting system of the army uniforms [2]. The solution to this problem proves that the automated sorting process can increase the ability to manage tasks and reduce the average time of moving the workpiece in the system significantly. [3] shows that Delta robots can identify products by the physical characteristics but cannot prove that the product sorting system can be sorted correctly and faster than the human labor system.

The DES technique is not only applicable for manufacturing settings but can be applied to other settings, such as the traffic modeling and out-patient clinics [4], [5] and [6]. [7], [8] and [9] model the health examination centers in hospitals with DES to experiment with strategies for reducing the waiting time of the patients.

3. HDD Sorting Process

Our HDD sorting process sorts 6 product families. Each product family consists of many stock keeping units (SKU). The current layout of the HDD process is shown in Figure. 1 and Figure. 2 shows the proposed conveyor system

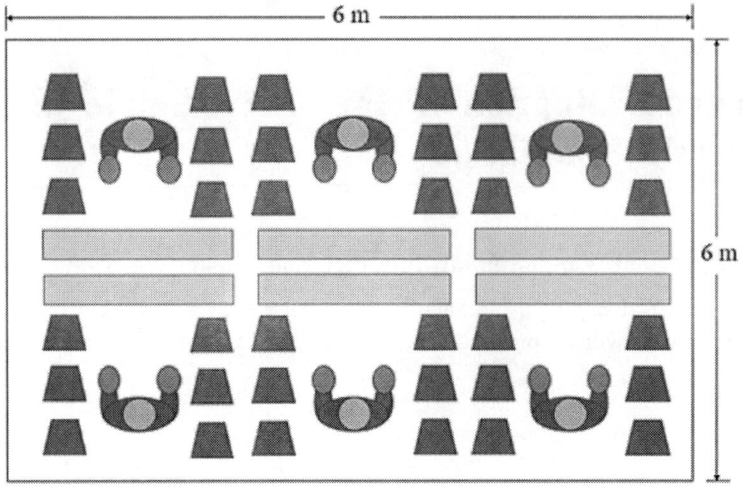

Figure 1. Current layout of the HDD sorting process.

At the present, the sorting process requires 6-12 people per shift. HDDs are transported by trolleys which have two sides and can stock 60 HDDs per side. The sorting department is located on the 1st floor, and it is preceded by the testing department on the 2nd floor. The factory manager is interested in using an automated loop conveyor system for the sorting process because it is space-saving and probably more accurate than humans.

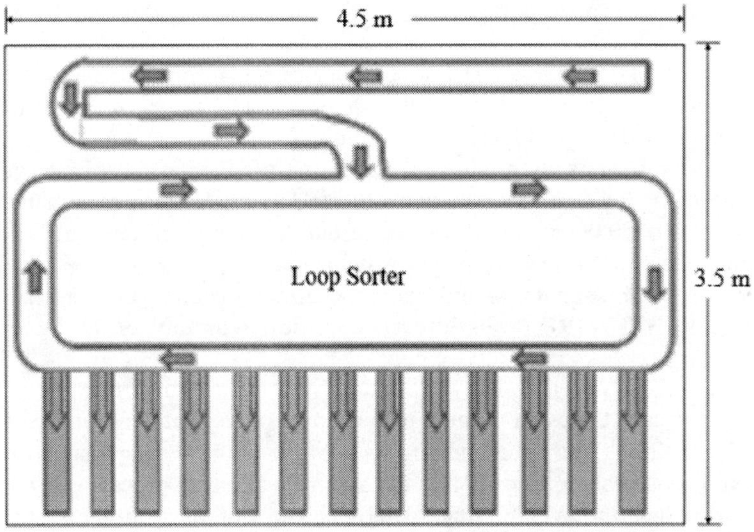

Figure 2. Proposed layout of the HDD sorting process.

The simulation model is built on ARENA. First, the data is collected and fit with statistical distributions. The DES model of the current system is constructed and verified. Then, the current sorting system is modified to the automated sorting system. The numerical performance of the two systems are then compared.

IOP Conf. Series: Materials Science and Engineering **639** (2019) 012005 doi:10.1088/1757-899X/639/1/012005

3.1. Data Collection and Analysis

To simulate the operation of the sorting department, the following data is required: the data of the arrival rates from the testing department, the transportation time delay to the sorting department, and the waiting time to stack 120 HDDs in a trolley at the tester. Table 1 and 2 shows the probability distributions for all random processes.

Table 1. Interarrival times at the sorting department

Process	Probability distributions	Sample size
	Minutes	Days
Type A interarrival time	LOGN(0.31, 0.348)	26
Type B interarrival time	0.04 + WEIB(0.0244, 3.61)	26
Type C interarrival time	0.2 + 0.12 * BETA(1.2, 1.19)	26
Type D interarrival time	WEIB(7.16, 0.431)	26
Type E interarrival time	0.05 + LOGN(0.0172, 0.0109)	26
Type F interarrival time	0.01 + 0.02 * BETA(24.3, 16.5)	26

Table 2. Process and transportation times at the sorting department

Process	Probability distributions	Sample size
	Minutes	Days
Transportation time from testing to sorting Dept.	52 + GAMM(23, 0.713)	30
Transportation time from sorting to packing Dept.	52 + WEIB(29.3, 3.21)	30
Process time to sorting type of product	TRIA(319, 411, 428)	30
Process time to sorting SKU of product	407 + 68 * BETA(0.577, 0.631)	30

The simulation model is validated by the two-sample t-test of the work in process (WIP) between the simulation data and actual data in Minitab Software. Each data needs to verify the distribution because two-sample t-test can test only normal distribution. In the Figure 4 and 5, they are demonstrated that the simulation data have P-Value = 0.062 and actual data have P-Value = 0.340 respectively. Both data are greater than 0.05. Therefore, we can decide that they are the normal distribution.

As can be seen from the above verification, we can use two-sample t-test to validate the simulation model. The result of validation is shown in the Figure 3. The simulation data and actual data are not statistically different because P-Value = 0.802 is greater than 0.05.

Two-Sample T-Test and CI: Simulation, System

```
Two-sample T for Simulation vs System

             N  Mean  StDev  SE Mean
Simulation  30  9606    397       72
System      30  9441   3553      649


Difference = μ (Simulation) - μ (System)
Estimate for difference:  165
95% CI for difference:  (-1170, 1500)
T-Test of difference = 0 (vs ≠): T-Value = 0.25  P-Value = 0.802  DF = 29
```

Figure 3. Two-Sample T-Test of the work in process report

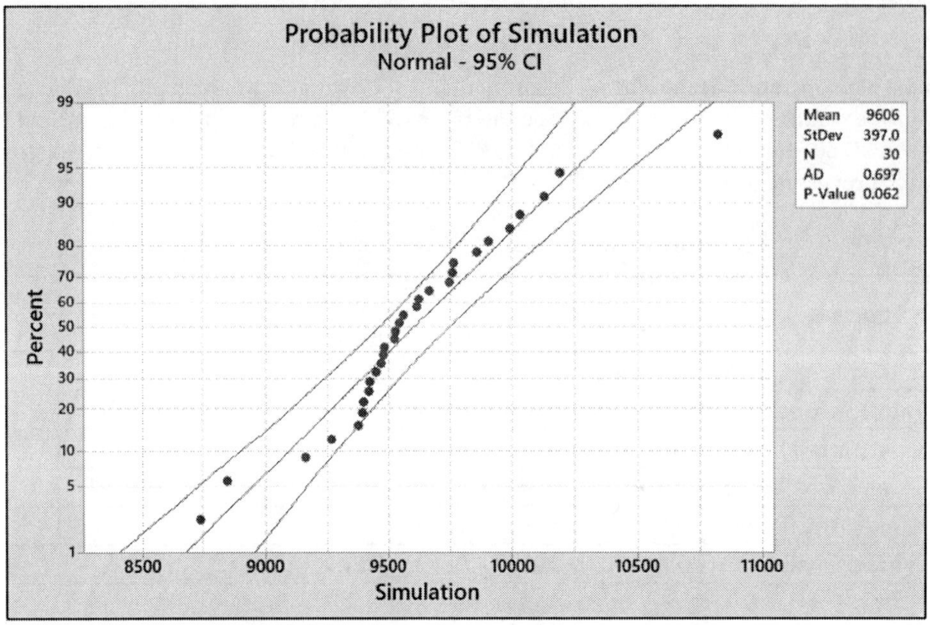

Figure 4. Probability Plot of Simulation Data

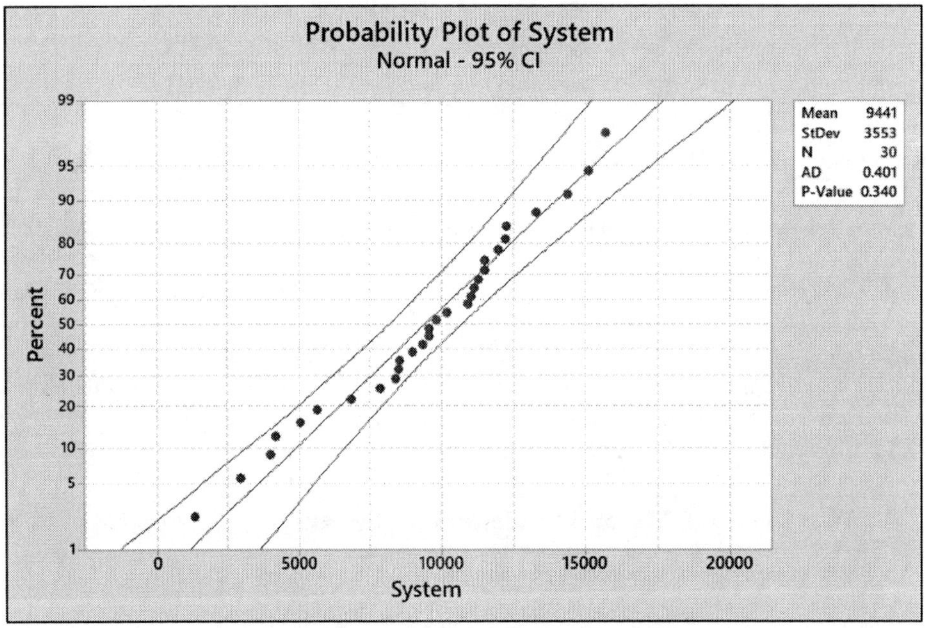

Figure 5. Probability Plot of Actual Data

3.2. DES Modeling

Arena logic models of the current sorting system and the loop conveyor system are shown in Figure 6 and 7. The HDD arrival processes are modeled with the Create modules. The primary sorters and secondary sorters are represented by the Process modules which provide the cycle time distribution information. Six staffs are responsible for sorting the mixed SKUs to identical SKUs.

We assume that machine failures are negligible. The automatic sorting system consists of one-way entrance and twelve plus one way exits. Because there are approximately 8 SKUs per product family,

12 exits are sufficient. The remaining exit are for some specific SKUs. The conveyor's speed is another important factor; the appropriate belt speed is about 2 m/s.

The simulation model of the automated sorting system has the Station and the Access modules which can model twelve plus one exits. Each exit has a Decide module to check the products and their SKUs before being sent to the packing department

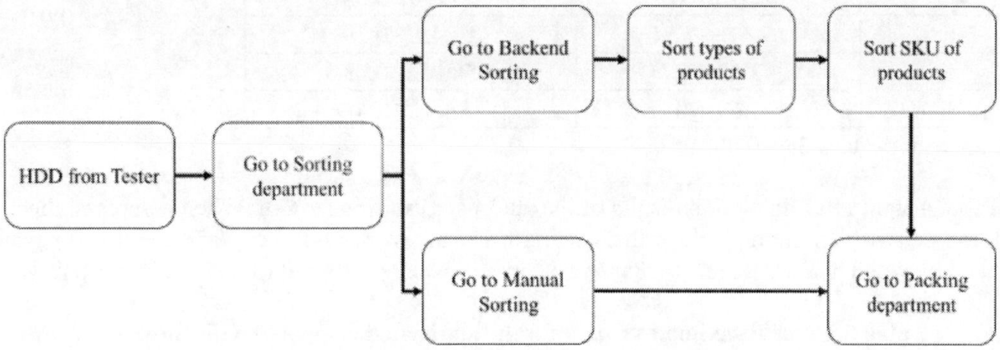

Figure 6. Arena logic models of the current sorting system

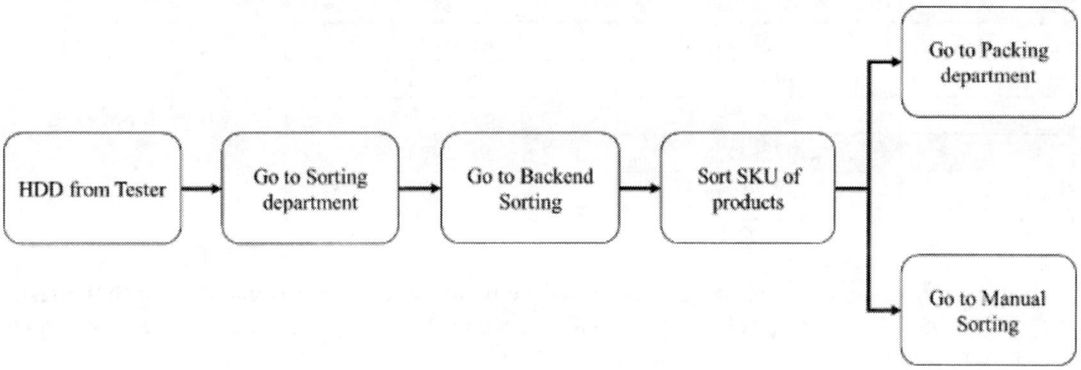

Figure 7. Arena logic models of the loop conveyor system

4. Simulation Results

The simulation model of the current sorting system is validated by comparing the work in process of the actual system and that of from simulation. The number of replications is 500. The sorting department works 2 shifts per day. The simulation results in Table 3 show that the current system is very congested; the average work in process is more than 9,000 HDDs, and the average time from the testing department to packing department is about 150 minutes.

Table 3. 95% CI of simulated results in terms of KPIs for the current sorting department

Product	Waiting time	Work-in-process	Time in system
	Minutes	Minutes	Minutes
A	127.59 ± 0.26	460.41 ± 0.83	150.35 ± 0.21
B	98.96 ± 0.16	1807.36 ± 2.64	
C	121.17 ± 0.29	516.81 ± 0.71	
D	69.43 ± 1.63	18.11 ± 0.55	
E	99.51 ± 0.16	1676.70 ± 2.44	
F	94.27 ± 0.16	4976.48 ± 7.31	

Table 4. 95% CI of simulated results in terms of KPIs for automated system

Product	Waiting time	Work-in-process	Time in system
	Minutes	Minutes	Minutes
A	7.11 ± 0.01	26.55 ± 0.03	4.74 ± 0.01
B	2.53 ± β	60.81 ± 0.04	
C	6.26 ± 0.01	29.04 ± 0.03	
D	87.41 ± 7.54	12.48 ± 0.29	
E	2.62 ± β	58.85 ± 0.03	
F	1.69 ± β	138.47 ± 0.09	

β: Half width is less than 0.00

Table 4 shows the simulation results of the automated sorting system. When compared these results with the current system in Table 5, the waiting time, the work-in-process, and the time in system are reduced by 74.78%, 85.09% and 96.85% respectively when compared to the current system.

Table 5. Percentage improvement for the automated compared with current system

Product	Waiting time	Work-in-process	Time in system
A	94.43%	94.24%	96.85%
B	97.45%	96.64%	
C	94.84%	94.38%	
D	-33.62%	31.56%	
E	97.37%	96.49%	
F	98.21%	97.22%	
Average	74.78%	85.09%	

However, the reason of Product D has a negative percentage of improvement because the result of the production order of Product D is significantly less than other production orders. Therefore, when we change the sorting system, it can increase the waiting time of Product D.

Lastly, the automated sorting system can improve the performance of the current system, it requires a new conveyors and management software. The factory manager must make a decision if the investment in the new system is worth the savings.

5. Conclusions

We present simulation models of the HDD manual sorting process and its new automatic system. The overall results of automated sorting model simulation are better than the current system. The waiting time, the work-in-process and the time in system is reduced by 74.78%, 85.09% and 96.85% respectively.

Acknowledgment

The authors would like to thank you Asst. Prof. Chana Raksiri and Phatcharee Toghaw Thongrattana, Ph.D. for their support of this work.

References

[1] THANSETTAKIJ MULTIMEDIA, *WD occupy to Thailand "world hard disk hub"!! Close the factory in Malaysia - Thailand BOI guaranteed.* 2018 [cited 2018 4 September]; Available from: http://www.thansettakij.com/content/301767.

[2] Peixoto Robson, Dias Luís, Maria Sameiro Carvalho, Pereira Guilherme, and Geraldes Carla A S, *An automated warehouse design validation using discrete simulation, in International*

Conference on Intelligent Transportation Systems (ITSC). 2016: Windsor Oceanico Hotel, Rio de Janeiro, Brazil. p. 199-204.

[3] Zhang Wenchang, Mei Jiangping, and Ding Yabin, *Design and Development of a High Speed Sorting System Based on Machine Vision Guiding*. Physics Procedia, 2012. 25: p. 1955-1965.

[4] Salimifard Khodakaram and Ansari Mehdi, *Modeling and Simulation of Urban Traffic Signals*. International Journal of Modeling and Optimization, 2013: p. 172-175.

[5] Kamrani Mohsen, Abadi Sayyed Mohsen Hashemi Esmaeil, and Golroudbary Saeed Rahimpour, *Traffic simulation of two adjacent unsignalized T-junctions during rush hours using Arena software*. Simulation Modelling Practice and Theory, 2014: p. 167-179.

[6] Khokhuntod Niramon, *Improving Kasetsart University Bus Routes By Simulation Modeling and Analysis*, Master's degree of Industrial Engineering. 2014, Kasetsart University.

[7] Jacobson H Sheldon, Hall N Shane, and Swisher R James, *Discrete-Event Simulation of Health Care Systems*, Master's degree of Mechanical and Industrial Engineering. 2006, University of Illinois at Urbana-Champaign.

[8] Wongsammacheep Thanon, *The Discrete-Event Simulation Model of a Health Screening Center for Improving Patient's Waiting Time*, Master's degree of Industrial Engineering. 2012, Kasetsart University.

[9] Al-Araidah O, Boran A, and Wahsheh A, *Reducing Delay in Healthcare Delivery at Outpatients Clinics Using Discrete Event Simulation*. International Journal of Simulation Modelling, 2012. 11(4): p. 185-195.

Development of Automated Vertically Stacked Hard Disk Drive Sorting Configurations Through Simulation Modeling

B Akbudalee, J Pichitlamken and S Suebsankul

Department of Industrial Engineering, Faculty of Engineering, Kasetsart University, 50 Ngamwongwan Road, Chatuchak, Bangkok 10900, THAILAND

budsarin.akb@ku.th, juta.p@ku.ac.th and suwiwat@yahoo.com

Abstract. This paper focuses on discrete-event simulation (DES) model of a vertically stacked HDD sorting system using the Arena software. We propose an approach decreasing the throughput time of sorting department that are inefficient. We find that the automated system can decrease the average waiting time, the work-in-process and the average total time in system by 59.24%, 59.87%, and 72.67%, respectively, compared with the current system.

1. Introduction

In 2019, Thailand's export market of hard disk drives (HDDs) continuously grows. KBank Research estimates that the overall HDDs exported in 2019 can reach around 252-262 million units, an increase of 3.6-7.8 percent from 2018 and equivalent to 12,576 -13,089 million USD [1].

The company in this case study is a HDD manufacturer. The current HDD sorting process uses manual labor. Many problems include insufficient working hours and too many HDD transfers so the HDD manufacturer wants to use vertical alignment technology to sort HDDs immediately after they leave the testers.

We develop a simulation model of the vertically stacked HDD process in Arena (Rockwell software) as a proof of concept before actually deploying it in the sorting department.

2. Literature review

Simulation model is a tool for process improvement to analyze alternatives before actual implementation, to study the system in a computer program, and to determine the guidelines for decision making to solve the problems without wasting time and cost of real experimentation.

Simulation is "the process of designing a model of a real system and conducting experiments with this model for the purpose either of understanding the behavior of the system or of evaluating various strategies for the operation of the system" [2]. The Discrete-Event Simulation (DES) model is "the operation of a system as a discrete sequence of events in time. Each event occurs at a particular instant in time and marks a change of state in the system" [3].

Simulation models are constructed for many problems, such as in manufacturing settings and traffic problems. For the manufacturing industry problems, [4] use Arena software to find the bottleneck, calculate the productivity and determine the requirement units. The result shows better efficiency of optimal productivity and utilization. In [5], the Arena simulation model helps to understand and analyze the bottlenecks and the solutions to improve system.

3. Methodology

The procedure in this research consists of 5 steps as follows.

3.1. Study the process

The HDD production process under this case study is shown in figure 1.

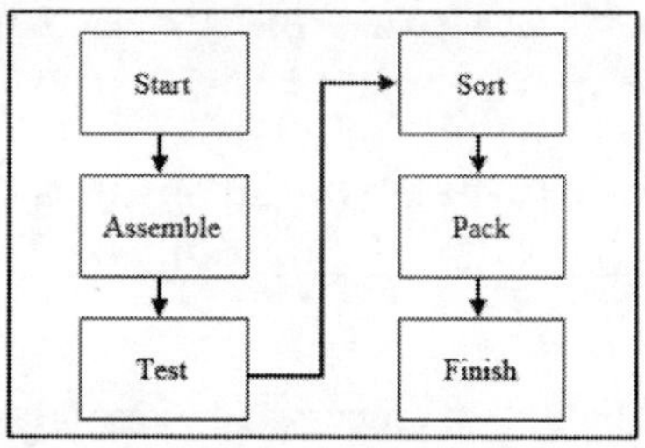

Figure 1. The HDD production process.

The process starts with HDD assembly, and they are tested in the tester. When HDDs leave testers, they are mixed, and they must be sorted before being packed.

Our HDD sorting process sorts 6 product families. Each product family consists of many stock keeping units (SKU). When testing finishes on the 2nd floor, the HDDs are packed in trolleys and transported via elevators to the sorting department on the 1st floor. Each trolley has 2 sides and can stock 60 HDDs per side.

The current process of sorting uses human labor and barcode technology to sort products. There are 2 6-worker shifts per day. About 80,000 HDDs can be sorted per day. The demand of HDDs fluctuate a lot; it can reach up to 150,000 HDDs per day so the manufacturer needs to hire up to 12 workers per shift. Due to this demand variability and high volume, the plant manager wants to change the manual sorting to the automated system to increase the productivity and lower cost.

The manager is interested in using the vertically stacked sorting system because it is space saving and probably more accurate than humans, and it requires less manual handling. The proposed layout of the HDD sorting process is shown in figure 2.

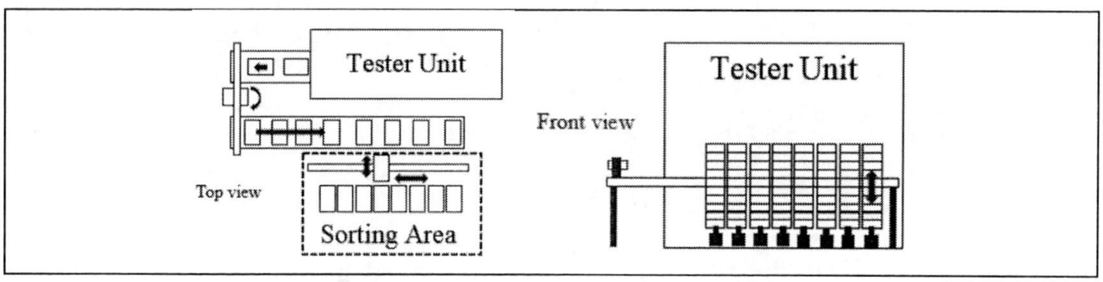

Figure 2. The proposed layout of the HDD sorting process.

In this paper, we construct a simulation model on Arena to simulate the vertically stacked sorting process.

IOP Conf. Series: Materials Science and Engineering **639** (2019) 012006 doi:10.1088/1757-899X/639/1/012006

3.2. Data collection

To simulate the operation of the sorting department, the following data is required: the arrival rates from the testing department, the transportation time to the sorting department, and the waiting time to stack 120 HDDs in a trolley at the tester. tables 1 and table 2 shows the probability distributions for all random processes.

Table 1. The probability distributions of interarrival time to the sorting process.[a]

Process	Probability distributions (Minutes)	Sample size (Day)
Product A interarrival time	LOGN(0.31, 0.348)	26
Product B interarrival time	0.04 + 0.04 * BETA(4.78, 3.93)	26
Product C interarrival time	0.2 + 0.12 * BETA(1.2, 1.19)	26
Product D interarrival time	WEIB(7.16, 0.431)	26
Product E interarrival time	0.05 + LOGN(0.0172, 0.0109)	26
Product F interarrival time	0.01 + WEIB(0.0127, 7.31)	26

[a]Surached Wongnibunkit(2019)

Table 2. The probability distributions of process and transportation times in the sorting process.[a]

Process	Probability distributions (Minutes)	Sample size (Day)
Transportation time from testing to sorting	52 + GAMM(23, 0.713)	30
Transportation time from sorting to packing	52 + WEIB(29.3, 3.21)	30
Process time to sorting type of products	TRIA(319, 411, 428)	30
Process time to sorting SKU of products	407 + 68 * BETA(0.577, 0.631)	30

[a]Surached Wongnibunkit(2019)

3.3. Developing a simulation model of the current system

Figure 3 shows the process flow diagram on Arena.

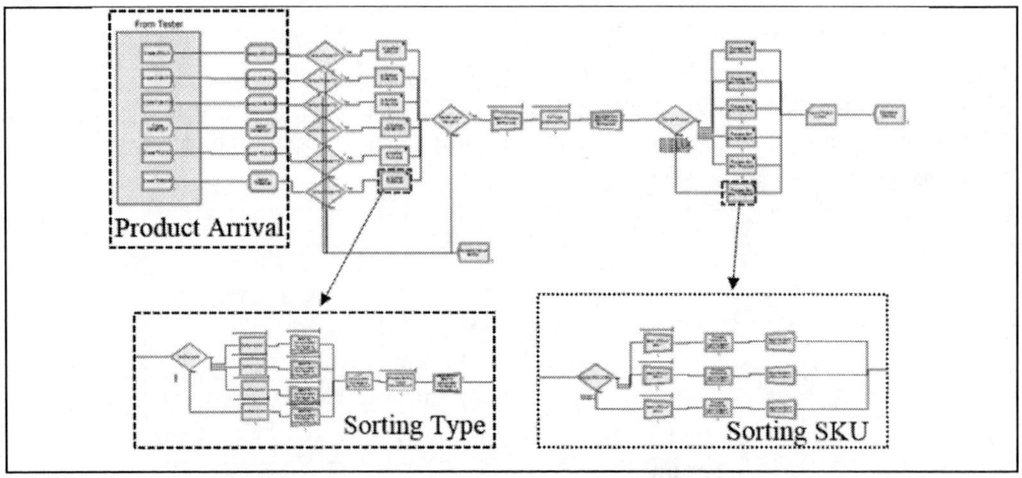

Figure 3. Arena model structure of current sorting department.

In a simulation model, Create Modules are starting points for Entities which are HDDs. Sorting types and sorting SKUs are represented by the Process modules.

IOP Conf. Series: Materials Science and Engineering **639** (2019) 012006 doi:10.1088/1757-899X/639/1/012006

3.4. Verification and validation of the model
The simulation model is validated by the two-sample T-test of the WIP from the simulation and that from actual data is shown in figure 4. They are not statistically different because p-Value of this test is greater than 0.05.

```
Two-Sample T-Test and CI: Simu, Real

Two-sample T for Simu vs Real

       N   Mean   StDev  SE Mean
Simu  30   9606     397       72
Real  30   9453    3561      650

Difference = μ (Simu) - μ (Real)
Estimate for difference:  153
95% CI for difference:  (-1185, 1491)
T-Test of difference = 0 (vs ≠): T-Value = 0.23  P-Value = 0.816  DF = 29
```

Figure 4. The result of Two-Sample T-Test.

3.5. Developing a simulation model of the vertically stacked system
In the new system, the sorting process is moved to be in front of the testers on the 2nd floor. The automatic sorting system consists of 8 stackers, each of which can contain 8 HDDs of the same SKU. Eight stackers are chosen because each tester has at most 8 SKUs product mix.
The process starts with HDDs from the testers, and barcodes are scanned. Cartesian robots pick and place in onto conveyers, moving HDDs to HDD shelves, and the stacker move it down 1 slot. When HDDs leave stackers, they are loaded into trolleys and transported to be packed. The simulation model of the automated sorting system is shown in figure 5.

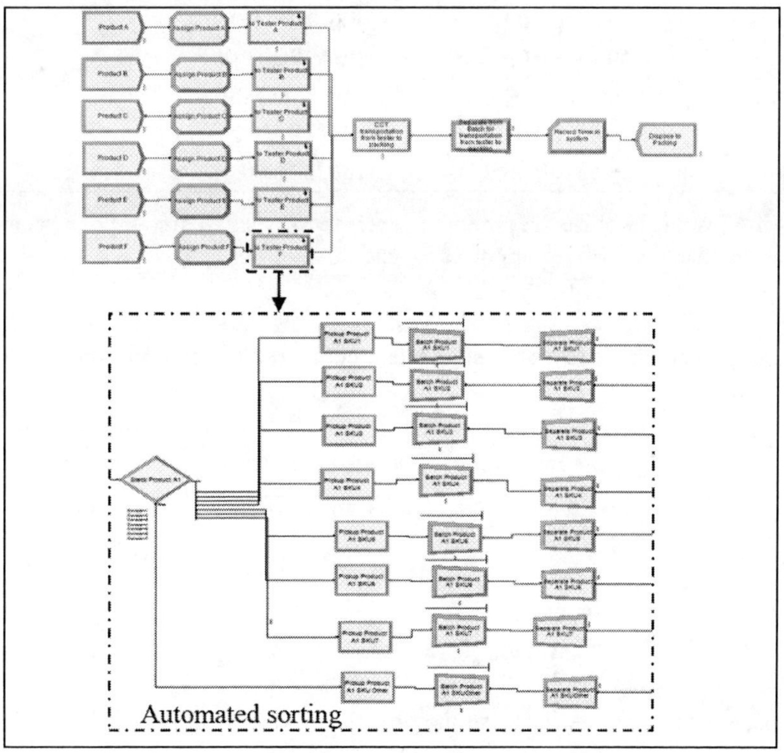

Figure 5. Arena model structure of automated vertical stack sorting

IOP Conf. Series: Materials Science and Engineering **639** (2019) 012006 doi:10.1088/1757-899X/639/1/012006

4. Simulation results

The key performance indices (KPIs) of the current sorting system model simulation are shown in table 3. The number of replications is 500, and the replication length is 24 hours.

Table 3. The 95% confidence intervals of the simulation KPIs of for the current sorting department

Product Family	Waiting time (Minute)	Work-in-process (Piece)	Time in system (Minute)
Product A	127.33 ± 0.27	460.13 ± 0.88	150.35 ± 0.21
Product B	98.80 ± 0.18	1804.11 ± 2.86	
Product C	121.13 ± 0.30	516.40 ± 0.72	
Product D	69.00 ± 1.73	18.09 ± 0.51	
Product E	99.39 ± 0.18	1674.67 ± 2.65	
Product F	94.11 ± 0.17	4968.17 ± 7.95	

The KPIs of the automated sorting system is shown in table 4. The comparison results of 3 KPIs are calculated into percentages as shown in table 5.

Table 4. The 95% confidence intervals of the simulation KPIs of the automated sorting system

Product Family	Waiting time (Minute)	Work-in-process (Piece)	Time in system (Minute)
Product A	59.38 ± 0.17	225.67 ± 0.38	41.09 ± 0.01
Product B	39.62 ± 0.02	689.12 ± 0.26	
Product C	49.73 ± 0.06	204.29 ± 0.19	
Product D	Not showing results[a]	40.47 ± 1.09	
Product E	39.05 ± 0.02	629.18 ± 0.26	
Product F	34.58 ± 0.02	1801.20 ± 0.68	

[a]Not showing results because the product D enters a very small amount of system compared to other products and therefore cannot be logged out.

Table 5. The percentage reduction results when compared the current system with automated sorting system.

Product Family	Waiting time	Work-in-process	Time in system
Product A	53.37%	50.96%	72.67%
Product B	59.90%	61.80%	
Product C	58.94%	60.44%	
Product D	Can't compare[a]	-50.64%	
Product E	60.17%	62.43%	
Product F	63.26%	63.75%	

[a]Can't compare because there is no waiting time the product D in automated sorting system.

Table 5 shows the percent improvement from the manual system to the automated sorting system. The result shows the reduction of the average waiting time, the work-in-process, the average total time in system by 59.24%, 59.87% and 72.67% respectively.

5. Conclusions

Our simulation study of the automated vertically stacked sorter shows that this automated system can increase the efficiency of the sorting process which reduces the waiting time, the work-in-process, and the time in system by 59.24%, 59.87% and 72.67%, respectively.

References

[1] KASIKORN Research Center. Thai HDD exports in 2019 are poised to top 250 million units following the rebound in the global computer market and a string of production relocations to Thailand. 2018 [cited 2018 26 December]; Available from: https://kasikornresearch.com/EN/analysis/k-econ/business/Pages/z2949.aspx.

[2] Shannon, R.E., 1975. Systems Simulation – The Art and Science, Prentice-Hall.

[3] Stewart Robinson (2004). Simulation – The practice of model development and use. Wiley.

[4] Neeraj RR, Nithin RP, Niranjhan P, Sumesh A, Thenarasu M. Modelling and simulation of discrete manufacturing industry. Materials Today: Proceedings. 2018;5(11, Part 3):24971-83.

[5] Kamrani, M., S.M. Hashemi Esmaeil Abadi, and S. Rahimpour Golroudbary, Traffic simulation of two adjacent unsignalized T-junctions during rush hours using Arena software. Simulation Modelling Practice and Theory, 2014. 49: p. 167-179.

Acknowledgments

The author would like to express the appreciation and deep gratitude to Prof. Phatcharee Toghaw Thongrattana, Asst. Prof. Chana Raksiri, and the senior plant manager for their support of this work.

Characterization of Physical Properties and Tensile Strength of Chitosan/Fibroin Composite via Freeze Drying Method

P Sripapha[1], N Kapao[2], and A Chaijaruwanich[2*]

[1] Biomedical Engineering Institute (BMEI), Chiang Mai University, Chiang Mai, 50200, Thailand.
[2] Advanced Manufacturing Technology Research Center (AMTech), Department of Industrial Engineering, Faculty of Engineering, Chiang Mai University, Chiang Mai, 50200, Thailand.

Email address: anirut@eng.cmu.ac.th

Abstract. Chitosan and silk fibroin composite are natural biopolymers widely used in biomedical engineering applications. This work studied the properties of a chitosan and silk fibroin composite were prepared by freeze drying method. The characterizations were operated by fourier-transform infrared spectroscopy, scanning electron microscopy, and x-ray diffraction. In addition, the tensile strength was obtained for various compositions. The results showed that the freeze-dried composite was of high porosity with 3D interconnected pores that are suitable for tissue engineering. In addition, both Fourier-transform infrared spectroscopy and x-ray diffraction indicated that the composite materials were water soluble.

Keywords: Chitosan, Silk fibroin, Biopolymer, Freeze-drying

1. Introduction
Chitosan (CS) is a polysaccharide consisting of random deacetylated and acetylated units (2-amino-2-deoxy-β-D-glucopyranose, and N-deacetylated, respectively) which are β-(1→4)-linked. Chitosan can be found in various natural products from marine sources [1,2]. Chitosan shows strong biological properties such as being antibacterial, biocompatible with cells, biodegradable, and able to absorb humidity [3]. These properties have led to its wide-spread use in various applications [4]. Silk fibroin (SF) is a linear polypeptide that can be extracted from the *Bombyx mori* silkworm. Natural silk fibroin is composed of various amino acid types such as glycine, alanine, and serine, which affect the solubility in water [5]. In addition, the properties of silk fibroin depend on the proportion of silk I and silk II, which are both secondary structure molecular conformations. Silk I is an amorphous structure that is water soluble, whereas silk II is more stable with higher crystallinity, which results in the water insolubility property [6,7]. In biomedical engineering, silk fibroin is used in controlled release [8], membranes for regeneration [9], and other high efficiency applications [10]. In this research, we focus on estimation of the physical and mechanical properties of CS/SF composites fabricated via freeze drying for application to various fields of tissue engineering.

2. Materials and Methodology

2.1 Materials
Chitosan was extracted from squid pens and silk fibroin was prepared from silk cocoons produced by *Bombyx mori* silk worm using the method presented in [11]. Initially, fibroin was dissolved into 99 ml of distilled water overnight then 1 ml of acetic acid (RCI Labscan Ltd., Australia) and chitosan were added to the fibroin solution and shaken gently. For tensile testing, 0.0025 ml of diluted

glutaraldehyde (Loba Chemie Pvt Ltd., India) was added to the solution for comparison of the strength property. A clear, homogeneous solution was obtained after dissolving CS/SF for 24 h. The ratios for blending CS/SF were 75/25, 50/50, and 25/75 by wt% (marked as 75CS/SF, 50CS/SF, and 25CS/SF, respectively) for 1 g of solid loading.

2.2 Freeze drying process
CS/SF solutions measuring 5 ml were drained into 24-well plates for the characterization of physical properties. For tensile testing, 50 ml of solution were transferred into a 70x100 mm acrylic mold. The pre-freeze and freeze stages were conducted at -20 and -80°C, respectively for 24 h. Then all CS/SF solutions were freeze dried at -50°C in a vacuum for 48 h.

2.3 Characterization
Fourier-transform infrared spectroscopy (FTIR) spectra were recorded using a NICOLET 6700 (United States) and collected over the range 4000-400 cm^{-1} with an accumulation of 64 scans and resolution of 4 cm^{-1}. X-ray diffraction (XRD) patterns were obtained using a Rigaku D/teX 250 (Japan) with CuK radiation in steps of 0.01° from 10 to 60° (2-theta angle). Freeze-dried CS/SF specimens were sectioned with a razor blade in the horizontal plane into 10x10x0.5 mm^3pieces then coated with a gold layer. Scanning electron microscopy (SEM) images were obtained using a JSM 5910 LV (Japan). Porosity and pore sizes were estimated using Image J from various SEM images. For tensile strength testing, a freeze-dried CS/SF sheet was cut into 30x80x2 mm^3 pieces and tested by the universal testing machine, Instron 5542 (United Kingdom), with a 500 N load and head cross-speed of 0.05 mm/s at room temperature and humidity lower than 40% RH.

3. Characterization and testing

3.1 General properties of specimen
All freeze-dried CS/SF specimens are shown in Figure. 1. Each of the specimens are white-colored sponges which cannot be identified by the naked eye. Glutaraldehyde is added when forming silk fibroin and was not used in these experiments except for comparison in the tensile strength test. Chitosan acts as a network and binding agent for silk fibroin in the CS/SF composites. Hence, 75CS/SF, and 50CS/SF are formed with smooth surfaces and can be cut into various shapes. However, at low chitosan concentration (25CS/SF), binding of the material was not as strong after freeze drying. Therefore, it collapsed easily and the porosity, pore size, and mechanical strength could not be estimated.

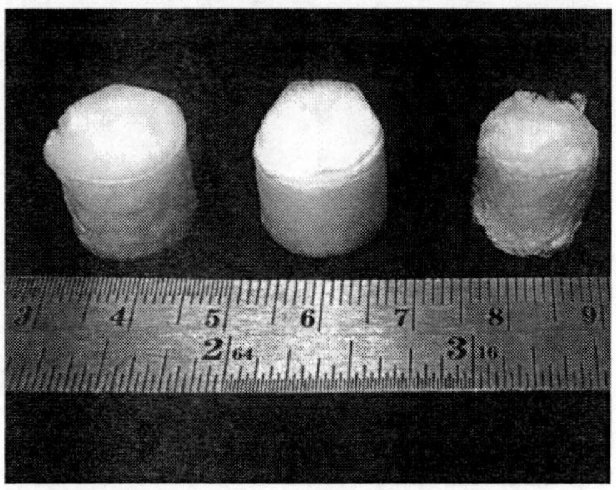

Figure. 1 Freeze-dried CS/SF specimens.

3.2 Fourier-transform infrared spectroscopy

The FTIR spectra of chitosan, silk fibroin, and CS/SF are shown in Figure. 2. The chitosan spectrum Figure. 2(A) shows bands at 3370 and 3236 cm^{-1} that represent hydrogen bonding, corresponding with the stretching of O-H and N-H respectively. The polysaccharide characteristic is indicated by the C-H bond at 2981 and 2879 cm^{-1} which correspond with symmetric and asymmetric stretching respectively. The functional group of N-acetyl is found at 1633 cm^{-1} (C=O stretching of amide I) and 1307 cm^{-1} (C-N stretching of amide III). However, amide II (N-H) could not be observed at 1150 cm^{-1} due to overlapping between peaks at close wavelengths. Amine at 1567 cm^{-1}corresponds with the bending of N-H. Both CH_2 and CH_3 are observed at 1412 and 1373 cm^{-1} with bending and symmetrical deformations, respectively. The bending vibration of hydroxide is shown at 1226 cm^{-1}. The C=O bond is subject to asymmetric stretching at 1121 cm^{-1}.

For the silk fibroin spectrum shown in Figure. 2(E), bands at 1639 and 1519 cm^{-1}correspond with amide I (C=O stretching) and amide II (N-H deformation and C–N stretching). The important bands in the range 1500 to 1700 cm^{-1} are associated with amides in silk fibroin, where 1600 to 1700 cm^{-1} corresponds with amide I and 1500 to 1600 cm^{-1}corresponds with amide II. The absorption in these regions is due to the peptide backbones of the amides. Regarding the analysis of the secondary structures of silk fibroin, the difference between silk I and silk II will be indicated by peaks in these bands. Characteristics of silk I with amide I and II peaks are observed at 1640 to 1660 $cm^{-1}$and 1535 to 1542 cm^{-1}, respectively. For silk II, there is a small shift from the silk I values. Specifically, amide I shifts from 1610 to 1630 $cm^{-1}$and amide II shifts from 1510 to 1520 cm^{-1} [12]. In this study, both amide peaks are shown at 1639 and 1538 cm^{-1} and correspond with silk I. All CS/SF specimen spectra have the same peak positions and their intensities are related to the content of pristine chitosan and silk fibroin.

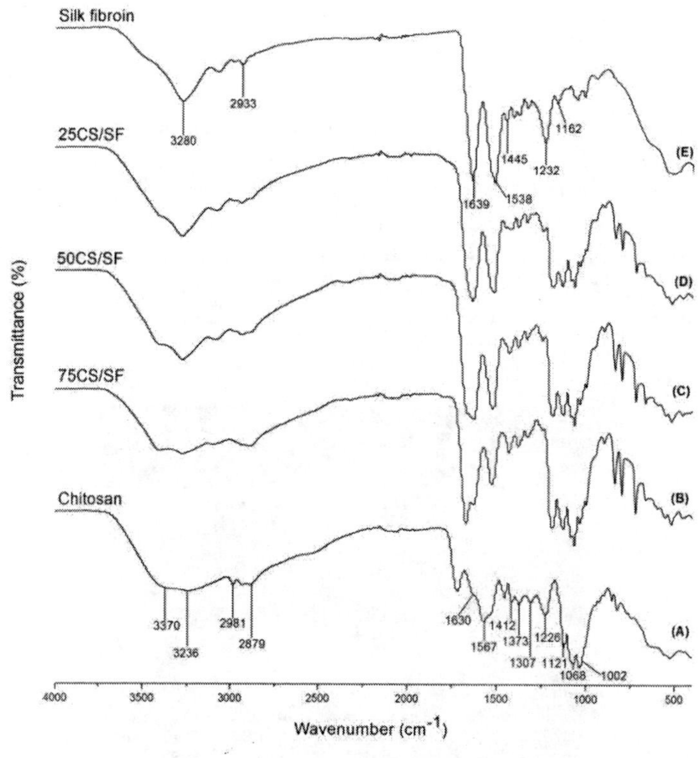

Figure. 2 FTIR spectra: (A) chitosan, (B) 75CS/SF, (C) 50CS/SF, (D) 25CS/SF, and (E) silk fibroin.

3.3 X-ray diffraction

Figure. 3 shows XRD patterns for all specimens. Both chitosan and silk fibroin show broad diffraction at 21.1°and 23.2°respectively due to their amorphous structure. The XRD pattern of Chitosan was reported in [13]. XRD patterns of silk fibroin correspond to the silk I crystalline peak [14], however, there can be various degrees of broadness in the pattern that depend on the solution for degumming [15], hence, the silk fibroin peak in Figure. 3 shows a large, broad diffraction when compared with another solution. XRD patterns of freeze-dried CS/SF are preserved when compared with the pristine diffractogram. Due to the overlap of chitosan and silk fibroin peaks, individual patterns could not be identified. However, a broad diffraction band confirms the large silk fibroin content, whereas more sharp diffraction is associated with chitosan.

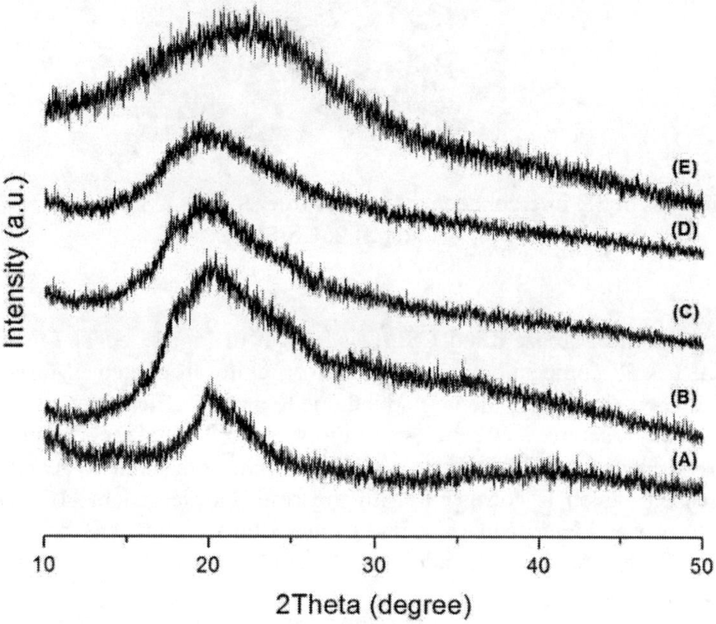

Figure. 3 XRD patters: (A) chitosan, (B) 75CS/SF, (C) 50CS/SF, (D) 25CS/SF, and (E) silk fibroin.

3.4 Scanning electron microscopy

SEM images of the horizontal-section of CS/SF are shown in Figure. 4. Three-dimensional interconnected pores are observed with different pore size and porosity. Without added glutaraldehyde for silk fibroin forming, chitosan serves as a network and binding agent in silk fibroin for the CS/SF composites as can be seen in Figure. 4 for the 75CS/SF and 50CS/SF. However, in the case of low chitosan content (25CS/SF), it can be seen that the capacity for binding is very low. Consequently, the freeze-dried CS/ SF composite was formed with weak bonds as shown in Figure. 4. For this composition the porosity, pore size, and mechanical strength could not be estimated. Table. 1 presents the pore characteristics of freeze-dried CS/SF. Careful investigation and observation of the pore structure of 75CS/SF in Figure. 4(A) shows smooth three-dimensional interconnected pores created by removing crystalline ice from the solution. On the other hand, 50CS/SF in Figure. 4(B) shows a pore structure similar to plates that are gently linked to each other due to less strong binding as a consequence of reduced chitosan content. The porosity was seen to decrease from 46.89 (75CS/SF) to 23.57 (50CS/SF) percent (the porosity in the 25CS/SF structure could not be estimated). On the other hand, pore size increased from 21-123 to 113-194 μm for 75CS/SF and 50CS/SF respectively.

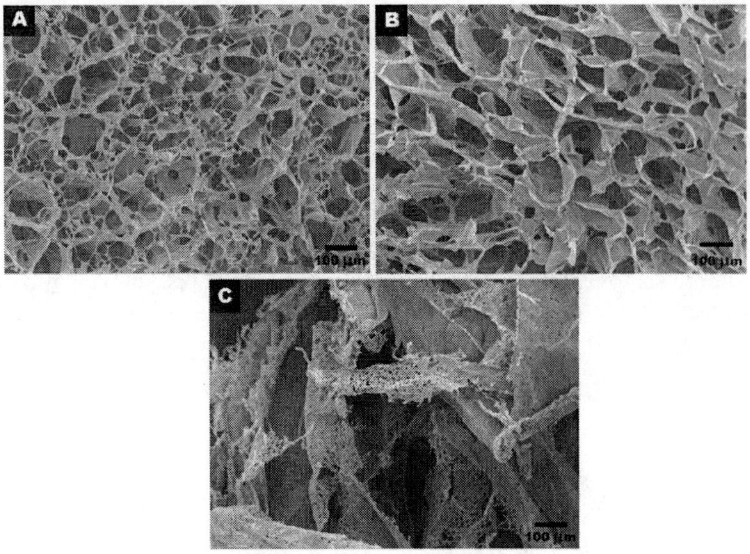

Figure. 4 SEM images of 3D interconnected pores of freeze-dried CS/SF: (A) 75CS/SF, (B) 50CS/SF, and (C) 25CS/SF.

3.5 Tensile strength

The mechanical strength of freeze-dried CS/SF is shown in Figure. 5. The results indicate that the tensile strength of CS/SF composites are dependent on chitosan content. Moreover, the mechanical strength increased when glutaraldehyde was added due to a more efficient combination of chitosan and fibroin into the CS/SF structure. Nevertheless, without glutaraldehyde, chitosan is the only substance which affects the strength of CS/SF. Hence, when the chitosan content decreased, the strength of CS/SF also decreased, which is consistent with the results observed in [16]. However, the tensile strength in this study was lower than [17], due to an electrospinning method shows nanofibers and pores in various CS/SF structures. In addition, freeze-dried method shows potential for massive production with appropriate properties.

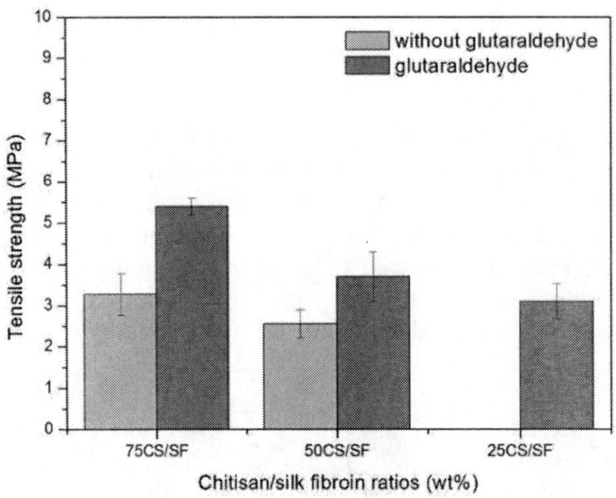

Figure. 5 tensile strength of freeze-dried CS/SF compared with and without added glutaraldehyde.

Table. 1 Pore characteristics of freeze-dried CS/SF

CS/SF (wt%)	Porosity (%)	Pore size (μm)
75/25	46.89	21-123
50/50	23.57	113-194
25/75	n/a	n/a

4. Conclusion

Freeze-dried CS/SF comprising of high porosity with 3D interconnected pores that are suitable for tissue engineering were fabricated. The results of FTIR and XRD indicated that the materials were comprised of highly amorphous structures and chemical functional groups that are water soluble. The mechanical strength was positively correlated with the chitosan content. Moreover, glutaraldehyde was seen to increase the tensile strength due to its ability to combine chitosan and fibroin into the CS/SF structure.

5. Acknowledgements

This research was financially supported by National Research Council of Thailand (NRCT). Technical advice from Biomedical Engineering Institute (BMEI) for the freeze-drying process.

References

[1] Lee Y, Kim H W and Kim Y H B 2018 *Food Sci. Biotechnol.* **27** 461
[2] Younes I and Rinaudo M 2015 *Mar. Drugs.* **13(3)** 1133
[3] Bashar M M and Khan M A 2013 *J. Polym. Environ.* **21(1)** 181
[4] Devlieghere F, Vermeulen A and Debevere J 2004 *Food. Microbiol.* **21** 703
[5] Altman G H, Diaz F, Jakuba C, Calabro T, Horan R L, Chen J, Lu H, Richmond J and Kaplan D L 2003 *Biomaterials.* **24** 401
[6] Marsh R E, Corey R B and Pauling L 1955 *Biochim. Biophys. Acta.* **16** 1
[7] Wang Y Q H, Wei K, Yang Y, Zheng R Y, Kim I S and Zhang K Q 2017 *Int. J. Mol. Sci.* **18(3)** 237
[8] Asakura T, Okushita K and Williamson M P 2015 *Macromolecules.* **48(8)** 2345
[9] Zhang S, Li J, Yin Z, Zhang X, Kundu S C and Lu S 2015 *J. Appl. Polym. Sci.* **132(32)** 1
[10] Koh L D, Cheng Y, Teng C P, Khin Y W, Loh X J, Tee S Y, Low M, Ye E, Yu H D, Zhang Y W and Han M Y 2015 Prog Polym Sci. **46** 86
[11] Rockwood D N, Preda R C, Yücel T, Wang X, Lovett M L and Kaplan D L 2011 *Nat. Protoc.* **6(10)** 1612
[12] Lu Q, Wang X, Lu S, Li M, Kaplan D L and Zhu H 2011 *Biomaterials.* **32(4)** 1059
[13] Hasegawa M, Isogai A, Onabe F, Usuda M and Atalla R H 1992 *J. Appl. Polym. Sci.* **45(11)**, 1873
[14] Jinfa M and Baoqi Z 2012 *J. Appl. Polym. Sci.* **125(3)** 2148
[15] Zhang H, Li L L, Dai F Y, Zhang H H, Ni B, Zhou W, Yang X and Wu Y Z 2012, *J. Transl. Med.* **10** 117
[16] Gobin A S, Froude V E and Mathur A B 2005 *J. Biomed. Mater. Res. A.* **4(3)** 465
[17] Cai Z X, Mo X M, Zhang K H, Fan L P, Yin A L, He C L and Wang H S 2010 *Int. J. Mol. Sci.* **11** 3529

Influence of pre-stretch ratio on the electrical actuation performance of VHB elastomer

Zhen-Qiang Song[1], Kazuhiro Ohyama[2] and Shijie Zhu[1]

[1] Department of Intelligent Mechanical Engineering, Fukuoka Institute of Technology, Japan

[2] Department of Electrical Engineering, Fukuoka Institute of Technology, Japan

E-mail: zhu@fit.ac.jp

Abstract. In this work, the influences of pre-stretch ratio on the electrical actuation performance of dielectric elastomer actuator (VHB 4905) are systematically investigated based on theoretical calculation and experimental tests. As the pre-stretch ratio increases, the maximum actuation strain increases at the initial phase, while decreases as pre-stretch ratio further increases. The peak value of actuation strain is attained when the electromechanical instability is just refrained by the pre-stretch, which is experimentally measured to be 3.25, with the maximum actuation strain of 90%.

1. Introduction

Dielectric elastomer (DE), as one kind of electroactive polymers (EAPs), shows great capability to change its dimensions under the electronic stimulus, which makes it potentially applied as soft actuators and robots [1]. The basic operational unit of dielectric elastomer actuator (DEA) comprises a soft elastomer sandwiched between compliant electrodes on both sides. When voltage bias is applied on the electrodes, the electrically-induced Maxwell stress will compress the elastomer membrane along the thickness direction, which leads to the area extension and thickness contraction, and the electrical energy is converted into the mechanical movement in this process. It has been demonstrated that the actuation strain and the specific actuation pressure of the EAP are similar to those of natural animal muscles [1], which is promisingly utilized as artificial muscles in the medical and industrial domains.

Recently, numerous studies have conducted by researchers aiming at improving the actuation performance of DEA, such as developing new materials with higher electrical permittivity and higher electrical breakdown field [3]. Pelrine et al. [4] reported that by simply pre-stretching the VHB acrylic elastomer membrane, the electrical actuation strain was significantly escalated up to 100%. It was found that pre-stretching of DE membrane is able to increase the electrical breakdown field of acrylic elastomer dramatically [5]. Additionally, the electromechanical instability (EMI) of the DEA can be effectively avoided by the pre-stretching [6]. Therefore, the pre-stretching strategy was widely adopted by many researchers [7]. However, it is noted that over pre-stretching of the membrane will induce significant stiffening of the elastomer [8], which deteriorates the actuation performance of DEA. Therefore, the optimum pre-stretch ratio of DEA should be a trade-off of multiple limiting factors. In this work, the influence of the pre-stretching on the actuation behavior of DEA was systematically investigated with VHB 4905 acrylic elastomer, and the optimum pre-stretch ratio was pinpointed based on theoretical calculation and experimental testing.

2. Experimental methods

2.1. Mechanical tests

VHB 4905 acrylic elastomer (3M) with a thickness of 0.5 mm was utilized in this work. In order to establish the constitutive equation of the DEA, the equi-biaxial tension experiment was conducted on a customized test bench. The circular specimen with a diameter of 50 mm was prepared and loaded with 18 clips distributed along the circumferential direction. The clips were connected with a linear motor (GLM20AP, THK Co., Ltd.) through an assembly of steel wires and pulleys, which was controlled by the LabVIEW software, as schematically shown in figure 1(a). The stretching force was measured with a load cell (U3B1-20K-B, Minebea Co., Ltd). A camera was mounted over the VHB membrane to record the stretch ratio in real time during loading process. The stretching velocity of the linear motor was set to be 2 mm/s, with an average strain rate of 0.1/s for the specimen.

2.2. Electrical actuation test

Circular specimens were equi-biaxially pre-stretched to various ratios before being fixed on a rigid frame with an inner diameter of 100 mm, as schematically shown in figure 1(b). The pre-stretch ratio is defined as the ratio of deformed diameter to the undeformed diameter of the DE film, which is selected to be 1, 1.5, 2.25, 2.85, 3, 3.25, 3.5, 4, 4.5, 5 and 5.5 in this work. An active region with a diameter of 5 mm was coated with conductive carbon grease as compliant electrodes. The voltage increase ramp rate was set to be 20 V/s, provided by a high voltage power supply (Model HAR-30P1, Matsusada Precision Inc). A camera was mounted over the membrane to record the deformation during the actuation process.

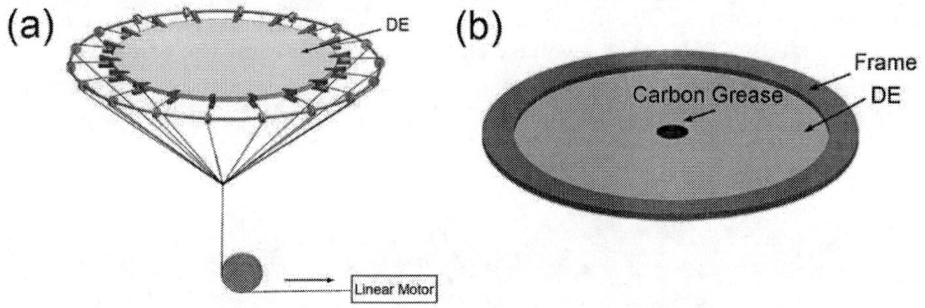

Figure 1. The schematics for the (a) equi-biaxial tension loading and (b) electrical actuation test.

3. Results and discussion

3.1. Equi-biaxial tension

The relationship between the nominal stress and stretch ratio for the VHB 4905 elastomer under equi-biaxial tension loading is shown in figure 2. Yeoh model is utilized to characterize the mechanical response of the elastomer, which is given as [8]:

$$\sigma(\lambda)=2(\lambda-\lambda^{-5})\left[C_1+2C_2(2\lambda^2+\lambda^{-4}-3)+3C_3(2\lambda^2+\lambda^{-4}-3)^2\right], \tag{1}$$

where σ is the nominal stress, λ is the stretch ratio, C_1, C_2 and C_3 are hyperelastic material parameters determined by fitting the experimental data, which are 0.0488 MPa, -2.646×10^{-4} MPa and 2.734×10^{-6} MPa, respectively.

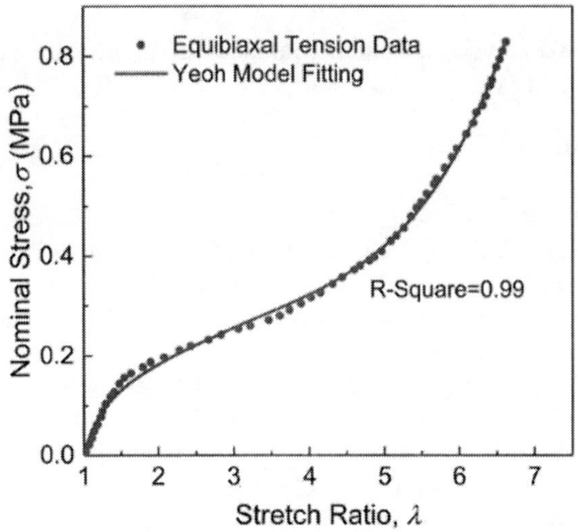

Figure 2. The relationship between nominal stress and stretch ratio under equi-biaxial tension.

3.2 Electromechanical results

The stress-stretch relationship for the dielectric elastomer film subjected to a combination of external force and voltage is expressed as [9]:

$$\frac{P}{\pi DH}+\varepsilon_0\varepsilon_r\left(\frac{\Phi}{H}\right)^2\lambda^3=\sigma(\lambda), \tag{2}$$

where P is the external force for pre-stretching the membrane, D and H is the initial diameter and thickness of the membrane, ε_0 is the free-space permittivity (8.85×10^{-12} F/m), ε_r is the relative dielectric constant (4.03 for VHB 4905), Φ is the voltage applied on the DE film. The variation of stretch ratio as a function of applied voltage under different pre-stretch ratios (λ_P) can be calculated based on the equation (1) and equation (2), as is shown in figure 3.

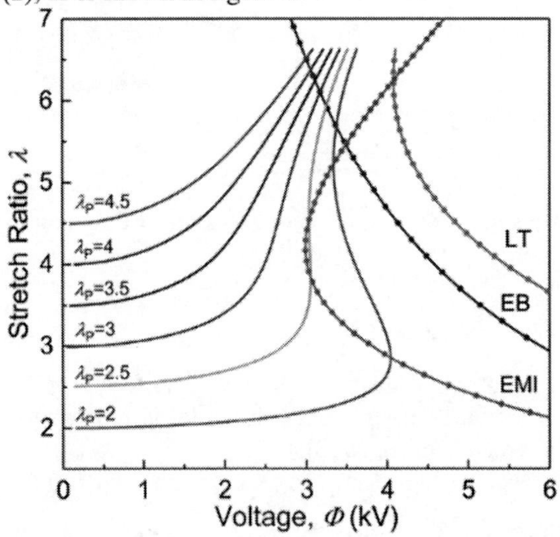

Figure 3. The theoretical λ-Φ relationship under various pre-stretch ratios from 2 to 4.5.

Because the DEA is essentially a deformable capacitor, the charges stored on the DE film, Q, are given as:

$$Q = \frac{\varepsilon_0 \varepsilon_r \lambda^4}{H} \Phi . \tag{3}$$

For the dielectric elastomer, over high voltage will lead to the electrical breakdown (EB) of the material, which is given as:

$$\Phi = E_{EB} H \lambda^{-2} , \tag{4}$$

where E_{EB} is the electrical breakdown field. It is noted that the E_{EB} of VHB elastomer is related to the stretch of the membrane, which is E_{EB}=30.6$\lambda^{1.13}$ (MV/m) [9]. Combining equation (3) and equation (4), the EB boundary line can be numerically derived, as shown in figure 3.

As voltage increases, the deformation of the active region is solely supported by the Maxwell stress, and further increase of the voltage will lead to the compressive stress in the planar direction of the membrane, which results in the wrinkles of the DE film and damages the function of the actuator. This failure mode, also known as loss of tension (LT), can be solved by setting P=0 in equation (2), which is shown in figure 3.

When voltage is applied, the elastomer membrane thins down because of the Maxwell stress, which results in even higher electric field to further deform the elastomer at the same voltage. This kind of positive feedback may cause extremely high electrical field and bring about electrical breakdown, which is known as electromechanical instability (EMI) failure. The corresponding boundary line for EMI failure can be calculated as dΦ/dλ=0 with equation (2), which is given as [9]:

$$3\varepsilon_0 \varepsilon_r \left(\frac{\Phi}{H} \right)^2 \lambda^2 = \frac{d\sigma(\lambda)}{d\lambda} , \tag{5}$$

and the numerical calculation results combining equation (3) and equation (5) are shown in figure 3.

In figure 3, The intersection points of the actuation λ-Φ plots with failure limiting plots determine the maximum actuation strain that can be attained, which is defined as $\gamma_{max} = 100\% \times (\lambda_{max} - \lambda_P)/\lambda_P$, where λ_{max} is the stretch ratio at the intersection points. From figure 3, it can be seen that as the pre-stretch ratio increases, the EMI failure can be gradually avoided, and EB failure becomes the only determining factor of γ_{max}. The theoretical values of γ_{max} as a function of pre-stretch ratio are plotted in figure 4. It can be seen that discontinuity emerges at the point of λ_P=2.6, where EMI failure is just refrained, and γ_{max} reaches the peak value of 132%. The experimental data were also presented in figure 4, and it can be seen that the theoretical calculation qualitatively follows the trend of experimental results. The optimum pre-stretch ratio is experimentally measured to be about 3.25, with the γ_{max} of 90%. The experimental and theoretical values of breakdown voltage at various pre-stretch ratios are shown in figure 5. It can be seen that the theoretical results matches experiments well, and the breakdown voltage decreases with the increase of the pre-stretch ratio.

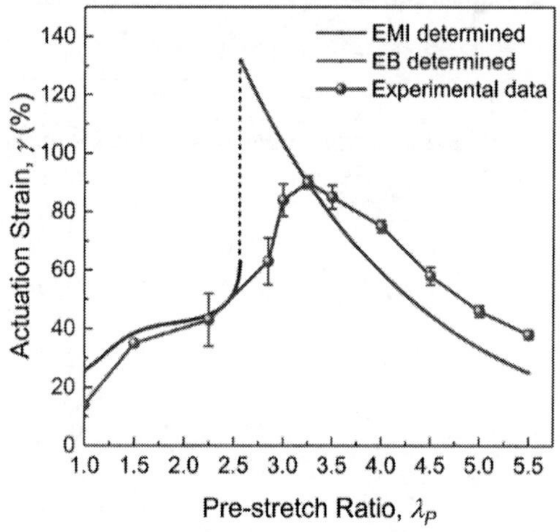

Figure 4. The actuation strain as a function of pre-stretch ratio.

It is noted that in figure 4 and figure 5, even though the basic features of experimental actuation tests are reproduced by the theoretical calculation, obvious errors exist between the experimental data and theoretical results. Two reasons may account for this kind of mismatch. Firstly, the theoretical analysis in this work neglects the viscoelasticity of the VHB elastomer, which has significant impacts on the electromechanical response of the actuator. Secondly, the influence of the electrode (carbon grease) is not considered. During the stretching process of the membrane, the electrodes coated on the membrane also endures large deformation, therefore the viscosity and electrical conductivity of the electrodes should dramatically affect the performance of the actuator. Even though the theoretical calculation in this work is simply based on an ideal dielectric elastomer model, it still sheds light on the performance optimization of DEA.

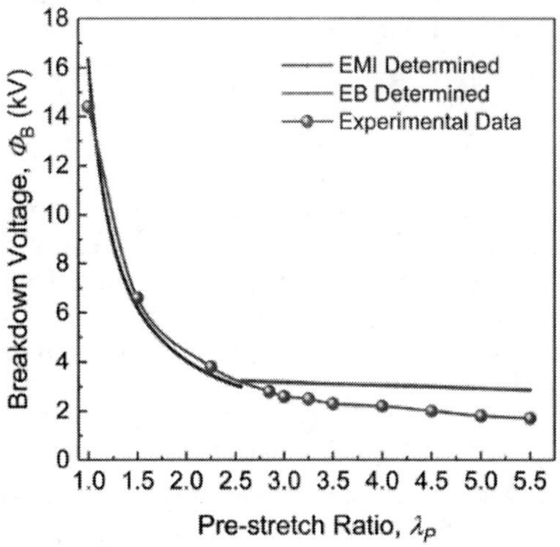

Figure 5. The breakdown voltage as a function of pre-stretch ratio.

4. Conclusions

Based on the theoretical calculation and experimental tests, the maximum actuation strain of DEA (VHB 4905) increases with the pre-stretch ratio at the initial stage, while decreases when the pre-stretch ratio is over large. The optimum pre-stretch ratio is reached when the EMI failure is just refrained, which is experimentally measured to be about 3.25 with the maximum actuation strain of 90%.

Acknowledgments

This work was supported by the Strategic Research Foundation Grant-Aided Project for Private University from Ministry of Education, Culture, Sports, Science and Technology, Japan.

References

[1] Shian S, Bertoldi K, Clarke D R 2015 *Adv. Mater.* **27** 6814-9
[2] Wax S G, Sands R R 1999 *Proceedings of the SPIE International Symposium on Smart Structures and Materials: Electro-Active Polymer Actuators and Devices* pp 2-10
[3] Opris D M, Molberg M, Walder C, Ko Y S, Fischer B, Nüesch F A 2011 *Adv. Funct. Mater.* **21** 3531-9
[4] Pelrine R, Kornbluh R, Pei Q, Joseph J 2000 *Science* **287** 836-9
[5] Huang J, Shian S, Diebold R M, Suo Z, Clarke D R 2012 *Appl. Phys. Lett.* **101** 122905
[6] Koh S J A, Li T, Zhou J, Zhao X, Hong W, Zhu J, Suo Z 2011 *Journal of Polymer Science Part B: Polymer Physics* **49** 504-15
[7] Bozlar M, Punckt C, Korkut S, Zhu J, Chiang Foo C, Suo Z, Aksay I A 2012 *Appl. Phys. Lett.* **101** 091907
[8] Ward I M, Sweeney J 2012 *Mechanical properties of solid polymers* (John Wiley & Sons)
[9] Koh S J A, Keplinger C, Li T, Bauer S, Suo Z 2011 *IEEE/ASME Trans. Mechatron.* **16** 33-41

IOP Conf. Series: Materials Science and Engineering **639** (2019) 012009 doi:10.1088/1757-899X/639/1/012009

Sustainable activity of engineering education and research of association for advanced education in Asia

Toshio Wakabayashi[1,2], Noriyuki Komine[1], Hidekazu Tsuji[1], Shigenori Tomiyama[1], Tadashi Ariga[1], Yoshiaki Moriya[1] and Toshio Iijima[1]

[1] Association for advanced education in Asis
3-8-13 Atsugi City Kanagawa 243-0037 Japan

[2] wakaba@rg7.so-net.ne.jp

Abstract. In this paper we describe activities of members in the association for advanced education in Asia (AAEA). It is a specified non-profit organization in Japan. The members are in activity at National University of Laos, Laos, King Mongkut's Institute of Technology Ladkrabang, Thailand and universities in Vietnam, Malaysia and others. They have collaborated with their counterparts such us professors and students on specialized education and research. A lot of activities are useful and effective for our sustainable fields.

1. Introduction

In 2006, an association for advanced education in Asia (AAEA) was established as a specified non-profit organization in Kanagawa, Japan. The mission is to build the creative and sustainable society through collaborative activities concerning education and research. We set up (1) to do collaboration with research and events, (2) to do human resources development by educational collaboration, (3) to support international exchange between Japan and Asian countries etc.

In this paper our activities are summarized in each area. In section 2, the robot contest which has been done in Laos and in section 3, the collaborative researches and activities in Laos and Thailand are described. In section 4, the research activities which have been performed in Vietnam and Malaysia are mentioned. Their outputs and problems are discussed in detail. They are useful and effective activities for collaborations. The effort is continued in order to realize a sustainable institute in the countries.

2. Robot contest in Laos

2.1. Purpose of the robot contest in Laos A subsection

The first robot contest in Lao was held in Vientiane by organizing by the Faculty of Engineering of National University Laos (NUOL) in 2002. The international robot contest was held on 2003, 2004, 2005 and 2006 respectively, together with Japan and Thailand teams. The 14th robot contest 2018 was held in the Sokpaluang campus of NUO in Vientiane. The Laos robot contests provide three purposes as the International interchange of students, the Human resources of creative specialists, and the acquisition of robot technology.

The international robot contest provide in student interchange with the university student of Laos, the vocational school student and the Laos neighbourhood students in Asian countries. The students will be able to study for high knowledge and technology by making a robot under the rule of the robot

IOP Conf. Series: Materials Science and Engineering **639** (2019) 012009 doi:10.1088/1757-899X/639/1/012009

contest. The education by using robot technology not only provides in conventional technical knowledge but also grows the creative thinking technology. In addition, a mainly microcomputer, electronics, machine, electricity, and information technology are indispensable for teaching robot technology. The students have been challenging the programming of microcomputer with interesting in engineering, science and technology. This is considered as an education of PBL with integrating the skill of computer engineering, solving problem, and team working.

2.2. Purpose of the robot contest workshop

The robot committee, the Faculty of Engineering of NUOL organized a robot contest workshop on applied microcomputer to robot design for students from technical college in Lao including the Faculty of Engineering and the Faculty of Sciences of NUOL in collaboration with the NPO/ AAEA. During the workshop, the participants learned about fundamental knowledge of electrical circuits, transistor circuits and motor control circuits with driving microcomputer for robot design from Japanese robot expert at the NPO/AAEA. The Japanese robot expert presents the embedded technology with combined the software engineering and hardware engineering by using practical robot. They have learned various techniques fact that they faced to use microcomputer equipment in producing their robots at the workshop.

2.3. Achievement and the expected results

It is one of achievement that the students have obtained various abilities and ingenious robot technology in the robot contest match [1]. The robot contest workshop in collaboration with NPO/AAEA is expected as an education of PBL with integrating the skill of computer engineering, problem solving, and team working for students of technical college in Lao including the Faculty of Engineering and the Faculty of Sciences of NUOL in Laos.

2.4. Robot contest 2018 in Lao

The contest is a game that an autonomous robot carries one colour box to the target position with cooperating manual robot, and to compete for a score according to the rule of the robot contest 2018. In the contest, the students of participated team prepared a manual robot and an autonomous robot to participate the robot contest. The student is operating the manual robot to carrying a cube box with colour to an autonomous robot. The autonomous robot should be automatic carrying to target colour box stand by tracing the black line, and the robot should be setting the target box stand by detecting a colour. The student team will get a score if the autonomous robot will be carrying to target colour box stand successfully. In the contest, two fields are set up as shown in figure 1 and 13 teams in each field participated.

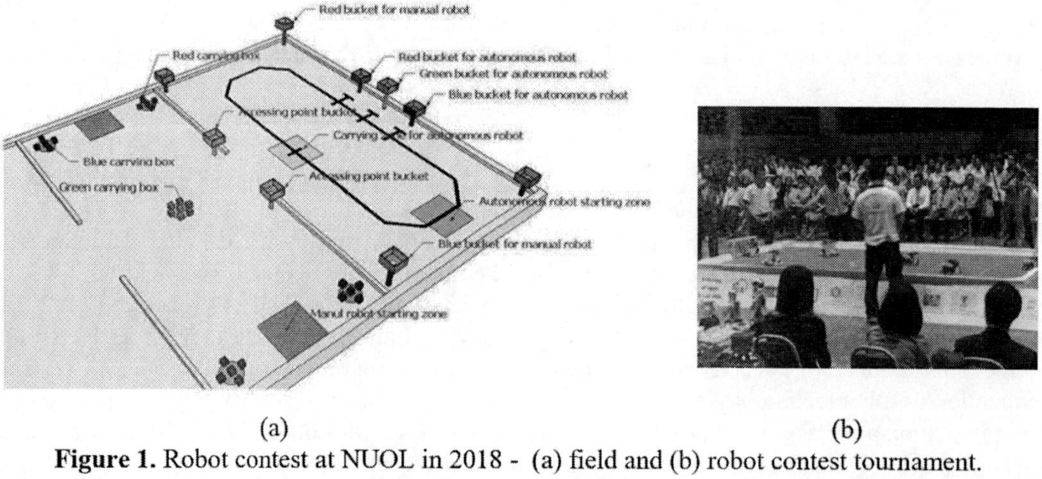

(a) (b)

Figure 1. Robot contest at NUOL in 2018 - (a) field and (b) robot contest tournament.

IOP Conf. Series: Materials Science and Engineering **639** (2019) 012009 doi:10.1088/1757-899X/639/1/012009

3. Collaborative research and activity in Laos and Thailand

3.1. Development of a renewable energy source and eco car in Laos

It is important to develop a renewable energy source and a new type of car in order to build a sustainable society. Especially, in Laos petroleum is not produced and its price very expensive compared with other economic price. Research on production of biomass gas from tree branch and corn core has been done as a challenge topic and succeeded to make small experimental plant. This result will be represented in near future.

Research on design and manufacture of electric vehicle is taken notice by government because Laos has much electric energy resource by hydro many power stations and petroleum is not produced in Laos as mention above. It is very important to bringing up human resources in university to develop electric vehicle in own country. Four departments, namely mechanical, electric, electronic and information ones cooperated, and first electric vehicle was designed and manufactured in Laos as shown in figure 2 [2]. On the other hand research on development and simulation of electric tricycle was performed in KMITL because exhaust gas from the tricycle (TukTuk) is one of the air pollution causes in Bangkok and big cities in Thailand [3].

Research on application of microbubble to purification of waste water .is recently interested one in the world [4]. Fortunately manufacturing the experimental apparatus and nozzle which is important part for generation microbubble are not difficult in workshop of the university. NUOL started the application research of the microbubble to purification of waste water in 2016, and KMITL did application of it to washing fruit surface in 2017. They are shown in figure.3. Associate professor and lecturer of NUOL will present obtained result in this conference.

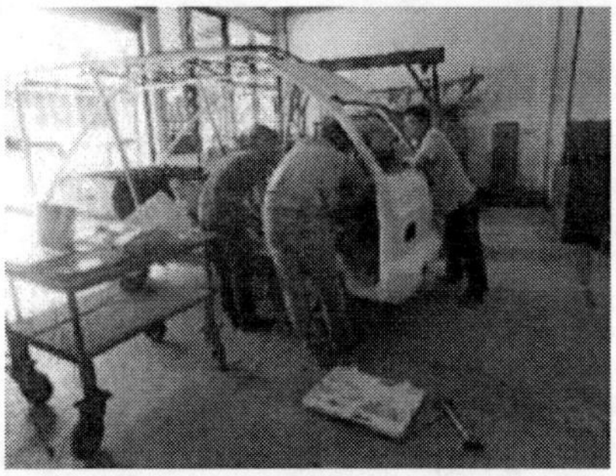

Figure 2. Manufacturing electric vehicle in NUOL

(a)

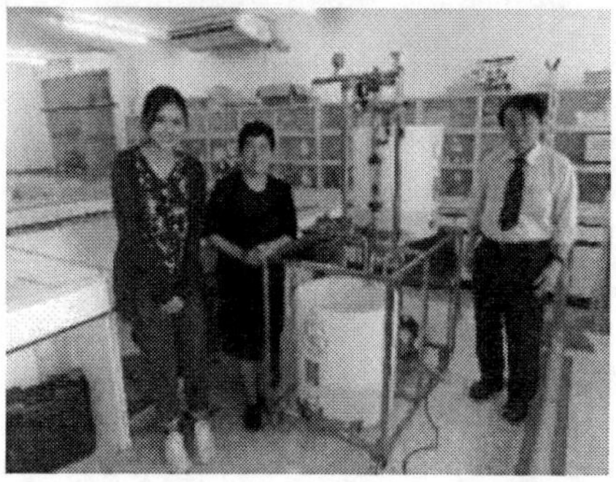

(b)

Figure 3. Experimental apparatus of microbubble – (a) NUOL and (b) KMITL.

Figure 4. Research discussion.

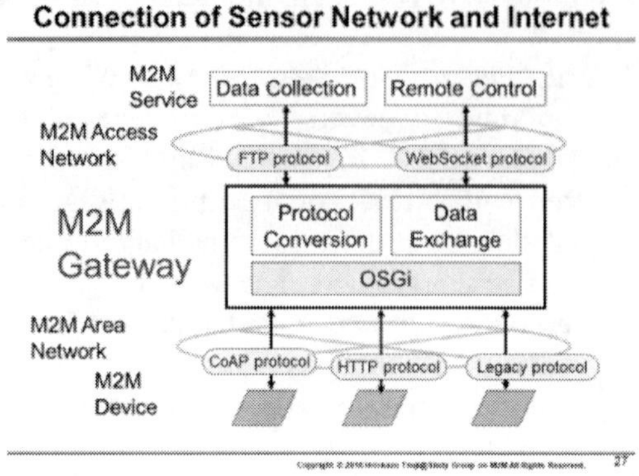

Figure 5. Special lecture content: Connection of sensor network and internet.

3.2. Research on microwave propagation

In Laos and Thailand, humidity is very high and temperature also very high. Therefore, the water vapor of the air increases. The radio waves of a microwave band decrease in proportion to water vapour content. The relation between the satellite TV receiving by a small aperture antenna and the water vapour content has been discussed experimentally and theoretically since 2002 [5].[6]. The figure 4 shows the scene of research discussion. Research on modelling and measurement of airplane flutter phenomenon on television signal has been done in Thailand. The airplane flutter phenomenon was considered theoretically and experimentally [7]. As research on propagation needs the human power, research funding and measuring devices in order to do, such research is not done actively in Laos and Thailand.

3.3. Lecture and research on M2M/IoT

Special lecture on Introduction to M2M/IoT was done by Dr. H. Tsuji at NUOL on Nov. 1, 2016. Many students and the faculty members attended and there were many questions and answers. Also, he had the special lecture on M2M/IoT at KMITL on Nov. 9, 2016. The faculty members and students attended the lecture (figure 5 for content) and many questions answers were made. Many attendees were interesting in research on M2M/IoT at both faculties of engineering. After the lecture research discussions were done and some have been continued through the internet. In near future, some papers will be submitted at the conference.

4. Research activity in Vietnam and Malaysia

4.1.Research on low power wireless monitoring system

Shrimps are very sensitive with water condition changing. Shrimps can get disease if changing is not processed, even though the changing is small in a short time. VASEP (Vietnam Association of Seafood Exporters and Producers) reports that the successful harvest probability is only 66%. Therefore, a system replacing human always monitors water quality can reduce risks for farmers. The objective of the research is to design a wireless monitoring system that monitors 24/7 shrimp water quality and warns farmer abnormal conditions. The overall architecture of the system is shown in figure 6. The experiment in the sites is shown in figure 7.

It has been performed to develop an integrated environment including data management, web server, web client application, smart phone application to connect farmers, technical engineers, companies, government together. Farmers can share their water condition and learn from others. Engineers and companies can access monitoring data, give farmer advice or recommend. Base on statistical analysis

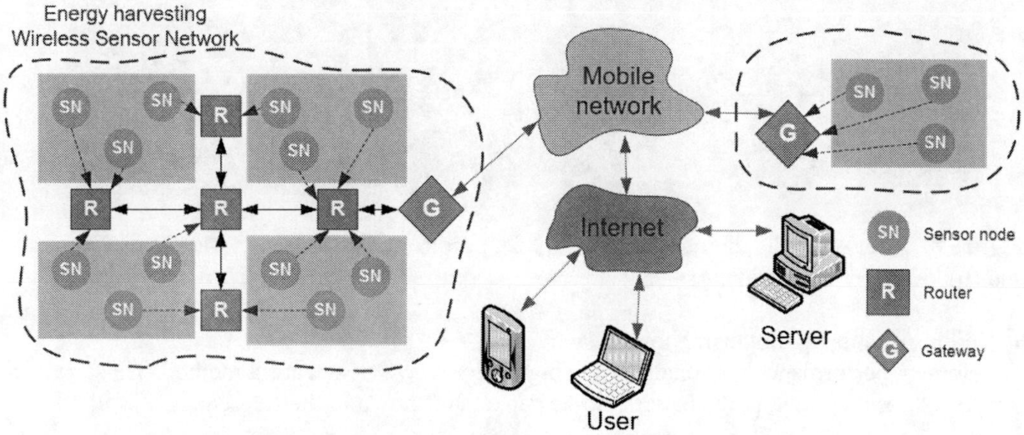

Figure 6. Overall architecture of shrimp pool water quality monitoring system

(a) (b)

Figure 7. Experiment in the sites – (a) Can Gio site and (b) Soc Trang site

of monitoring data, the government has an overview, can make a plan and has some directional guidelines The research will be able to sustain itself in future by conducting research and deployment for companies with different places and multiple shrimp ponds in the south of Vietnam due to the ability to increase their shrimp production

4.2. Collaborative research at universities in Malaysia

Research on the material Science and joining science has been done at the universities such as Universiti Malaya(UM), Universiti Malaya Perlis(UMP), Universiti Kebangsaan Malaysia(UKM), and Universiti Sains Malaysia(USM). Main research themes are as follows;

- Joining technology of various ceramics (Al2O3, AlN)
  Ceramics (aluminum oxide, aluminum nitride) and copper alloy are joined using special filler metal (porous Cu,+Ag+Cu+Ti). Then, the joining interface is analyzed to evaluate the joining property.
- New type of semiconductor cooling device
  Combining copper plate and porous metal (Cu,Ni) to produce a new type of semiconductor cooling device
- Joining of aluminum alloy by laser heat source
  Aluminum alloy is joined (brazed) using the fiber laser.
- Addition of special metal (porous metal) to solder joint
  Construct a new joining interface for joining strength and heat resistance of soldered joints
- Diamond particle joining technology
  The diamond particles are fixed on the stainless steel plate by using a special joining material.

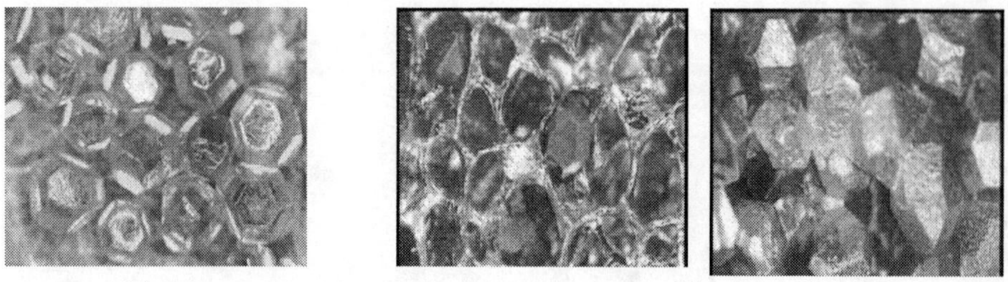

(a) (b)

Figure 8. Fixation of the diamond particle by Cu porous metal (brazing) - (a) Before joining and (b) After joining (brazing)

- Joining of copper alloy using microwave
  Joining performance was examined using a microwave as a heating method. Those research results are reported in the papers. Some papers are shown in the references [8], [9]

4.3. Collaborative research and lectures in MJIIT

MJIIT was established by Malaysia and Japanese governments in Kuala Lumpur in 2011 and was a faculty of Universiti Technologi Malasia (UTM). Vision is leading in cutting edge technology education and research Also missions are (1) providing Japanese style engineering education blended with Malaysia distinctiveness for sustainable industry and society and (2) leading in academic and research excellence in Electronics, Precision, Environmental & Green Engineering and Management of Technology. Most noteworthy is education and research based on both Malaysia and Japanese styles and international development of human resources in ASEAN.

MJIIT is constituted of 12 programs of PG and UG in niche areas. Departments are (1) Electronics System Engineering(UG, PG), (2) Mechanical Precision Engineering (UG,PG), (3) Chemical Process Engineering (UG), (4) Environmental Engineering and Green Technology (PG) and (5) Management of Technology (PG). Table 1 is numbers of students, and graduates in 2018. T. Wakabayashi was a professor from 2011 to 2014.

MJIIT had collaborative research with KMITL and the title is a study on planar antennas elder care systems and communication networks in sustainable society. The outputs were four papers and five symposium papers [10], [11], [12]. The central frequency of the band notch characteristic is obtained by changing the slit position, length and width. The figure 9 shows the return loss characteristics of the UWB planar antenna (.L_d=21, L_g=24, W=21, L_1=16, L_2=12.8, F_1=0.4, h_c=0.035, h_s=1 mm, ε_r=4.2, *e*=0.6,) Other results of the collaborative research are subscribed in the paper [13].

Table 1. Numbers of students and graduation

	students	graduates
UG	783	365
PG	455	167
total	1238	532

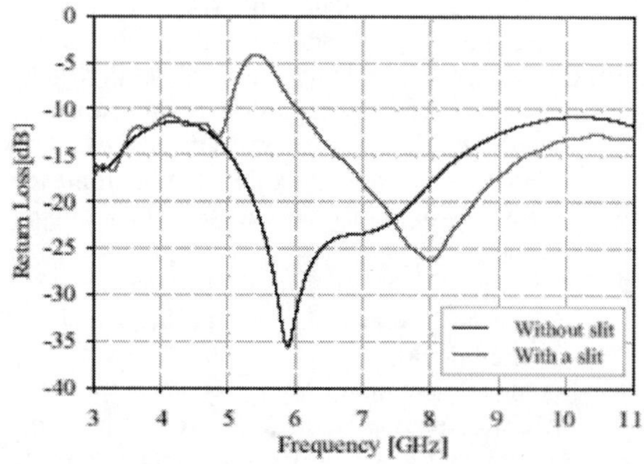

Figure 9. Return loss of UWB planar antennas

5. Conclusion

Association for Advanced Education in Asia(AAEA) is a small specialized non-profit organization with 12 members. Prof. Dr. T. IIJIMA is the president of NPO/AAEA. The members have been independently collaborated with their counterparts with their own policy to realize sustainable society since 2006. Those collaborative research and activities will be continued in order to achieve the missions of the association.

Finally, N. Komine, T. Iijima, Y. Moriya, H. Tsuji, S. Tomiyama, T. Ariga and T. Wakabayashi describe the parts of 2, 3A, 3B, 3C, 4A, 4b and 4c, respectively. This is summarized by T. Wakabayashi.

6. References

[1]"Home news Friday Aug.19, 2016", Vientiane Times

[2] Nakamura T et al. 2009 Development and Performance of an Electric Trike with In-Wheel Motors, *Proceedings of the 24th International Electric Vehicle Symposium and Exposition* (Stavanger, Norway) Paper No. 5190523 pp.1-6

[3] Nakajima T et al.2009 A Study on Driving Performance of an Electric Trike *Proceeding of Japan Society on Simulation and Technology* (Tokyo) . pp. 108-111

[4] Hediki T 2014 Micro and Nanobubble, Pan Stanford Publishing Pte. Ltd.

[5]Satahach P,.Southisombath K, Mangnomek S, Ikeda S, Moriya Y and Takagi K 2000 The Relationship between Satellite Broadcasting Signals and Rain Characteristics in Laos Proc 2000 *Asia-Pacific Symposium on Broadcasting and Communications*(Thailand) .

[6] Southisombath K, Wakabayashi T and Moriya Y 2005 Observation of Water Vapor Attenuation at Ku Band in Tropical Region *IEICE Trans.* **E88-B** 6 pp 2446-2448

[7] Wongkeeratikul A, Supnithi P, Noppanakeepong S, Leelaruji N and Hemmakorn N 208 Modeling and Measurement of Airplane Flutter Phenomena on TV Broadcasting Signal *IEEE Transactions on Broadcasting,* **54** 2 pp173-181.

[8] Wongkeeratikul A, Supnithi P, Noppanakeepong S, Leelaruji N and Hemmakorn N 2008 Modeling and Measurement of Airplane Flutter Phenomena on TV Broadcasting Signal *IEEE Transactions on Broadcasting* **54** No.2 pp. 173-181

[9] Ariga T ,et al 2017 Influence of BA4047 filler addition through Mamdani fuzzy logic optimization for double-sided T-joint welding of aluminum alloys using low-power fiber laser *The Int.Journal of Advanced Manufacturing Technology* (Springer–London)

[10] Arig T et al 2017 A novel method of brazing Cu/Cu-7.0Ni-9.3Sn-6.3P/Cu using microwave – Hybrid –heizung *Materialwissenschaft und Werkstofftechnik,* issue 3 .pp 1-7

[11] Vongsack S, Lamultree S, Osklang P, Phongcharoenpanich C, Kosulvit S , Hamamoto K, and and Wakabayashi T 2012 A UWB Bidirectional Rectangular Ring Antenna Fed byCDM with a Rod and Ridges for Constant Beam Direction; *HPCIJ* **2012** ID691214 12pages

[12] Tangjitjetsada M, Rakluea P, Anantrasirichai N, Benjangkaprasert C and Wakabayashi T 2014 CPW-fed Compact Thin-Film UWB Antenna with Dual Band Notched Characteristics *,IEEJ Trans.* 9 pp421-426 .

[13] Fatimah S, Wakabayashi T and Abd Raihim M K 2014 Simulated Characteristics of UWB Planar Antennas Characterized by a Slit on an Elliptical Element and Ground Conductor on a Dielectric Substrate *Frontier of Applied Plasma Technology* **7** 2 pp.77-82

IOP Conf. Series: Materials Science and Engineering **639** (2019) 012010 doi:10.1088/1757-899X/639/1/012010

A comparison of using Euclidean and road distance to solve facility location problems

J Pittayarugsarit and Y Song

Department of System Management, Fukuoka Institute of Technology, Fukuoka, Japan

mjm18201@bene.fit.jp and song@fit.ac.jp

Abstract. To solve any facility location problem, the distance between facility locations and demand nodes is a crucial factor. In previous studies, the distance used most frequently is the Euclidean distance, despite that the road distance should be used when we travel by car or on foot. In this paper, we focus on the difference of the results by using the Euclidean distance and road distance to solve facility location problems. We apply the data of Fukuoka City to the p-median model and the p-center model, and compare the optimal solutions with Euclidean distance and road distance. Numerical results suggest that wide difference exists in some cases.

1. Introduction

Facility location decisions are an important factor in strategic planning for achieving commercial success and competitive advantages. Location decisions also arise in a wide range of public and private sectors. Since the cost of property holding and facility construction are high, facility location or relocation projects are a long-term investment. Thus, poor location decisions may lead to increased costs and decreased competitiveness. In other words, the success or failure of any facility depends in part on the location that was chosen. The decision makers must choose facility location that will not only perform well according to the current circumstance, but will continue to be profitable throughout the facility's lifetime, even as environmental factors, populations, and market trends change [1]. Hence, for finding beneficial or optimal facility locations, the facility location problem is raised.

For solving the facility location problem, the distance between facility locations and a demand node (e.g., residential communities, industrial estates) is one of the crucial factors. Currently, the distance can be obtained by measuring the length of the Euclidean distance and the road distance between locations. Note that the road distance is always greater than or equal to the Euclidean distance. The Euclidean distance is a length of a straight line between two locations, and the road distance is the length of the road from one location to another location that is used in real life, i.e. the exact traveling distance by car. Even so, on related works, concerning the location problem, the Euclidean distance is used most frequently [2].

For that reason, this study concentrates on the different results obtained between using the Euclidean distance and the road distance to solve the facility location problem. In this paper, the P–median and P–center problem, classical types of the facility location problem, are considered to solve the problem. This study uses Gurobi optimizer to demonstrate the simulation results.

The rest of this paper is divided in three main parts as follows: Section 2 illustrates the methods and formulation of the p-median and p-center problem. Then, the simulation results tested in reallocation are presented in Section 3. Finally, the conclusion and future work are discussed in Section 4.

2. Problem formulation

The *p*-median and *p*-center problem, both have the problem of selecting the specified number (p) of optimal location from various candidate locations to locate a facility without considering the area that the facility can cover. However, the method of these two problems is different.

In order to formulate these problems, we set $i \in \mathrm{I}$ to represent a demand node, and the facility can be located at candidate location $j \in \mathrm{J}$. Introducing the following notations:

c_{ij} = Distance from demand node i to facility location j

p = Number of facilities to locate

$$x_{ij} = \begin{cases} 1, \text{ if demands at node } i \text{ are served by the facility at site } j \\ 0, \text{ otherwise} \end{cases}$$

$$y_j = \begin{cases} 1, \text{ if the facility is located at candidate location } j \\ 0, \text{ otherwise} \end{cases}$$

2.1. *p-median problem formulation*

The method to select the optimal location of the median problem is by minimizing the total distance from each demand node to the nearest facility which can serve those demands.

With the notations mentioned above, the P–median problem model can be formulated as follows:

$$\text{minimize} \quad \sum_{i \in I} \sum_{j \in J} c_{ij} x_{ij} \tag{1}$$

$$\text{subject to} \quad \sum_{j \in J} x_{ij} = 1 \qquad \forall i \in I \tag{2}$$

$$\sum_{j \in J} y_j = p \tag{3}$$

$$x_{ij} \leq y_j \qquad \forall i \in I; j \in J \tag{4}$$

$$x_{ij} \in \{0,1\} \qquad \forall i \in I; j \in J \tag{5}$$

$$y_j \in \{0,1\} \qquad \forall j \in J \tag{6}$$

The objective function (1) minimizes the total distance between each demand node and the facility which it serves. Constraint (2) states that all of the demand at node i must be served by the facility at some candidate site j. Constraint (3) ensures that only p facilities can be located. Constraint (4) confines that demands at node i cannot be served by the facility at node j unless the facility is located at candidate site j. Constraints (5) and (6) are the binary constraints.

2.2. *p-center problem formulation*

The center problem selects the location by minimizing the maximum distance which is the distance from a demand node to the nearest facility.

With the notations mentioned above, the P–center problem model may be formulated as follows:

$$\text{minimize} \quad Z \tag{7}$$

$$\text{subject to} \quad \sum_{j \in J} x_{ij} = 1 \qquad \forall i \in I \tag{8}$$

$$\sum_{j \in J} y_j = p \tag{9}$$

$$x_{ij} \leq y_j \qquad \forall i \in I; j \in J \tag{10}$$

$$\sum_{j \in J} x_{ij} c_{ij} \leq Z \qquad \forall i \in I \tag{11}$$

$$x_{ij} \in \{0,1\} \qquad \forall i \in I; j \in J \tag{12}$$

$$y_j \in \{0,1\} \qquad \forall j \in J \tag{13}$$

Equation (7), the objective function is to minimize the maximum distance between the demand node and the nearest facility (Z). Constraint (8) ensures that all of the demands at node i must be served by facility at some candidate site j. Constraint (9) confines that only p facilities can be located. Constraint (10) stipulates that demands at node i cannot be served by a facility at node j unless the facility is located at candidate site j. Constraint (11) states that the distance between any pair of demand node i and facility at site j which can served must be less than or equal to the maximum distance between the demand node and the nearest facility (Z) Constraints (12) and (13) are the binary constraints.

3. Numerical results.

This section presents two simulations of using the road distance and Euclidean distance applied to the p-median and p-center problems. The optimal solutions of the simulation were performed with Gurobi Optimizer.

The simulations test with a real location in Japan. The first simulation test with Chuo Ward, Fukuoka City in Fukuoka Prefecture, Japan, by considered 12 primary schools as demand nodes and 22 gas stations as candidates of facility locations. The second simulation tests with a larger area, Fukuoka City in Fukuoka Prefecture, with 146 primary schools and 178 gas stations as demand nodes and candidates of facility locations as well. The road and Euclidean distance data are obtained from Google Map [3] and Benricho website [4].

To indicate how the result using the road distance and using the Euclidean distance is different numerically, we considered following these steps:

(1) Obtain an optimal location using the road and Euclidean distance, denoted by R and E, respectively.
(2) Optimal location of using road and Euclidean distances, denoted by $V(R)$ and $V(E)$.
(3) Compute the objective function value of Euclidean distance's optimal location but using road distance, denoted to $V_R(E)$.
(4) Calculate the percentage difference (%Diff) between $V(R)$ and $V_R(E)$.

Note that the objective function value of using road distance is equal to $V(R)$.

The simulation results of applied various lengths in the p–median problem and the p-center problem is shown in table 1 for Chuo Ward area, and table 2 for Fukuoka City area. The example of located facilities and its covered demand node in Chuo Ward is shown in figure 1 for the p-median problem result, and figure 2 for the p-center problem result, where it was determined that $p = 3$.

Table 1. The result of applied the p-median and p-center problem with Chuo Ward.

p	p-median problem				p-center problem			
	Optimal Value (km)		$V_R(E)$ (km)	%Diff.	Optimal Value (km)		$V_R(E)$ (km)	%Diff
	$V(R)$	$V(E)$			$V(R)$	$V(E)$		
1	25.25	19.6	27.35	8.30%	3.80	2.50	3.90	2.60%
2	17	12.9	18.15	6.80%	2.60	2.00	3.40	30.80%
3	14	10.4	15.45	10.40%	1.67	1.40	2.30	35.30%

Table 2. The result of applied the p-median and p-center problem with Fukuoka city.

p	p-median problem				p-center problem			
	Optimal Value (km)		$V_R(E)$ (km)	%Diff.	Optimal Value (km)		$V_R(E)$ (km)	%Diff
	$V(R)$	$V(E)$			$V(R)$	$V(E)$		
1	1103.11	882.91	1103.11	0.00%	25.07	13.86	29.98	19.57%
2	852.31	662.79	868.71	1.92%	16.15	11.37	34.32	112.49%
3	634.79	492.62	671.25	5.74%	13.73	8.87	49.35	259.35%
4	572.04	439.59	609.98	6.63%	10.16	7.81	49.35	385.66%
5	526.78	397.02	581.41	10.37%	8.64	7.14	25.44	194.43%

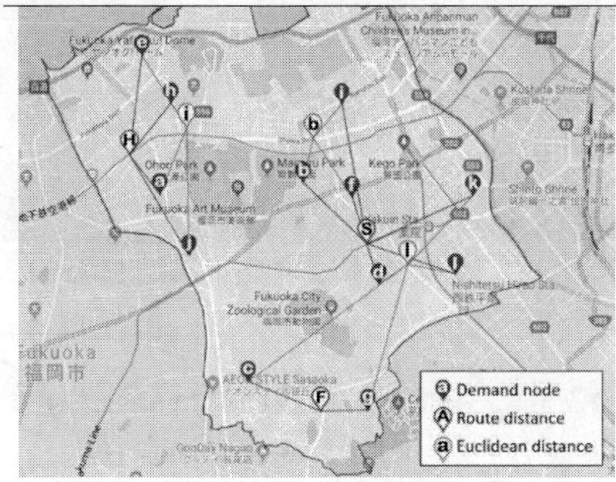

Figure 1. The result of 3-median problem.

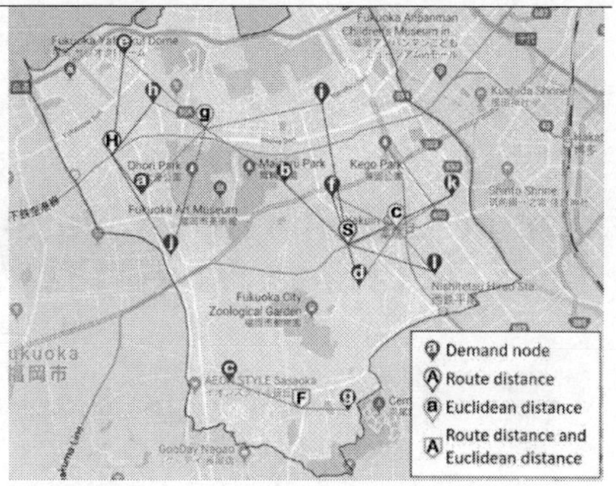

Figure 2. The result of 3-center problem.

From the results, the percentage difference of using road distance and using Euclidean distance is rather high. Moreover, it not only causes a difference in optimal location but also causes the number of the covered demand nodes of each facility too. To describe the reason why this %Diff is rather high, we decided to compute to the correlation coefficient and the average of %Diff of all road distances and Euclidean distances as shown in table 3. The correlation coefficients are positive and close to 1, meaning that there is high correlation between the two types of distances. That is, if the road distance gets larger the Euclidean distance will get larger too. Furthermore, from the average of %Diff of the Euclidean and road distance, it is indicated that the length of road distance in these simulations, are greater than the length of Euclidean distances by an average of 24.6% for Chuo Ward area, and 21.3% for Fukuoka city area.

Table 3. The correlation coefficient and the average difference percentage of the road and Euclidean distances in various area.

	Chuo Ward	Fukuoka city
Correlation Coefficient	0.975	0.945
Average %Diff.	24.6%	21.3%

4. Conclusion and future work

This paper compared the differences in optimal solutions of facility location problems by using the road distance and the Euclidean distance. We applied the data of Fukuoka City to two classical facility location models: the p-median and p-center problems. The numerical results have shown that the percentage difference between using road distance and Euclidean distance is rather high in some cases, especially for the p-center problem. For future work, we intend to apply this idea to check other location models.

References

[1] Owen S H and Daskin M S 1998 Strategic facility location: A review *European Journal of Operational Research* **111** 423–47
[2] Liu L and Zhou J 2016 A road-like demand location problem based on squared-Euclidean distance *2016 International Conference on Logistics Informatics and Service Sciences* 16673386
[3] Google 2019 Google map Online: https://www.google.co.jp
[4] Everybody's knowledge 2019 Measuring the distance between two points Online: https://www.benricho.org
[5] Daskin M S 2013 *Network and Discrete Location 2nd Edition* (New Jersey: John Wiley & Sons)
[6] Kubo M, Pedroso J P, Muramatsu M and Rais A 2016 *Mathematical Optimization: Solving Problems using Gurobi and Python 1st Edition* (Japan: Kindaikagaku)

IOP Conf. Series: Materials Science and Engineering **639** (2019) 012011 doi:10.1088/1757-899X/639/1/012011

Factors Affecting Decision to Select Government R&D Support Projects for SMEs

Rattanasak Wannakul[1], Supachai Pathumnakul[1] and Sirawadee Arunyanart[1,*]

[1] Supply Chain and Logistics Systems Research Unit, Department of Industrial Engineering, Faculty of Engineering, Khon Kaen University, Khon Kaen, Thailand

sirawadee@kku.ac.th

Abstract. Nowadays, Thailand industry is still lacking technological and scientific application for making competitive capacity and value-added products. For this reason, the government needs to initiate special projects for supporting industrial research and development (R&D). In order to gain the highest benefit of the government budget, the process of selecting projects to be supported is considered important since it has an effect on the future prosperity of the country. Multiple evaluation factors, which may be conflicted by nature, are required for selecting suitable projects among several alternatives. The objective of this research is to study the factors that should be used in appraising the government R&D support projects for SMEs. Analytical Hierarchy Process (AHP) is applied to determine 3 main factors which consist of 9 sub-factors used for selecting suitable projects in order to receive funding from the government. The result shows that the important weight of the 3 main factors; entrepreneur, researcher, and project details are 0.565, 0.235, and 0.200 respectively.

1. Introduction

Understanding of the role of innovation in the process of small and medium-sized enterprises (SMEs) has initiated to attract research consideration over the last few years [1]. Scientific collaboration network is an academic network where hubs are researcher and spokes are co-authorships. This linkage is powerful participation that innovates research environment [2]. The triple helix model of research capacity development is based on the synergy between academia, industry, and governments for economic growth and social development [3]. Accordingly, SMEs has been an important mechanism of novel economies and many scholars mention that SMEs is a crucial factor of sustainable economic growth [4]. Nowadays, SMEs has shifted to a modern concept of “Innovative Entrepreneur” [5]. The Innovative Entrepreneur has become one of the key driving forces for the development and it is very crucial for the economic growth of any countries. The purpose of accomplishing circular economic, industrial development, provincial growth, employment creation, sustainable production, and advancement of new technology have regularly depended on entrepreneurial development [6].

In this regard, several countries like Thailand have initialized some policies to support innovative Entrepreneur. Thailand 4.0 policies is an economic model that aims to unlock the country from several economic challenges resulting from past economic development models. These challenges include “a middle-income trap”, “an inequality trap”, and “an imbalanced trap” [7]. Industrial Research and Technology Capacity Development Program (IRTC) is a Thai government funding program constructed to expedite the research and development projects for SMEs in Thailand. Entrepreneurs who are implementing and developing process improvements are the primary prospects to receive research subsidies through IRTC, however, uplifting sociological and economic impact technology adoption projects that result in new potentialities are also considered.

Since the government must spend a lot of money on plenty of R&D projects; and, some projects need to disburse high investment, project selection will balance between risk and benefit which occur during the process of each project. At present, the selection of projects is expertly worked by an internal and external committee, however, the opinion of this expertise is not an evidence base and it is difficult to assess a variety of R&D projects without bias. Analytical Hierarchy Process (AHP) is an effective tool for dealing with complex decision making and may assist the decision maker to set priorities and make the best decision [8]. Many scholars used the AHP to choose R&D projects in the enterprise sector; nevertheless, the selection process of government-granted R&D project is less discussed [9].

This paper represents an AHP model for determining factors affecting government-funded project selection. Consideration of decision maker is applied to perform the pairwise comparisons of the qualitative performance data related to Project details, Entrepreneur, and Researcher that should be considered during Government R&D Support Projects for SMEs selection. Decision maker alike compared factors with each other to generate the hierarchical decision structure. The paper is organized as follows. Section 2 presents the literature on R&D project selection and the application of AHP in the classified decision factor. Section 3 discourses the methodology applied. Results are represented in Section 4. Section 5 concludes the research with the discussion section.

2. Literature review

2.1. Background: R&D project selection in Thailand

As a result of the available resource constraints, and the diversity of R&D project alternatives, funding the correct project is significant and spend a lot of time [10]. In Thailand, some funding government agencies such as regional science park have responsibility to subsidize projects for SMEs. The procedure for developing project is as follow. Firstly, funding agency visits the industrial sector to evaluate the information and problems needed to be solved or target to be achieved. In second step, funding agency selects the researcher from university or research institutes to deeply diagnose the technical problems. Lastly, industrial sector and researcher jointly develop project proposal according to problems and industrial demand. As previously mention that decision making in selection of R&D Projects that will be supported by government agencies is one of the most momentousness processes because R&D Projects is an activity that consumes large investment budget. For this reason, funding agencies have to give serious consideration and trade-off between benefits and risk that will arise from projects. The selection process is done by a panel of experts from inside and outside the funding agency. On the contrary diversity of expertise's nomination committee and decision criteria that are used to consider cannot be clearly specified. It is difficult to evaluate without prejudice and risking unfairness in considering the project.

2.2. The AHP Method

The AHP has been acknowledged as one of the solid and elastic multi-criteria decision-making (MCDM) tools for dealing with complex problems [11]. The method demands the decision committee to contribute consideration regarding the interrelated emphasis of either criterion which is then employed in designation a preference for either selection alternative. The AHP method comprises of four processes include construction of the decision hierarchy, formulation of the pairwise comparison matrix, computation of the weights of criteria factor, and computation of the consistency ratio [12].

Process 1: Hierarchical decision making is constructed by analyzing the selection purpose, decision criteria and feasible alternatives. Each level consists of main criteria, sub-criteria, and alternatives could be expanded to the hierarchy for declare the selection objective in information.

Process 2: Pairwise comparison matrices are formed after constructing the decision hierarchy. Decision committees correlate the emphasis of the decision criteria in matches conforming to the objective of the selection goal adopting the nine-point rating score invented by Saaty [13].

Saaty's rating score is described in Table 1. If there are various decision maker in the selection process, their considerations in the pairwise comparison matrices would be summarized applying the geometric mean.

Process 3: Following completing the pairwise comparison matrices, the model has been computed the weights of criteria factor. To calculate the weights, each factor in the pairwise comparison matrix is divided to summarize of its column. After that, the mathematics mean of each row is computed. The values achieved from this computation superseded the weights of the criteria factor.

Process 4: The final process in AHP model is to compute the Consistency Ratio (CR) in either pairwise comparison matrix. CR determines how invariably the considerations of decision committee are. If CR is underneath 0.1, accordingly the consideration of the decision makers can be approved as unvarying. In other words, if CR is more than 0.1, the decision makers are appealed to re-judgment the pairwise comparisons continuously the considerations turn into steady result [14].

3. Methodology

In the AHP hierarchy model for selecting government R&D support projects for SMEs, the objective should be to select the projects that have the capability for achievement by initiating worth. Entrepreneur characteristics, researcher characteristics, and project details factors are the three main criteria that are adopted in the selection of government R&D support projects for SMEs. These criteria are generally subdivided into individual sub-criteria. In this research, the entrepreneur characteristics criterion is segregated into budget, staff, and market. Budget grades are depended on the entrepreneur's authorized capital. Staff grading is performed depending on in-house R&D staff that estimated in accordance with the organization diagram and the task appearances. Market are evaluated according to the availability in the targeted market. The researcher characteristics criterion is segregated into the experience, education level, and availability. Experience scores are depended on preceding research and experiences. Education scoring is performed depending on the degree of certificate that was enclosed to the proposal. Availability is evaluated according to their teaching and research workload. The project details factors criterion is subdivided into business impact, plan consistency and risk. Business impact criteria demonstrate the consequence and significance of an impact to the business if the disruption was to occur to a system, process, and other business operations. Plan consistency grades are based on reasonable of a work plan and actual time. The risk is evaluated by assessing probability and impact should consistently be performed. All these factors originate from literature review then the decision makers are interviewed for screening factors that affect the decision to select government R&D support projects for SMEs. Corresponding hierarchy is as represented in Figure 1.

Next process, sub-criteria beneath each main criterion are correlated to each pair of criteria by three decision makers. These decision makers have seen the project statistics from the information document that were provided to them. On that occasion, the comparison of the main factors has been achieved, and the AHP model has generated the local arrangement for these criteria.

Finally, these following steps are applied to score the criteria weights of three main criteria and nine sub criteria. Local and global weights are identified with the weighted score for the total selection problem. After that, the decision makers are interviewed for pairwise comparison to correlate the importance of the decision criteria. Table 2 demonstrates both the local and global weights for the whole criteria. Ranks column displays significant criteria among the whole criteria. The most significant criterion has been discovered as "market" and the least significant criterion is "risk". Entrepreneur characteristics criteria are distinguished when correlating with researcher characteristics and project details factors. "Experience" is the most significant criterion amidst the researcher characteristics. "Plan Consistency" is the most important criterion among the project details factors.

Table 1. Saaty's rating score.

Level of importance	Definition
1	Indicate the two factors are of same importance
3	Indicate the two factors the early is slight important than the latter
5	Indicate the two factors the early is obviously important than the latter
7	Indicate the two factors the early is more important than the latter
9	Indicate the two factors the early is extreme important than the latter
2-4-6-8	The median value of the above consideration

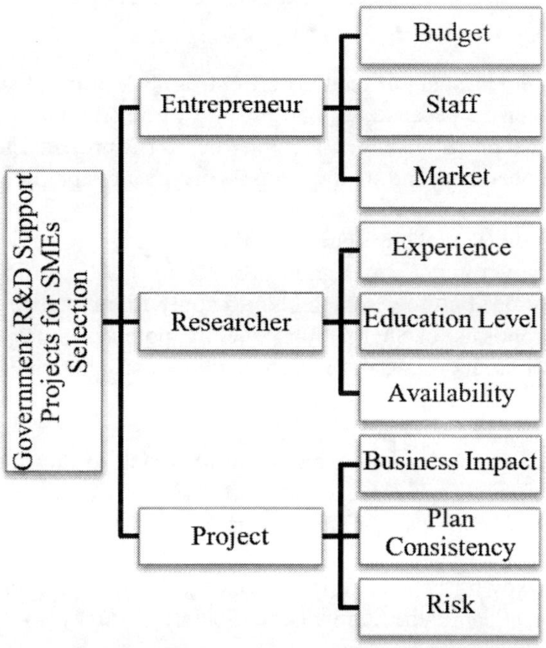

Figure 1. The hierarchical structure of the decision problem.

4. Results

Government R&D support projects for SMEs should be completely investigated before receiving a support decision. In this regard, it is meaningful to characterize the construction of purposes that produce an innovative entrepreneur project outstanding. In this research, the results have presented that the most significant factors in government R&D support project selection are the entrepreneur characteristics criteria: "Market", "Staff", and "Budget". The emphasis of the project details factors (Plan Consistency, Business Impact, Risk) do not importantly affect the whole score of the possible choice. The greatest values of these criteria factors enlarge the R&D project's potentiality of being profitable. "Market" criteria, the preceding criteria, is appraised according to opportunity in the prospect market. A threat of new entranced technologies and competitive rivalry are both discussed in the assessment. On this subject, it could be declared that the "market" that the project prospects are the paramount criteria that should be concerned. A dependable-constructed market diagnosis can be extremely dominant on the decision of decision committees. The parallel between the recommended technology and the market necessitates dexterity. In part of ranking the decision criteria, this paper uses the global weight of each criterion to prioritize the decision criteria that involve the government R&D funding projects selection.

Table 2. Criteria weights.

Criteria	Local Weights	Sub-criteria	Local Weights	Global Weights	Rank
Entrepreneur	0.565	Budget	0.329	0.186	2
		Staff	0.214	0.121	3
		Market	0.457	0.258	1
Researcher	0.235	Experience	0.413	0.097	4
		Education Level	0.345	0.081	6
		Availability	0.242	0.057	8
Project	0.200	Business Impact	0.328	0.066	7
		Plan Consistency	0.441	0.088	5
		Risk	0.231	0.046	9

5. Discussion

In this paper, "Risk" is the least significant criterion from the entire of sub-criteria. Nevertheless, risk criterion should be concerned because the hardest part of developing a custom risk appraisal feature requires diagnosis of the potential and effect on the R&D project should a risk incidence occur. The feasibility of a risk occurring and its impact on a project are applied in an organization as decision assistance.

Although this paper points out the weight of the three main criteria and nine sub-criteria that affecting the decision to choose government R&D grant projects for SMEs, it does not state how to deal with alternative project selection. Future research should apply the model to test with real data of several R&D funding project proposals for SMEs. Subsequently, potentially projects will be subsidized. It will be of a benefit for academia, industry, and governments, for sustainable development in countries.

References

[1] Acemoglu D 2003 Root Causes: A historical approach to the role of institutions in economic development *Finance and Development* pp 27-30

[2] Wolfgang G 2004 *Handbook of Quantitative Science and Technology Research* (Netherlands: Springer)

[3] Ranga M and Etzkowitz H 2013 Triple Helix systems: ananalytical framework for innovation policy and practice in the Knowledge Society *Industry and Higher Education* **27** no 3 pp 237–262

[4] Paul J and Shrivatava A 2016 Do young managers in a developing country have stronger entrepreneurial intentions? Theory and debate *International Business Review* **25** issue 6 pp 1197-1210

[5] Didem Z and Durmus U 2018 Assessment of techno-entrepreneurship projects by using Analytical Hierarchy Process (AHP) *Technology in Society* **54** pp 41-46

[6] Miller D and Garnsey E 2000 Entrepreneurs and technology diffusion: how diffusion research can benefit from a greater understanding of entrepreneurship *Technology in Society* **22** pp 445-465

[7] Jones C and Pimdee P 2017 Innovative ideas: Thailand 4.0 and the fourth industrial revolution," Asian International *Journal of Social Sciences* **17** issue 1 pp 4-32

[8] Bhushan N and Rai K 2004 *Strategic decision making: Applying the Analytic Hierarchy Process* (London: Springer)

[9] Ornek A S and Danyal Y 2015 Increased importance of entrepreneurship from entrepreneurship to techno-entrepreneurship (startup): provided supports and conveniences to techno-entrepreneurs in Turkey *Social and Behavioral Sciences* **195** pp 1146-1155

[10] Fayolle A 2007 *Handbook of Research in Entrepreneurship Education* (Cheltenham: Edward Elgar Publishing)

[11] Arunyanart S and Pruekthaisong S 2018 Selection of multi-criteria plant layout design by combining AHP and DEA methodologies *MATEC Web of Conferences* **192**
[12] Palcic I and Lalic B 2009 Analytical Hierarchy Process as a Tool for Selecting and Evaluating Projects *International Journal of Simulation Modelling* **8** pp 16-26
[13] Saaty T 1980 *The Analytic Hierarchy Process* (New York: McGraw Hill International)
[14] Polata G and Erayb E 2015 An integrated approach using AHP-ER to supplier selection in railway projects *Procedia Engineering* **123** pp 415-422

Acknowledgments
This research is supported by Faculty of Engineering, Khon Kaen University and the Supply Chain and Logistics Systems Research Unit, Department of Industrial Engineering, Faculty of Engineering, Khon Kaen University, Khon Kaen, Thailand.

International Freight Transportation Problems Under GMS-CBTA: Case Study of Thailand-Cambodia

Pratan Uttamapurna, Jaruwit Prabnasak, Ronnachai Tiyarattanachai
International College King Mongkut's Institute of Technology Ladkrabang, Bangkok, Thailand.

Email: Ut.pratan@gmail.com

Abstract. The governments of Thailand and Cambodia have enacted a policy under the GMS agreement to facilitate and increase trade and transport capacity. One of the key components is to improve land transport performance on the designated routes between Bangkok and Phnom Penh under the GMS-CBTA. However, the policy has not been considered effective as expected because licensed logistic companies still do not use that privilege. They also use the traditional method in which cargo is transhipped by local trucks at the borderline. In light of the problem, this study systematically investigated the root causes of the agreement ineffectiveness. Some recommendations for improving the performance of operation are given.

Keywords. Cross border, GMS-CBTA, Land transport, Thailand, Cambodia.

1. Introduction

Presently, the growth of border trade between Thailand and Cambodia is increasing every year, which now Thailand as much as 145,799.82 million baht/year in the year of 2018 [1]. The most favorite way to transport cargoes is land transportation. The governments of the two countries have implemented the policy under the ASEAN agreement to increase transport potential. The current major policy is to operate an agreement of land transport on the agreed routes between Thailand-Cambodia that called Greater Mekong Sub-Region Cross Border Transport Agreement or R1 route (GMS CBTA) [2].

According to the Transportation Institute (2013) [3], they studied the transport route from Aranyaprathet – Poipet – Phnom Penh, found that Aranyaprathet border delays frequently occur in customs clearance. The Poipet border has some delay in the CIQ process, and the road between Poipet - Phnom Penh is in poor conditions. Many transport companies found the problems about the misunderstanding of legal issues and bad coordination between the two countries make the agreement inefficiencies. Furthermore, Cambodia has the problem of high logistics cost when compared with those in other countries in the region and the world [4]. However, Cambodia government has tried to improve the logistics sector in the country and the rolling out of the Automated System for Customs Data (ASYCUDA). The goal of this improvement is to enable easier, cheaper, and faster transport of goods across borders into and from Cambodia [5].

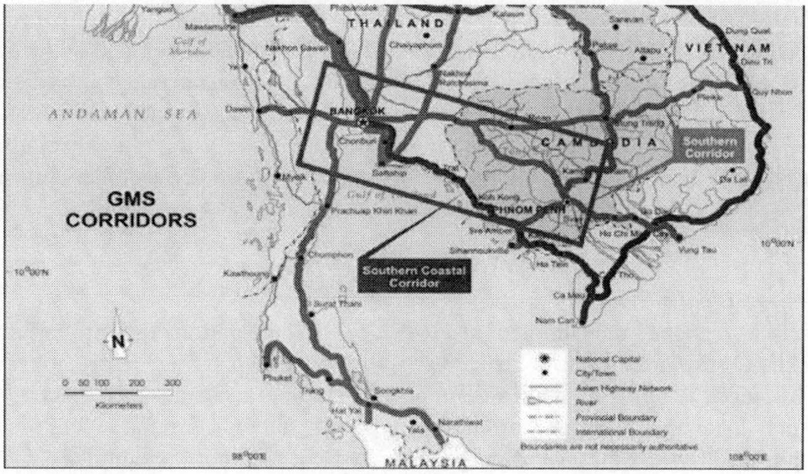

Figure 1. ASEAN agreement routes between Thailand-Cambodia (R1).

Even if the government of both countries try to develop cross-border transportation policy and also infrastructure along agreement route. But with above problems, therefore, this is interesting things for researcher to study, investigate the problems, and also explore advantage and disadvantage of GMS-CBTA to give information to transport operator, government of both countries, and people who interested to be improve and recheck before operate and do any new policy for the best benefit of Thai and Cambodian people. Likewise, many researchers have been studied land transportation on such route but still don't have any research about cross-border transportation under GMS CBTA.

2. Method

The method is divided into two main issues are to determine the framework of the study and conducting the research method within the framework. For the content of methodology has amplified as follow;

2.1. Frame Work

This research is studied and analyzed information on international transportation problems under GMS CBTA. Firstly, study current border transportation between Thailand-Cambodia at Aranyaprathet customs, Poipet customs and general method that used in freight transport. Secondly, study transportation data under GMS CBTA permit on such route.

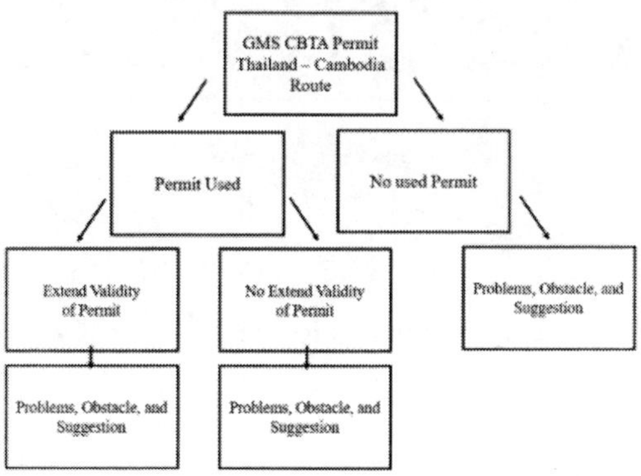

Figure 2. Procedure of questionaire design.

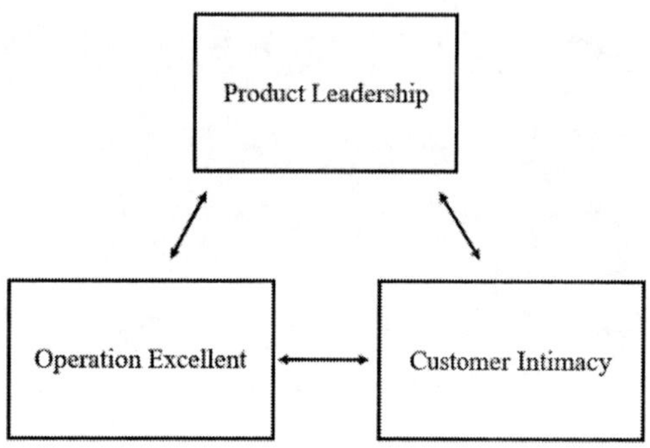

Figure 3. Value Discipline Model.

2.2. Research Methodology

For the method of research begins from study the previous research, literature review, review related theory, and review secondary data used in freight transport. Then, survey and collecting primary data that essentially used in transport under GMS CBTA permit to comparative advantage and disadvantage, analyze, and summarize. Otherwise, other methods for collecting data are questionnaire design, interview, and analysis data permit by following three side models of Value Discipline Model which composed of Product Leadership (PL), Operation Excellent (OE), and Customers Intimacy (CI) [6].

3. Results

3.1. Survey as Designated Route.

The researcher has designed the shipment as GMS CBTA route and make it properly with a customer in Thailand. The route started from the truck yard, then pick the cargoes before going to the border to do export clearance. After that, make an import clearance and transport cargo to destination in Cambodia. Total distance from truck yard until arrived destination is around 700 kilometers (one-way). This shipment uses 3 trucks to transport cargo. The scope of transport as Figure 4.

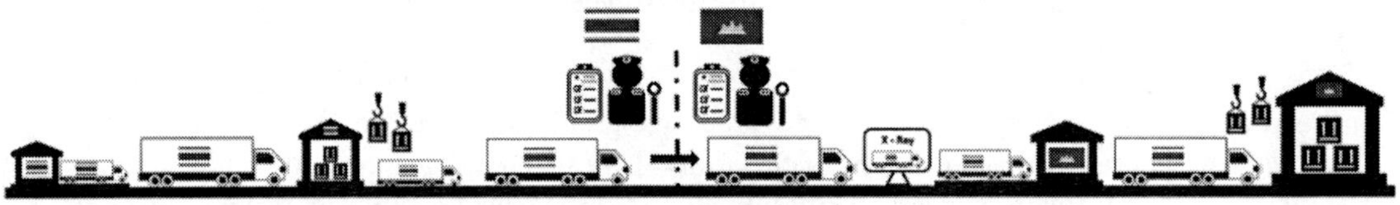

Figure 4. Step of transportation from Origin to Destination.

3.1.1. GMS Cross-border and customs clearance in Thailand site. Truck yard in Thailand is located at Klongluang, Pathum thani. Then trucks pick up the cargoes at Ratburana road before going to Aranyaprathet-Poipet border at Srakaew province to do export clearance. After finished export customs clearance process, these trucks are taking National No.5 road (NH5) of Cambodia which is connected with Aranyaprathet customs in Thailand site. Likewise, this road is also called AH1 as ASEAN highway name.

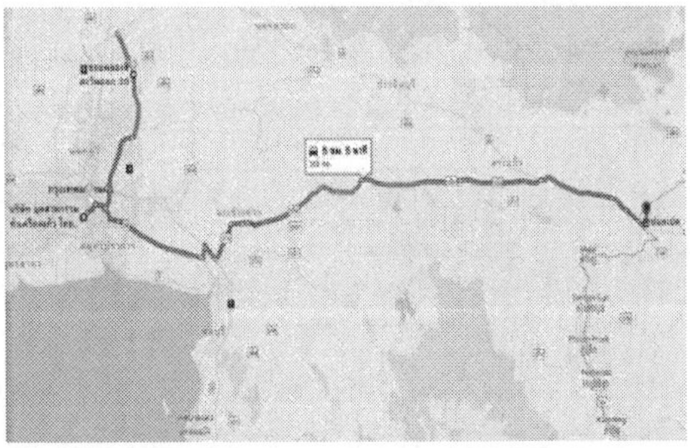

Figure 5. Designated Route of Survey (AH1) in Thailand.

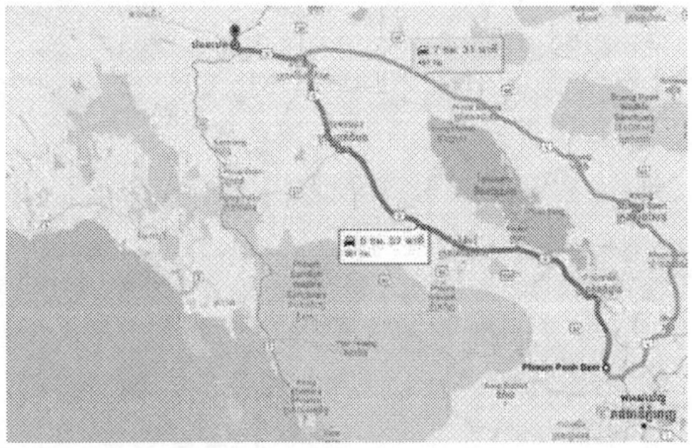

Figure 6. Designated Route of Survey (AH1) in Cambodia.

3.1.2. GMS Cross-border and customs clearance in Poipet, Cambodia. After trucks have arrived in Cambodia. Cambodian broker which are waiting at Poipet border collects the documents which customs of Thailand issued for the import process. Camcontrol is the department that checks cargo and calculates tax and duty which importer must pay to Cambodia government. Then customs office is issued and authorized Letter which means Thai trucks already get the permit to bring their cargoes to Cambodia.

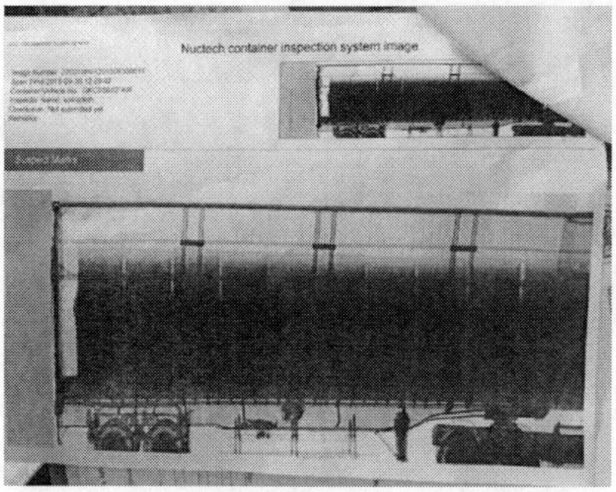

Figure 7. Document of X-Ray.

Figure 8. X-Ray at Container Scan Station.

After that, drivers drive the trucks to go to X-Ray at Container Scan Station of Poipet Customs and Excise office. Importantly, customs broker takes our GMS permit document which issued from DLT to checked and persistent with Poipet customs. Then they check company name and truck number in their list to be issued the license and give the permit to Thai truck for transport on NH5.

3.1.3. GMS CBTA Transportation on national No.5 road.

Figure 9. Unloading at factory in Kampong Chhnang Province.

After finished import customer clearance process and get the permitted for the trucks as GMS CBTA regulation, the driver can drive the truck through inside Cambodia more than temporary permit area or more than ten kilometers to deliver cargoes to customer's factory at Phnom Penh Beer Co., Ltd. This factory is located at Kampong Chhnang Province, Cambodia which far from Central of Phnom Penh around fifty Kilometers. However, even we have the permitted, but our Cambodian agency sends their staff goes with the trucks to escort the trucks along the way.

Table 1. Distance, transport and customs clearance cost

		Cost/truck	
Route	Distance (km)	Transportation[a]	Customs Clearance[b]
Thailand site	330	23,700 Baht / 10w truck	3,000 baht / 10w truck
		33,200 Baht / Trailer	3,000 baht / Trailer
Cambodia site	365	-	992 USD/ 10w truck
			992 USD/ Trailer

[a] Transport with GMS CBTA permit on both Thailand and Cambodia site.
[b] Included documents, customs clearance, escort in Cambodia.

3.1.4. Summarize of the survey. Even if the total distance between Thailand and Cambodia is not too different but in Cambodia site need 2 days for import clearance and transportation because the procedure of customs clearance is still manual. Moreover, mostly, road condition in Cambodia is still 2 lanes which make truck unable to speed. For transport cost, the expense of a ten-wheel truck is 23,700 baht/truck and trailer is 33,200 baht/trailer from origin to destination. Customs clearance cost in Thailand is 3,000 baht/truck/trailer which included any customs document, and customs clearance cost in Cambodia is 992 USD/truck/trailer which included any documents and escorts the trucks on the way as in figure 4.

3.2. Questionnaires.
The researcher chooses the data that importance and interesting from 50% of respondents. The GMS CBTA permit is used 66.7% from components and never use the permit 33.3% of components. All respondents or 100% of them still need to extend the validity of GMS CBTA permit.

Table 2. Level of the company that use and no use GMS CBTA permit from questionnaire.

		Problems of GMS CBTA	
Company	Level of usage (%)	Level of problems (%)	Main Problems[a]
Use Permit	67%	74.89%	Language, cargo insurance, truck insurance, expertise and risky of route, road condition, traffic condition, traffic law, opposite driving lane, unloading at destination, no backhaul, light in night time, communication with driver, and GPS tracking
No use permit	33%	46.39%	Cargo insurance, truck insurance, expertise and risky of the route, road condition, traffic condition, traffic law, opposite driving lane, maintenance, light in the night time, and communication with the driver

[a] These problems are in Cambodia site.

3.2.1. Problems level of GMS CBTA permit user. The average result of users is 3.74 point or 74.89%. This's mean the permit has many problems. However, the most problems from components are in Cambodia site composed shipping cost, language, cargo insurance, truck insurance, expertise and risky of route, road condition, traffic condition, traffic law, opposite driving lane, unloading at destination, no backhaul, light in night time, communication with driver, and GPS tracking. The level of these problems are between 3.51-4.50 point or 70-90%. This means they have many problems.

3.2.2. Problems level that why no use permit company. The average result is 2.32 point or 46.39%. This's mean few problems. The most problems that make these companies are not using permit is in Cambodia site composed of cargo insurance, truck insurance, expertise and risky of the route, road condition, traffic condition, traffic law, opposite driving lane, maintenance, light in the night time, and communication with the driver. The level of problem is between 2.51-3.50 point or 50 – 70 %. This means they have moderate problems.

3.3. Interview

The researcher has a good opportunity to interview staff from the Department of Land Transport (DLT), Mr.Therawech Nilkamheng. He said "Not too much operator uses their trucks to transport in this route as GMS CBTA. Maybe because of the problems of the standard of insurance, not good condition of the road, and traffic law in Cambodia."

The researcher has comments from companies who use the permit. The interesting comment is when they need to extend the validity of the permit, they must prepare many documents and difficult with unclear procedure for preparing. Moreover, the corruption and intimidation to Thai truck is the big problem. The truck must be escorted from Cambodian all the way to protect any problem which has more expense.

3.4. Analysis of data

The advantage of GMS CBTA permit can be concluded by triangular; 1) Product Leadership (PL). GMS CBTA permit can be used as an agreement and make an advantage. 2) Operation Excellent (OE). Reduce transportation timing *and* cost in Cambodia. 3)Customers Intimacy (CI). Trustable from the customer with 100% cargo taken care and reduce damage from move cargo to Cambodia truck.

4. Conclusions and Recommendations

4.1. Conclusions

The study found that the problems of GMS CBTA permit on Thailand-Cambodia are the road conditions, GPS tracking, truck repair, truck insurance, cargo insurance, language, corruption, and backhaul. Moreover, still found the problems of GMS CBTA permit from an operator that composed of the validity of the permit is a short time, permit extension is complicated, and coordination between two governments and support from government is not good enough.

4.2. Recommendations

GMS CBTA is the agreement of great-mekong country which tries to develop logistics in these regions. Thai government try to promote to private sectors to use this privilege. The most important problem is how to communicate between government to government with efficiency from management to operation worker. If these problems still not solve, the result of problems is not to the government, but the private sector must be losing the money and opportunity when they try to use the permit.

Otherwise, this research still not investigate and compare the exact cost for transportation between use or not use the permit. Therefore, this is interesting if getting study cost structure in the future.

IOP Conf. Series: Materials Science and Engineering **639** (2019) 012012 doi:10.1088/1757-899X/639/1/012012

References

[1] Department of Foreign Trade (DFT) 2019 *Phap Ruam Karn Kha Chai Dan kub Prathet Phuen Ban Malaysia-Myanmar-LAO PDR- Cambodia 2018*

[2] Asian Development Bank (ADB) 2011 *Greater Mekong Subregion Cross-Border Transport Facilitation Agreement: Instruments and drafting history*

[3] Department of Land Transport 2013 *The Manual of preparation for Thai company to ASEAN, the kingdom of Cambodia*

[4] Infrastructure and Regional Integration Technical Working Group (IRTWG) 2015 *Overview on Transport Infrastructure Sectors in Cambodia (5th Edition)*

[5] Srivastava P and U Kumar 2012 Jul *Trade and Trade Facilitation in the Greater Mekong Subregion*

[6] Treacy M and Wiersema F 1993 *Customer Intimacy and Other Value Disciplines* (Havard Business Review)

IOP Conf. Series: Materials Science and Engineering **639** (2019) 012013 doi:10.1088/1757-899X/639/1/012013

Gripper Design for Automatic Trimming System in Forging Process

S Jantanalach and S Vongbunyong

Institute of Field Robotics, King Mongkut's University of Technology Thonburi, 126 Pracha Uthit Rd., Bang Mod, Thung Khru, Bangkok, 10140, Thailand

Email address: sanan.j@mail.kmutt.ac.th, Supachai.von@fibo.kmutt.ac.th

Abstract— Labor shortage is one of the most common yet critical problems in industrial sectors. Using robotics and automation to replace human labor is one of the most effective solutions. In this research, robotic technology is implemented in the trimming process which is one of the key steps in faucet manufacturing. The system is designed to be able to deal with uncertainties of the shape of flash that are occurred due to the impression-die forging process. A gripper with specific jaws is designed to suit the workpiece. A vision system is used to identify the position and orientation of the workpieces. As a result, with the gripper and the vision system, the workpieces can be grasped, transferred, and loaded into the trimming machine at high success rate in comparison to human labor. In this paper, the design procedure of the gripper and performance of the gripper are emphasized.

1. Introduction

Forging is one of the most important manufacturing processes of mass production of metal parts. Forging is a process where metal is pressed, pounded or squeezed under great pressure into high strength parts known as "forgings". The process is normally performed at high temperature by preheating the metal part to a desired temperature before it is worked. It is important to note that the forging process is entirely different from the casting process, as metal used to make forged parts is never melted and poured. [1] The forging process can create parts that are stronger than those are manufactured by any other metal working process. Therefore, forgings are mostly used in applications that reliability and human safety are critical. Nowadays, three types of forging methods are used to mass production of the metal parts. Impression-die forging is one of the most reliable method. However, for the disadvantage, the forged workpieces will have flash, which is excess material. It needs to be removed by trimming process (See Figure 1).

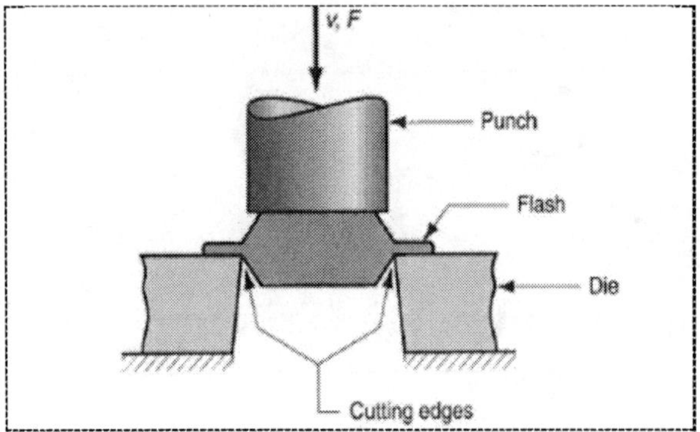

Figure 1. Trimming processes

Trimming process is a process for cutting the flash and remove it from the workpieces. A punch die is used to press on the workpieces through trimming die in order to remove the flash (See Figure 2).

Figure 2. (a) Forged Workpiece, (b) Trimming Die, (c) Trimming Machine, (d) Trimming Workpiece, (e) Finish Goods

Traditionally, trimming process requires human operators involved with loading workpieces into the trimming machine and machine operation. However, trimming process is harmful to the operators due to the sharpness of the flash that has been removed. In this research, a robotic system with specifically designed gripper and vision system is developed in order to replace the human operators. As a result, the performance and safety level of the system will increase.

This paper focuses on the design of the gripper that is appropriated with forging workpieces. The gripper is mounted on a robot arm's end-effector in order to grasp and handle the workpieces accurately. According to the uncertain shape of the flash, to design the gripper which is capable of manipulating workpieces with shape variations of the flash is a challenging problem.

A number of research works were conducted in the design and applications of robot gripper. In regard to the mechanism design, Krishnaraju *et al.* (2015) [2] designed three-fingered gripper and Martínez (2015) [3] designed a two-fingered gripper. Both prototypes are designed based on impactive gripper type that aim for grasping parts with various shape. However, modification of jaws are essentially required in order to use with the forging parts in this project. From the previous work conducted by the authors, Vongbunyong *et al.* (2018) [4] developed an automatic handling method for valves trimming system. The system consisted of a robot arm, a gripper, and vision system. The workpieces are simpler according to the shape of workpiece which is symmetry. As a result, the grasping position can be determined more easily. This research takes the concept from this previous work and improve the grasping performance by designing a more complex gripper.

This paper is organized as follows: Introduction and literature review (Section 1), Methodology (Section 2), gripper design (Section 3), experiments Results (Section 4), and Conclusion (Section 5),

IOP Conf. Series: Materials Science and Engineering **639** (2019) 012013 doi:10.1088/1757-899X/639/1/012013

2. Methodology

2.1. System Frameworks

The problem in handling the forging workpieces is resolved by using a robotic system. The gripper is designed to be capable of handling all workpieces with flash shape uncertainties. The gripper is one of the most important parts of the entire system as it directly contact with the workpiece while grasping (design procedure, see Section 3). The performance of the gripper and the vision system affects the system's performance.

In summary, the robotic system in this research consists of 5 primary modules as follows: (a) a gripper module, (b) industrial robot module, (c) vision system module, (d) trimming machine, and (e) conveyor belt. The robot controller of the industrial robot is the master of this system. Controller is connected with the slave modules, gripper and vision system modules with I/O and TCP/IP protocol, respectively. The system diagram is presented in Figure 3.

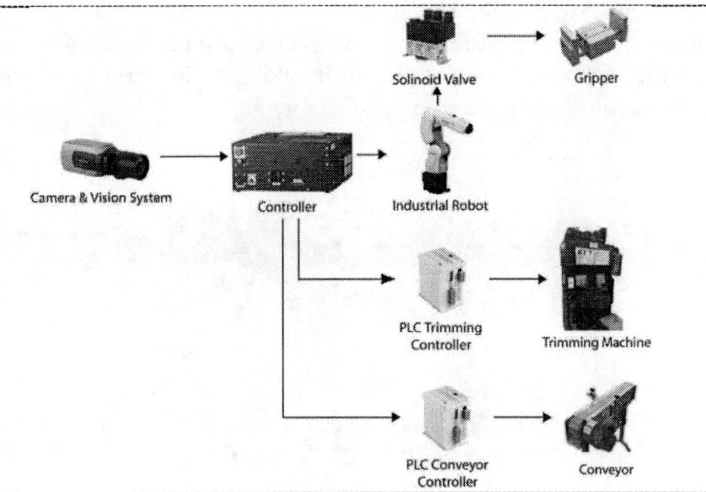

Figure 3. System Diagram

2.2. Primary Modules

(a) Gripper is a parallel air gripper (MHL2-10D2Z) [5] is used in this project because of its lightweight (430g) and long stroke (96 – 156 mm at close and open position) in order to support forged workpieces. The jaws of the gripper are made of metal (SCM4) for durability. The gripper connected with the end-effector the robot arm.

(b) 6-axis articulated robot (Nachi MZ04) [6]: this model is selected due to specification of the high speed movement, high accuracy with ±0.02 mm. position repeatability, workspace with 541 mm reach, and 4 kg payload.

(c) Smart camera (OMRON FQ2-S20100F) [7] has a built-in processor that can process images and communicate with the robot controller via a TCP/IP protocol. The vision system for detecting the workpieces are programmed by using FQ Touch Finder Simulation software. It will sent information about position of workpiece to robot controller.

(d) Trimming Machine (MecolPress) is able to generate force of 150 metric ton with capacity of 1,000 piece/hr. The 4 trimming dies on a rotating table of the machine can rotate to support the forged workpieces. This machine is controlled by the robot controller which is connected via a Programmable Logic Controller (PLC).

(e) Conveyor belt connected with the Programmable Logic Controller (PLC) and Robot controller connect to Conveyor belt through PLC same as Trimming Machine. The conveyor will feed workpieces to picking position at the end of conveyor with capacity of 1,000 piece/hr.

IOP Conf. Series: Materials Science and Engineering **639** (2019) 012013 doi:10.1088/1757-899X/639/1/012013

3. Gripper Design

3.1. Requirements and constraints
Requirements and constraints for designing the gripper are as follows. The workpiece is made of brass (JIS C3771) weight 125 g. The accuracy required for inserting the workpiece into the die is within ±2 mm. The area on the workpiece that can contact with jaws are holes and surface of body of workpiece. The area that should be avoided is the flash as the shape is uncertain, so that it is unreliable to design the jaw based on it. The forbidden zone is on one side of the workpiece as it needs to be placed on the die. To improve the robustness of the system against shape uncertainty, the gripper must be able to grasp the workpiece even the input position from the vision system is inaccurate (see Figure 4).

3.2 Design principle
The gripper used in this research is impactive gripper type (jaw and friction grippers). The impactive gripper is a mechanical gripper grasping objects by using the friction force acting between active surfaces of the object and gripper's jaw. On a contrary, the jaw grippers use the friction force together with the jaws' geometry to support the object [8].
The reason that impactive grippers are chosen over suction gripper because the limited surface area on the workpiece that the suction can be performed. In addition, the surface of workpiece is covered by graphite powder which can cause the suction pad clogging.

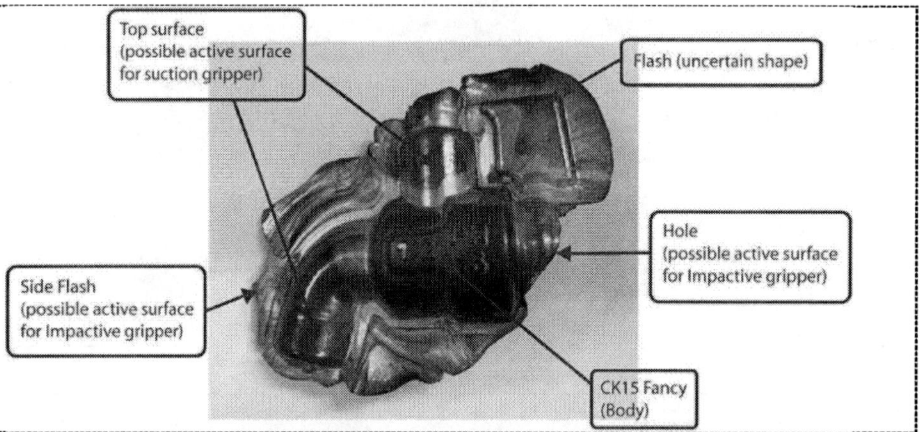

Figure 4. Analysis Possible active surface for impactive grippers and suction grippers.

3.3 Jaws design
Active surface is the area on the workpiece that will be contacted by the jaws. From the geometry analysis of the workpiece, possible active surfaces are a hole and side flash. Therefore, one jaw will be inserted in the hole and another jaw is for supporting the flash. The jaw design secure the position and orientation of the workpiece for moving to correct position of trimming die (see Figure 5). The cavity at the center of the gripper is designed for securing the position of workpiece. The workpiece will be enforced to stay in that position while the gripper is approaching and pressing on the workpiece (see Figure 6(c)).

The gripper is operated according to the steps shown in Figure 6. Firstly, (a) the jaws are open and (b) the gripper approaches the workpiece by moving down from the top. Next, (c) it will move down to contact with workpiece. Afterwards, when workpiece stay in correct position, (d) the jaws will be closed. As a result, one side of the jaw insert in to the hole and another side support the flash. After position of workpiece fully secured it is ready to be transferred to the target position.

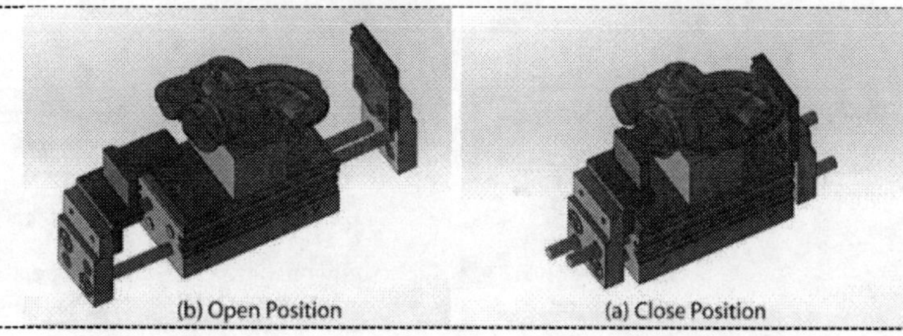

Figure 5. Impactive Gripper: Open and Close position. (Version 3.)

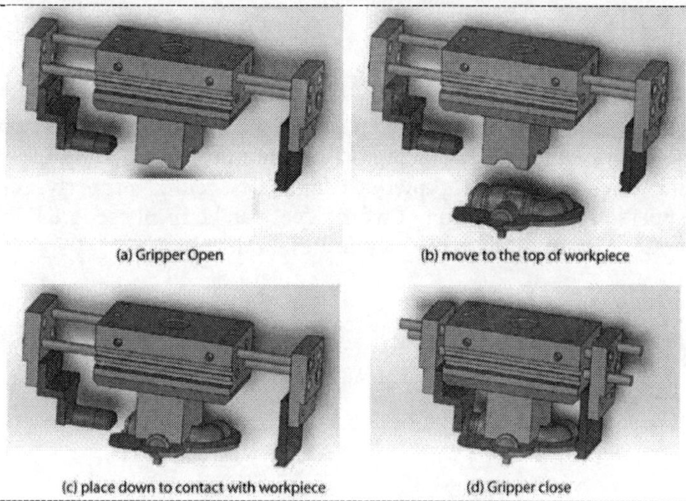

Figure 6. Step-by-step grasping

In this research, three variations of jaws sets (see Figure 7) are designed and tested in order to find the most effective design. From a preliminary test, a comparison between three designs is shown in Table 1.

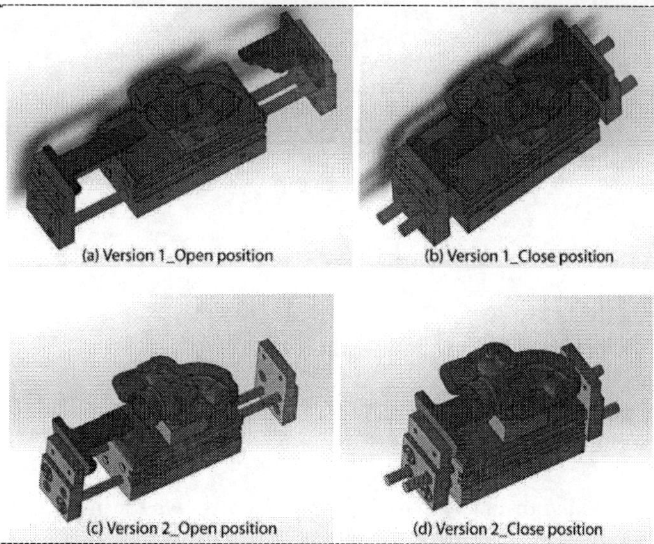

Figure 7. Impactive gripper jaws design version 1 and version 2

Table 1. Comparison performance of the secure position of gripper each version

Version	**1**	**2**	**3**
Technical	Jaw & Friction	Jaw & Friction	Jaw & Friction
Image			
Advantage	Secure position & Orientation	Use only one jaw gripper	Secure position & Orientation
Disadvantage	1. One side jaw interrupt workpiece 2. Use two side jaws	1. Unsure position & Orientation	1. Use two side jaws
Grasping test	8 from 10 pieces	5 from 10 pieces	10 from 10 pieces

According to the testing result, Ver.1 gripper can secure the position and orientation of workpieces but one jaw interrupts the workpieces while placing on the trimming die. Ver.2 gripper has only one jaw, so that the position and orientation of workpieces cannot be secured properly. Ver.3 exhibits the best performance as the position and orientation of workpieces can be firmly secured. It can grasp workpieces with 100% success rate. As a result, it is selected to further develop.

3.4 Force calculation

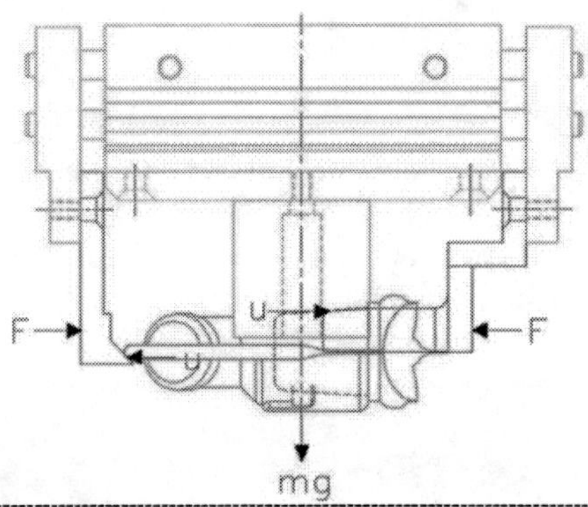

Figure 8. Calculation of Required Gripping Force

$$F = \frac{mg}{2\mu} a$$

Where

- F : Gripping force [N]
- μ : Friction coefficient between jaw and workpiece
- m : Workpiece mass [kg]
- g : Gravitational acceleration (= 9.8 m/s2)
- mg : Workpiece weight [N]

Gripping force is calculated with a margin of "a" = 4, which allows for impacts that occur during normal transportation, etc. [5]

IOP Conf. Series: Materials Science and Engineering **639** (2019) 012013 doi:10.1088/1757-899X/639/1/012013

4. Experiment

4.1 Experiment setup

The experiment was designed to evaluate the performance of the gripper, especially, in regard to the acceptable tolerance of the workpiece's position and orientation. The maximum acceptable tolerance needs to be identified as it will affect the performance of the entire system. The gripper test setup is shown in Figure 9.

The experiments were done in 4 parts: *Test-1* is to test the positional tolerance in x-y axes with 3 mm increment. *Test-2* is to test the rotational tolerance about z-axis with 5 degrees increment. (See Figure 10) *Test-3* is to test rotation of workpieces each angle and gripper grasp. (See figure 11.) *Test-4* is to test continuous grasping workpieces to place on trimming die. (For first and second test, researcher test at laboratory and Third and Fourth test, researcher test at on-site of the factory)

For *Test-1* and *Test-2*, The 10 sample workpieces were selected and the gripper grasped 3 times each. For *Test-3* and *Test-4*, The 100 sample of workpieces were randomly selected in order to exhibit shape variations (i.e. Thickness, contour) of the flash. The sample workpieces are shown in Figure 12.

(a)Gripper connected with industrial robot (b) Gripper testing with workpieces

Figure 9. Experimental working of gripper

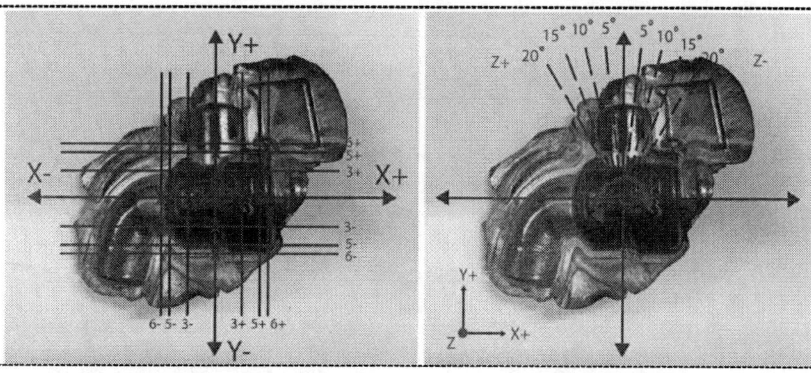

(a) Distance of moving (b) Distance of rotation

Figure 10. (a) Distance of moving in X axis and Y axis (b) Distance of rotation in Z axis

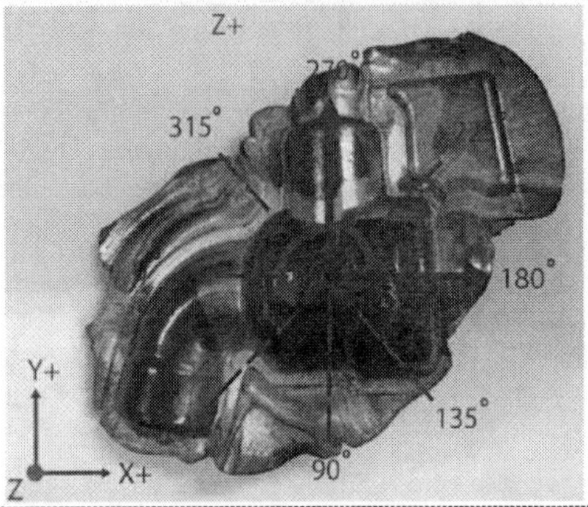

Figure 11. Rotation of workpieces in each angle.

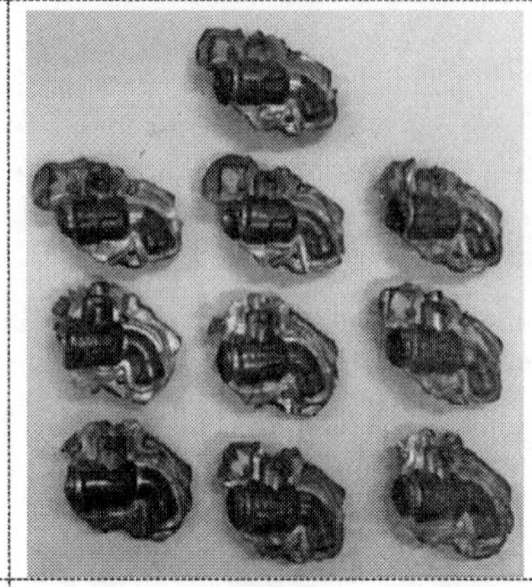

Figure 12. Sample workpieces that is used for experiment

In this experiment, test about precision of the gripper that can grasp workpieces in all angle even through rotation the workpieces in any direction in the same plane and in this test, researcher define the text ½ PN16 is on the top.(See figure 11.)

4.2 Experimental results

Table 2. Tolerance in X axis and Y axis. (Test-1)

Distance (mm.)	-6	-5	-3	0	3	5	6
Axis X	×	×	×	√	√	√	×
Axis Y	×	√	√	√	√	√	×

From the experiment in Test-1, the maximum error that the gripper can successfully grasp the workpiece in x-axis, y-axis are [0, 5] mm, [-5, 5] mm, respectively. If the position and orientation error of the gripper is higher than these tolerance levels, it failed to grasp the workpiece (see Table 2. and Figure 10 (a) for x-axis and y-axis)

Table 3. Tolerance of rotation in Z axis. (Test-2)

Distance (degree)	-20	-15	-10	-5	0	5	10	15	20	25
Axis Z	×	√	√	√	√	√	√	√	√	×

From the experiment in Test-2, the maximum error that the gripper can successfully grasp the workpiece in z-axis [-15, 20] degrees, respectively. If the orientation error of the gripper is higher than these tolerance levels, it failed to grasp the workpiece (see Table 3. and Figure 10(b) for z-axis.)

- Remark : √ = Gripper can grasp the workpiece
  × = Gripper cannot grasp the workpiece

Figure 13. Test the gripper on-site in the factory

For Test-3 and Test-4, the on-site test was conducted in the factory in order to observe the system while working under the actual environment as the conditions which are different from the lab condition. (See figure 13.)

Table 4. Results of Rotation of workpieces each angle and gripper grasp 100 times each (Test-3)

NO.	Angle	Quantity of workpiece that can grasp (Pcs.)	Quantity of workpiece that cannot grasp (Pcs.)	Remark
1	0	99	1	*1
2	45	98	2	*2
3	90	94	6	*3
4	135	95	5	*4
5	180	98	2	*5
6	225	98	2	*6
7	270	97	3	*7
8	315	98	2	*8
Average	-	**97.125%**	**2.875%**	-

From the experiment in Test-3, the gripper can grasp workpieces all of rotation of each angle before moving to trimming die. The gripper can grasp workpieces average 97.125% and the gripper cannot grasp workpieces average 2.875%

Table 5. Result of continuous grasping workpieces to place on trimming die (100 pcs. 3 time) (Test-4)

Test	Quantity of workpiece that can grasp (Pcs.)	Quantity of workpiece that cannot grasp (Pcs.)	Remark	Time (Sec)/Piece
1	97	3	*1	3.51
2	98	2	*2	3.42
3	96	4	*3	3.65
Average	**97%**	**3%**	-	**3.52**

From the experiment in Test-4, the gripper can continuous grasping workpieces to place on trimming die. (100 pcs. 3 time) The gripper can grasp workpieces average 97% and the gripper cannot grasp workpieces average 3%. The average time 3.52 Sec/Piece or about 1,022 Pcs. /Hr. which the objective of this project is 3.6 Sec/Piece or 1,000 Pcs. /Hr.

5. Conclusion

In this research, the impactive gripper used for grasping the workpieces and transferring to the trimming die in trimming machine was developed. This is integration of a gripper, an industrial robot arm, a vision system, a trimming machine, and a conveyor belt. The challenging issue is the uncertain shape of the flash of forged workpieces. The jaw gripper was designed specifically for solving problem of uncertain shape of workpieces and can grasp the workpieces transfer to the trimming die precisely. From the experiment, the gripper was proved to be able to grasp the workpieces with 100% success rate within the acceptable levels of tolerance. With respect to the on-site test, the gripper can grasp the workpieces with 97% success rate so the gripper can achieve the objective of this project.
For future work, the full system will be developed. The information of the workpieces will be provided by the conveyor system. Eventually, the system will be fully automated and will be implemented in the actual production line.

6. References

[1] Forging industry association, URL : www.forging.org/about#what
[2] Krishnaraju A, Ramkumar R, Lenin V R., "Design of Three Fingered Robot-Gripper Mechanism.," Mechanical Engineering, Mahendra Engineering College, Namakkal, India, 2015.
[3] Álvaro Meneses Martínez., "Mechanical design of a robot's gripper.," Warsaw University of Technology, Poland, 2015.
[4] S. Vongbunyong, P. Archeewawanich, . P. Akaratham, C. Martchan and . K. Thamrongaphichartkul, "Automatic Handling for Valves Trimming System.," in *ICOME2018*, Phuket,Thailand, 2018.
[5] SMC, Catalog-MHL2 Series, 2017
[6] Nachi-Fujikoshi-Corp., Standard specifications MZ04-01/MZ04E-01-CFD, 2017.
[7] Omron, User's Manual. Smart Camera FQ2-S/CH Series, 2013.
[8] Gareth J. Monkman, Stefan Hesse, Ralf Steinman, Henrik Schunk, "Robot Grippers" Wiley-VCH Verlag GmbH & Co., 2017

Acknowledgments
We would like to thank you Asahi Thai Alloy Co., Ltd. Thailand, for technical support, financial support, and the demonstration site.

IOP Conf. Series: Materials Science and Engineering **639** (2019) 012014 doi:10.1088/1757-899X/639/1/012014

Assembly of Modular Robot for Cleaning Various Length of Solar Panels

K Chailoet and E Pengwang

Institute of Field Robotics, King Mongkut's University of Technology Thonburi, 126 Pracha Uthit Rd., Bang Mod, Thung Khru, Bangkok, 10140, Thailand

Email address: kampanat.chailoet@mail.kmutt.ac.th, eakkachai@fibo.kmutt.ac.th

Abstract. This paper discusses about the design and feasibility study for building a modular robot for cleaning solar panels. The objective was to design a universal module for transmission and manipulation that can be used with different length of solar panels and different arrangements in solar panel. Since there are various configurations for arranging the solar panel, this robot can be modified using an advantage of modular robots. The principle of working is to clean a solar panel with water, spiral brush, and rubber sweeper. The modular concepts were validated for the solar panel with a range of 1 to 4 meters in both testing and field environments. The solar panels cleaning robot can clean dirt and dust on the panel and increase the power generation of the solar panels. The target applications for this design can be extended for a solar farm, solar rooftop, and floating solar panels.

1. Introduction

In Thailand, the total production of electrical energy is 168,656 GWh in 2014, while the total production of electric energy produced from solar energy 1,928.66 GWh [1]. However, there are many factors affect the efficiency in producing electricity from solar cells such as the environmental conditions including light intensity, daylight hours, the temperature of cell panels, dust, and bird droppings. The dust has also resulted on degradations of solar panels. In Thailand, air pollution and agricultural activities can cause lots of dust. A Large amount of dust can accumulate on the glass surface of the solar panel and reduce the performance in producing electricity. According to Natthakan's research, the dust accumulation will affect the performance of solar power plants and production efficiency ratio. In that investigation, the efficiency of the solar panel is decreased by 1.6-3% for dust accumulation in 1 month and 6-8% for dust accumulation in 2 months. However, after cleaning the dust on solar panels, the efficiency ratio increased by 10% [2].

Hence, the cleaning process is crucially important to maintain the production of solar panels. Cleaning methods with water is suitable for Thailand due to bird droppings and agricultural dust [3]. Mostly, a solar power plant in Thailand still use a manual cleaning that is high cost (approximately 300 USD for 1 MW power plant) and need a lot of labours to clean up. The most common method for cleaning solar panels is a telescopic pole and bathroom mop.

The concept of a modular solar panel cleaning robot is an automatic system that is easy to set-up, low cost, and applicable with various size of solar panels. After the design and prototype processes, the robots are aimed to reduce dust on solar panel and consequently increase the performance in producing electricity.

2. System overview

Robots are designed with a modular concept to serve the panels with a length of 1-4 meters. The main method of cleaning arrays of solar panels is a rotation of cleaning brush on the top surface of panels with a spray of water. The design consists of a universal module for transmission that can be used with various length of the brush and different design in solar panels. This robot can be modified using the advantage of a modular robot that can be combined and disassembled within a short period of time. Each module is 2 meters in length with a total weight of 25 kg. The design considerations for this modular robot are a simple design, easy to operate and set-up, adjustable, lightweight, and low cost.

2.1. Components

The prototype for the modular robot for cleaning solar panels is designed for typical length of solar panel. The model of the prototype is with 1 modular (2 meters) as shown in Fig. 1.

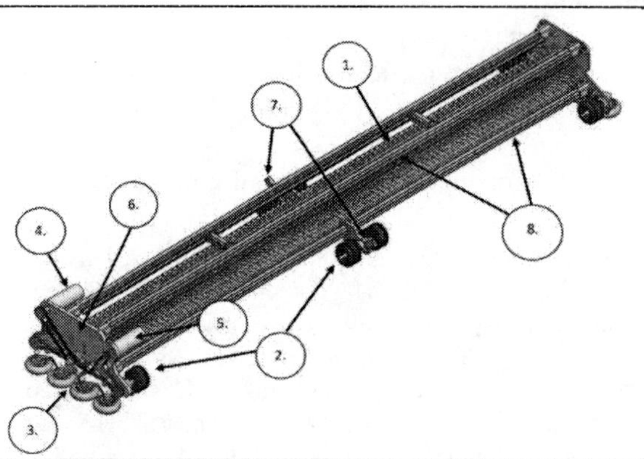

Figure 1. The components for solar panel cleaning robot (1. brush, 2. wheels, 3. support wheel, 4. brush motor, 5. motor of wheel driving, 6. side plate structure, 7. photoelectric sensors, and 8. aluminum pipe frame)

The components of automated robot for cleaning solar panels has a technical information and specifications as follows:

(1) Brush : Rotary brushes are mounted on the side plate structure with a length of about 2 meters.

(2) Wheels : Eight wheels, in conjunction with two axles, allow the robot to move smoothly and it also help in transport. These wheel will move on the top surface of the solar panels.

(3) Support wheel : A set of four supporting wheel on each side will move on the frame of the solar panels in order to ensure the position.

(4) Motor of brush driving : A 25 W DC motor (Zheng Motor: Model ZGX45RGG22.5i) with 250 rpm is used.

(5) Motor of wheel driving : A 25 W DC motor (Zheng Motor: Model ZGX45RGG36i) with 150 rpm is used.

(6) Side plate structure

(7) Photoelectric sensors : These sensors are used for edge-falling protection for solar panel cleaning robot, while working in an autonomous mode.

(8) Aluminium pipe frame : The structure is made of aluminium pipe frames, that are strong and lightweight.

2.2. Design and specification

This robot can be separate into 3 main systems which are (1) brush driving system (2) wheel driving system (3) adjustment system for the brush's compression distance. The detailed explanation is included in this section.

(1) Brush driving system: The brush driving system will be mounted a topmost part on the side-plate so that it is easy to adjust and disassembly. It consists of one DC motor for a nylon roller brush (spiral-shape) with a length of 194 centimeters, and a diameter of 15 centimeters as shown in Fig. 2.

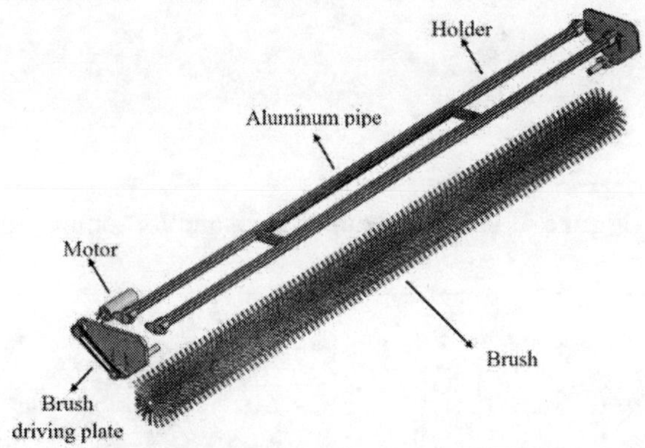

Figure 2. A module for brush system

This robot also has mini water sprinkler for wetting the solar panels while cleaning as shown in Fig. 3. The pipe is 2 centimeters in diameter and it is mounted on the top of aluminium pipes in brush driving system.

Figure 3. Water sprinkle system.

(2) Wheel driving system: The transmission system for wheel driving system can be separated into 3 main components which are (1) wheel driving that is connected to one motor with three-level of adjusted speed (0.6, 0.8 and 1.0 km/hr). These wheels are on the top surface of solar panels and are controlled by using encoders and PID, (2) supporting wheels on the side view that will move on the frame of solar panel, and (3) photoelectric sensors. The overview of the wheel driving system is shown in Fig. 4. The supporting wheel is shown in Fig. 5. A robot for solar panels cleaning in autonomous mode also has edge-falling protection. Hence, two photoelectric sensors are installed in a front and rear position near the middle supporting wheel with the orientation of 45° angle as shown in Fig. 6.

(3) Brush compression system: This system is designed to adjust the distance between the brush and the solar panels to ensure a complete surface contact as shown in Fig. 7. The brush compression system is in the triangle form with a brush driving system on the top of the wheel driving system. The adjustment system will attached both modules together by using 4 bolts. This can be adjusted for a range of 1 centimeter in a compression distance. The overall specification of the robot is shown in Table I.

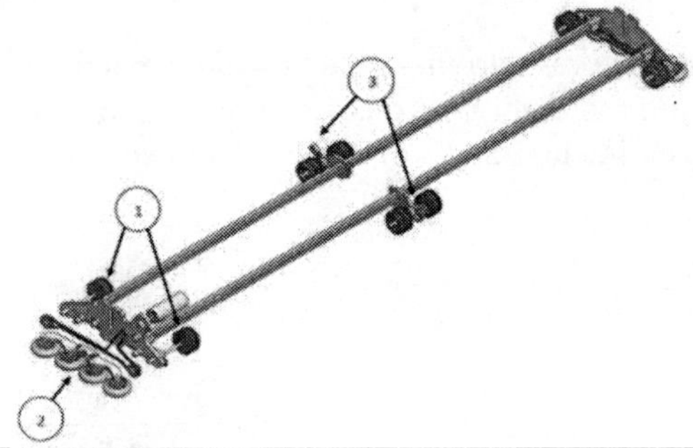

Figure 4. Wheel driving system and its components.

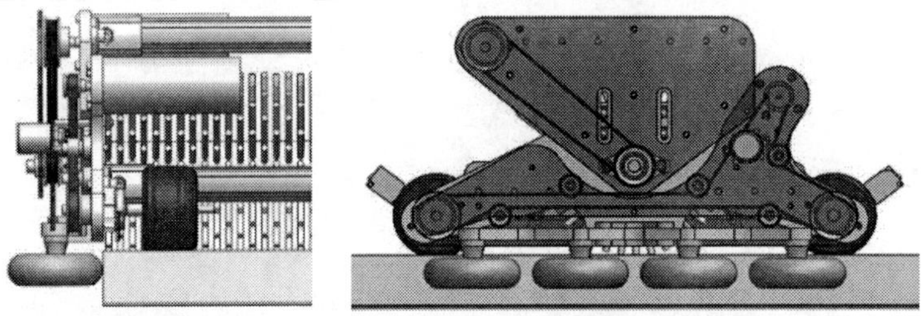

Figure **5**. Supporting wheels.

Figure **6**. Positions for sensor installation.

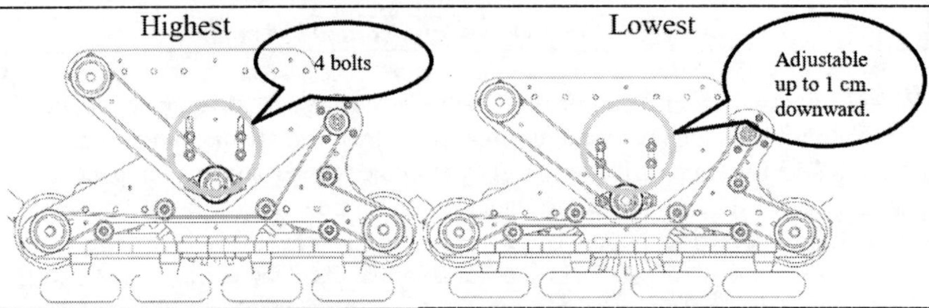

Figure 7. Brush adjustment and the assembly (side view).

IOP Conf. Series: Materials Science and Engineering **639** (2019) 012014 doi:10.1088/1757-899X/639/1/012014

Table 1. Summary of robot specification

	Robot Specifications	
1.	Length	200 cm
2.	Width	40 cm
3.	Weight	25 kg
4.	Battery working hours (20 Ah)	4-5 hr
5.	Maximum capacity for cleaning (1 module, Length: 2 meters)	2000 m^2/hr

Robot for cleaning solar panels is controlled by a microcontroller and protection sensor edge solar panels, when working in an automated system. The control diagram is shown in Fig. 8. This robot can be operated in both manual and automatic modes by selecting an option from a radio-frequency channel.

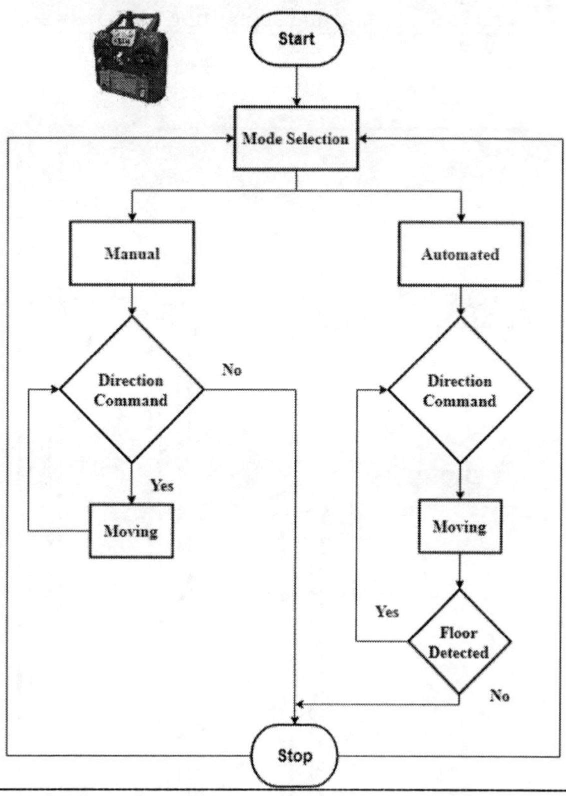

Figure 8. Flowchart diagram for solar panels cleaning robot.

In a manual mode, the robot can be controlled by using a joystick on a RF controller. The operator can easily control the direction of the cleaning and the beginning and ending time for the brush rolling. In such cases, the robot can go to the desired locations that need a re-cleaning. In an autonomous mode, the robot can be operated with the one-start button. The performance for the cleaning will continue until it reaches the other end of the strings, then it will return back to the original position on the other end.

2.3. Assembly of robot modules

Since there are various configurations for arranging the solar panel, this robot can be modified using an advantage of modular robots. Connection between each modular robot is 4 pins with a supporting plate between them as shown in Fig. 9. In this design of cleaning robot, they can be adjusted and connect to other module to obtain a desired length as shown in Fig. 10. Recently, the available range of total length is between 1 to 4 meters and it can be adjusted to match with customer's requirements in any length. The fabricated robot for 2-meters and 4-meters solar panels are shown in Fig. 11 and 12.

Figure 9. A connection method and supporting plate between each module.

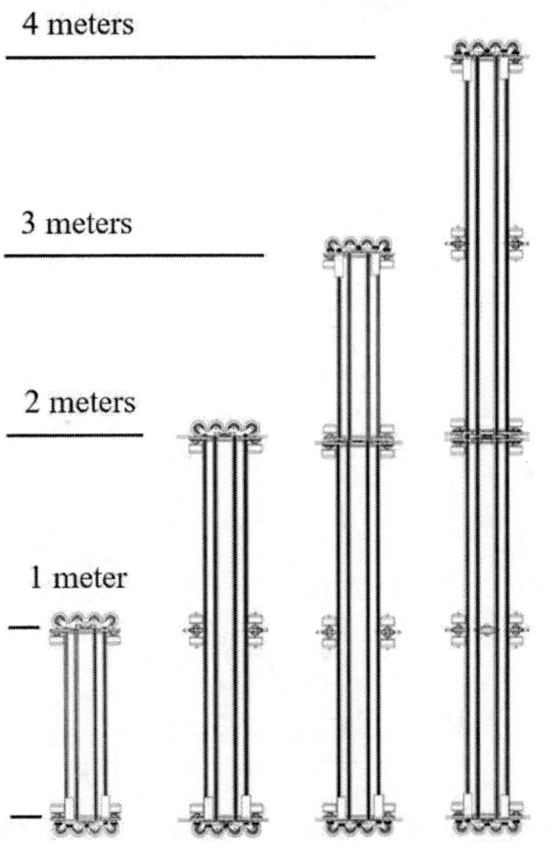

Figure 10. Example of cleaning robot for solar panels with a length of 1 to 4 meters.

(a) Robot length 2 meters.

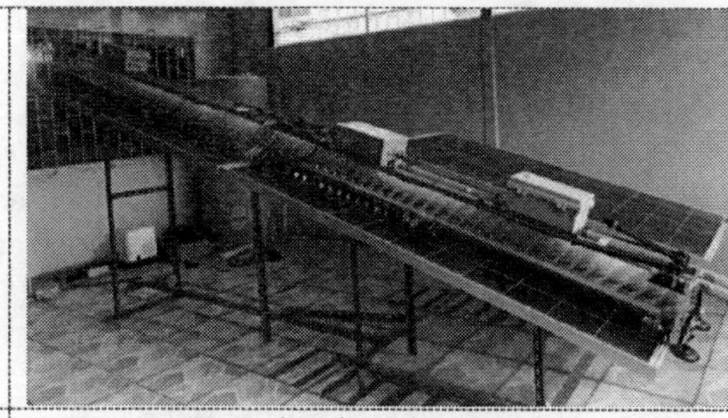

(b) Robot length 4 meters.

Figure 11. Robot for cleaning solar panels in laboratory environment with a length of (a) 2 meters and (b) 4 meters.

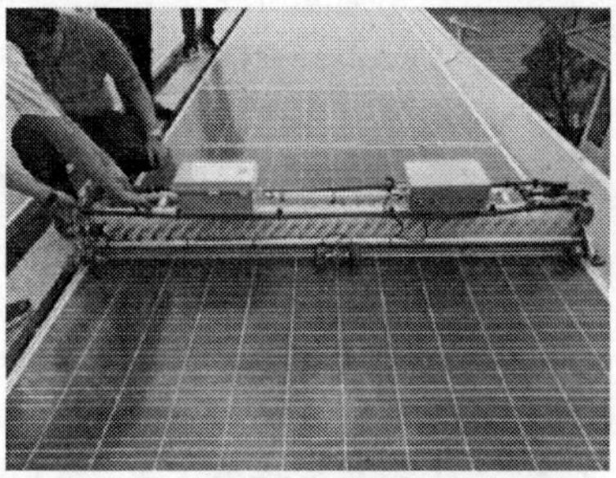

Side view

Top view

(c) Robot length 2 meters.

Side view

Top view

(d) Robot length 4 meters.

Figure 12. Robot for cleaning solar panels in laboratory environment with a length of (c) 2 meters and (d) 4 meters.

IOP Conf. Series: Materials Science and Engineering **639** (2019) 012014 doi:10.1088/1757-899X/639/1/012014

3. Experiment Results

There are three experiments on the performance for solar panel cleaning robot as follows.

3.1. Speed control

In this test, the speed of movement is designed to meet the 3 levels; slow, medium, and fast. The duty cycle of the PWM (pulse width modulation) is used to control the brush motor and drive motor. The values of PWM percentages and measured speed are shown in Table 2.

Table 2. Experiment results for speed control

	PWM (%)	Operate properly	Average time (sec) to travel in 50 m.	Measured speed (km/hr.)	Target Speed (km/hr.)
Slow	45	/	310	0.58	0.60
Medium	60	/	230	0.78	0.80
Fast	70	/	180	1.00	1.00

3.2. Solar panel gaps

In this experiment, the robot is evaluated to find the maximum values of the gap distance between the solar panel in each row. The performance is recorded and evaluated for the successful and failed cases. In this design, the maximum gap that the robot can pass is 3 centimeters. In case that the gap is less than 3 centimeters, all trials are successful. However, if this value for gap distance increase, the front wheel can be modified for larger diameter as well.

3.3. Edge-falling protection sensors

The performance of the solar panel cleaning robot in autonomous mode is enabled by using edge-falling protection. The distances between the robot and the solar panel edge are measured for different speeds of operations. From the results, edge-falling protection systems can work properly at slow and medium speeds, the distance between the robot and the solar panel edge is 2.54 and 2.34 centimeters respectively and the robot does not fall off the solar panel. However, the robot is tested with a high-speed level, the stopping distance is decreased to 2.04 centimeters, but the sensor can be adjusted to increase this stopping distance.

4. Results and discussion

Robot for cleaning solar panel is tested in power plants environments. In this test, the robot can clean better with water as shown in Fig. 13.

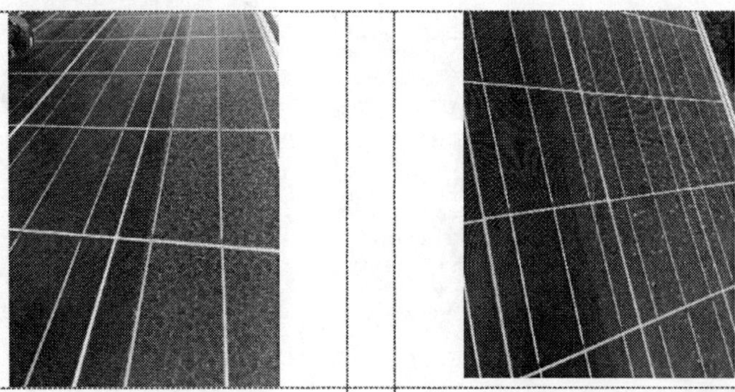

(e) Cleaning without water | (f) Cleaning without water

Figure 13. The experiment results from using robots to clean solar panels (e) without water (f) with water.

For the comparison of dust characteristics, Fig. 14 demonstrates the difference between the solar panel before and after the complete cleaning.

(g) Before the cleaning

(h) After the cleaning by using the 4 meters modular robot.

Figure 14. Dust characteristics on the solar panels (g) before the cleaning (h) after the cleaning by using the 4 meters modular robot.

For testing the efficiency of solar panels cleaning robot, the power plant is analyzed and compared between the two adjacent area with a number of 40 panels per set. The two group of data is collected for each set of panels; 1) Panel that is not cleaned (controlled set) 2) Panel that is cleaned by the robot. The overall electricity production of solar cells is compared for the day before and after the cleaning. As a result, shown in Fig.15, the solar panels after cleaning with robot can increase the electricity production approximately 3.40%. However, the efficiency of electricity production may increase if the dirt on the solar panel is very dirty.

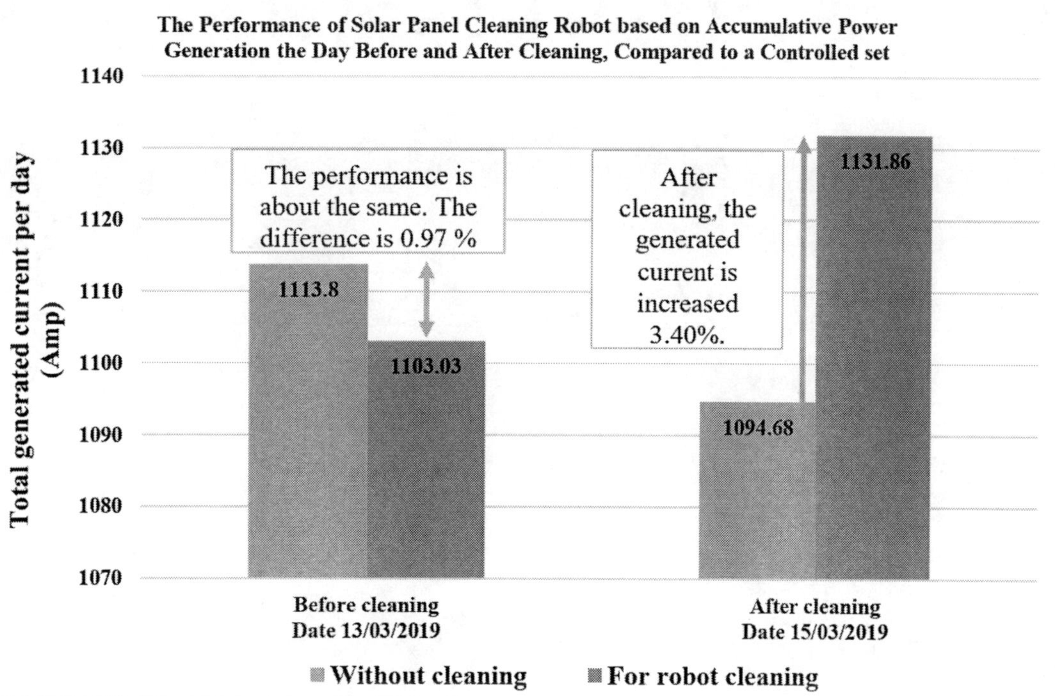

Figure 15. The comparison of the generated current Before and after cleaning with robot.

IOP Conf. Series: Materials Science and Engineering **639** (2019) 012014 doi:10.1088/1757-899X/639/1/012014

5. Conclusion

The modular robot for cleaning solar panels with various length of the brush is successfully demonstrated for a range of 1-4 meters of solar panels. The robot will clean with water sprinkles and Nylon brush that will not deconstruct the surface of solar panels. The module consists of the driving wheels and supporting wheels on the top and along the frame of solar panel respectively. The robot also has a fall-detection technique using photoelectric sensors. The characteristics of speed, gap distance, and stopping distance are examined for this design. The solar panels after cleaning with robot. A robot can clean reduce dirt and dust on the panel are generating electricity have increased.

6. References

[1] Solar power generation status report of Thailand 2014-2015, 2017, pp 13-17.
[2] N. Sakarapunthip, D. Chenvidhya, S. Chuangchote, K. Kirtikara, T. Chenvidhya and W. Onreabroy, "Effects of dust accumulation and module cleaning on performance ratio of solar rooftop system and solar power plants", Japanese J. Phys, 56, 2017.
[3] B. Liamphrachan, Ph. Thamkaew, K. Buayai and K. Kerdchuen, "The Study for Performance Ratio Increasing of Soiling Solar PV Systems", EENET J., vol. 1, 2017.

Acknowledgments

The authors are sincerely grateful to Electricity Generating Authority of Thailand (EGAT) and Thailand Research Fund (TRF) (Grant. No. RDG61D0007) for providing a research grant to support this work.

IOP Conf. Series: Materials Science and Engineering **639** (2019) 012015 doi:10.1088/1757-899X/639/1/012015

Prototyping a Fast and Precise Automation for Folding Vermicelli Process

K Kanjana-Ekkarin and E Pengwang

Institute of Field Robotics, King Mongkut's University of Technology Thonburi, 126 Pracha Uthit Rd., Bang Mod, Thung Khru, Bangkok, 10140, Thailand

Email address: kraingkrai.kan@mail.kmutt.ac.th, eakkachai@fibo.kmutt.ac.th

Abstract. This paper discusses about the design and experimental results for a precise automation system in folding vermicelli processes. The objective is to design a machine that can fold the soft and tangled flour sticks into the suitable shape that can be used in the drying box. The machine consists of four pneumatic actuators and microprocessors to perform that tasks automatically. Since there is a uncertainty on the weight of product, the effect of atmospheric humidity level, and the mixture of the vermicelli, we design the machine that can be adjusted, but still fast enough to replace a labor in this workflow. These designs are also tested with various folding technique. The spending time of the entire process is 23 seconds in order to complete the folding and dropping into testing box for the drying conveyor. The test result showed that only 0.5% of products are with defects in weight and shape.

1. Introduction

Vermicelli is the raw material which has the highest value, comparing with the same food type (such as noodle and pasta) [1]. Each year, Thai people consume vermicelli around 20,000 - 25,000 tons and export more than 3,000 tons [2]. Vermicelli is made of green bean and its process consists of 4 steps which are mixing, threshing flour, squeezing, and drying the vermicelli. From the industrial survey, there are the main 4 processes in drying and folding process that are time-consuming included (1) Cutting long line vermicelli into 34-37 cm (2) Weighting wet vermicelli products into 105 grams for 40-45 grams dried-products (3) Folding vermicelli into the packaging, and (4) Putting vermicelli in aluminium boxes in order to be used in drying process.

From the current survey, it was found that there are numerous staffs in vermicelli production in Thai industry [3]. The problem of using labour is difficult to control the factors such as speed, cutting accuracy, weighing, and dexterous in folding processes. Moreover, vermicelli is a food product which have to concern about food safety in terms of cleanliness such as matter of hair or microbials from staffs that can cause contaminations and microbiological infection. According to Thai Industrial Standards Institute, the standard of vermicelli (TIS. 444-2549) [4] has established with these criteria (a) a good vermicelli must have an appropriate ingredients (b) colour and size of the vermicelli are consistent (c) the product is clean, no physical contamination and scentless. (d) moisture should be below 12 percent by weight, and (e) the manufacturing process of vermicelli has to meet the industrial standards set [5]. These factors indicate that the vermicelli process is such a delicate and complicated that also required a high accuracy.

2. Overview of automation system

Automation system aims to reduce a number of staff in the production line. It is also help to increase the speed and accuracy of making product. However, there is still mandatory to have one staff to control 3-4 folding machines at the same time. The components of automation system before the drying process consist of 3 main parts; (1) Weighing and cutting system (2) Folding and transporting and (3) Conveying to conveyer belt as shown in Figure 1.

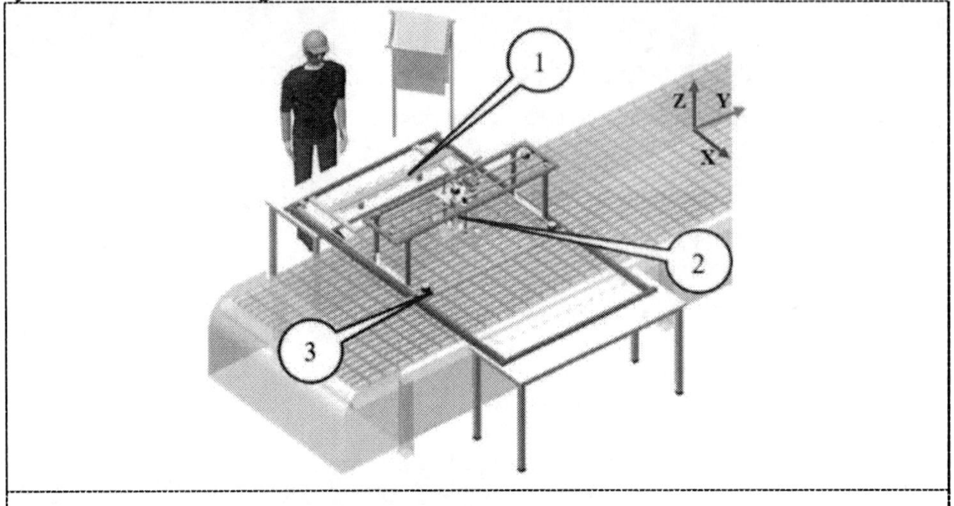

Figure 1. System overview for the folding machine; (1) Weighing and cutting system (2) Folding and transporting and (3) Conveying to conveyer belt.

The processes for the entire automation system consist of 5 steps as follows:

- A staff picks vermicelli from the raw material rack which has already been divided into a small section (Approximated length of 120 cm and width of 5 cm).
- A staff places vermicelli on the base of cutter which is equipped with a set of two force sensors in different height location to compute the individual weight at each point. This process is used to find the positions that can divide the same weight in each section of cutting. Because vermicelli normally has different density in each area and the middle part has the maximum thickness, this might result on the heaviest weight are in the middle if the plates have the same dimensions.
- Then a microcontroller is used to calculate the position of cutting. The cutting blade will move in a straight line driven by a linear actuator from the top of vermicelli. Hence, a cutting blade can cut a vermicelli into 3 pieces with similar weight.
- The vermicelli is then folded by pulling and pushing the plate that will be described in detailed in Section B.
- Folding and conveying the vermicelli will move into the position of the aluminum boxes. There will be a rectangular container of 22 chambers on a conveyer that moves with a constant speed of 0.36-0.45 m/s.

2.1. First Prototype of Vermicelli folding set

The prototype for weighting and cutting machine consists of three 100 x150 mm weighing pads, three force sensors, and supporting base for force sensors (Model: ARD2-2151. This weighing and cutting set was fabricated in order to demonstrate how much the weight changes for each force sensor before and after the cutting process. For the folding set, this designed system is aimed for fast and compact folding which can be installed in a limited area in production lines. This machine consists of pneumatic cylinder that can move in 3 axis (XYZ). Both systems are showed in Figure 2.

2.2. *Testing of First Prototype of vermicelli folding set*

The following procedures are taken in order to prove a concept of cutting blades on a series of cutting pads with force sensors.

- Use the 40-gram dry vermicelli, and soak for 5 minutes.
- Spread the vermicelli in a straight line (length~40 cm and width~5 cm).
- Place the vermicelli on the weighing set and record the values for each load cells.
- Separate the vermicelli by cutting blades and weight them again on each weighing plate. Then, each experiment is repeated for 3 times and the results after cutting are recorded and compared. From the results, the weights on each pad are has slightly changed, which are within the acceptable range of ± 2 grams for the wet vermicelli. The concept of folding vermicelli consists of four steps as shown in Figure. 3. The detailed explanations for each step are described below.

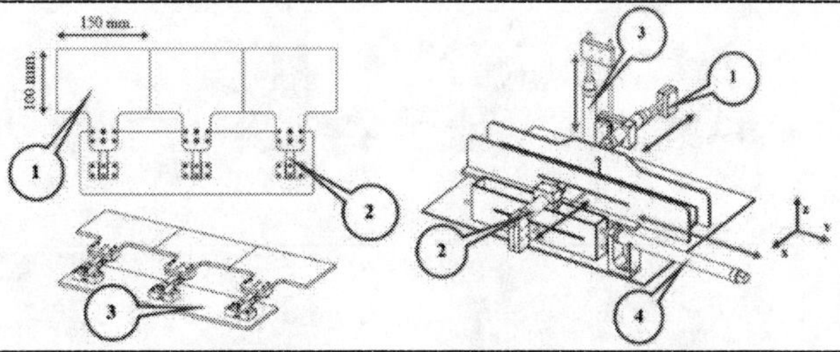

Figure 2. ***(Left)*** The components for the first prototype of vermicelli folding set; (1) a 100x150 mm weighing pad, (2) three force sensors, and (3) force sensors base. ***(Right)*** The first prototype for folding machine for vermicelli; (1) and (2) pneumatic cylinder in X axis, (3) pneumatic cylinder in Z axis, and (4) pneumatic cylinder in Y axis.

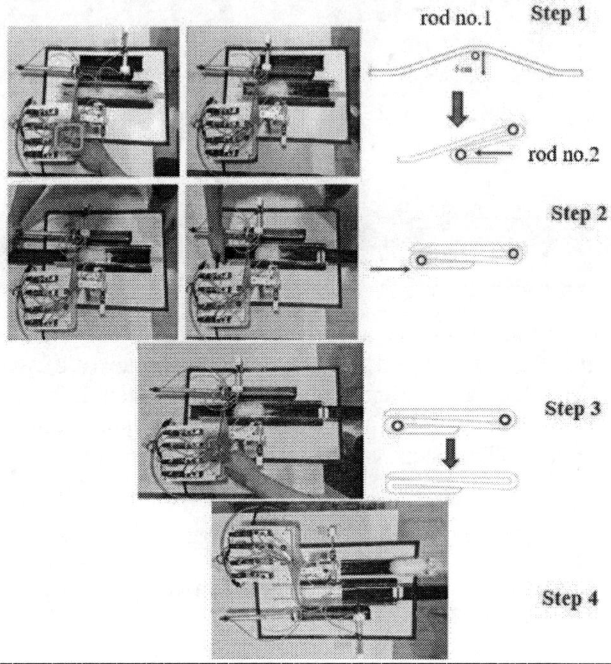

Figure 3. Vermicelli folding step; ***Step 1 :*** The steel rod no.1 is push underneath the center of the vermicelli and lift vermicelli up 5 cm. The steel rod no.2 is placed at the top of vermicelli and moved to the left. ***Step 2 :*** Folding the left end of the vermicelli by moving the pad to the right. The friction that occurs between the vermicelli and pad will fold the vermicelli down to the bottom. ***Step 3 :*** The rod no.1 and no.2 are removed to the initial positions. ***Step 4 :*** Then the vermicelli is removed from the folding set by pulling the container.

IOP Conf. Series: Materials Science and Engineering **639** (2019) 012015 doi:10.1088/1757-899X/639/1/012015

After folding the vermicelli with the folding set, the data has been collected by taking pictures from the top view and side view. The result showed in Table 1. From the results, the shape of the folded vermicelli is observed that there is some extending vermicelli in the red circles, which is called "mustache" and will affect the packaging process. This problem can be solved by cutting the ending tip of the vermicelli to the appropriate length before folding the vermicelli.

Table 1. Testing results of first prototype of Vermicelli folding set

No.	Top View	Side View
1	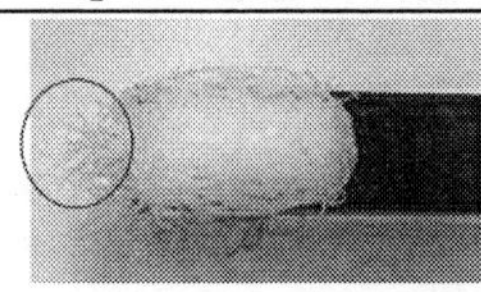	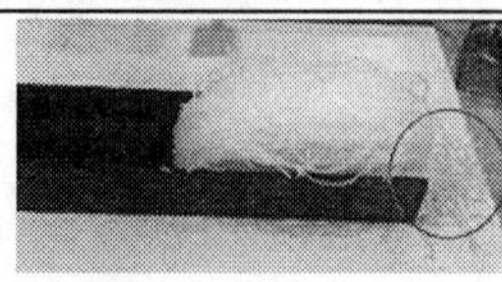
2	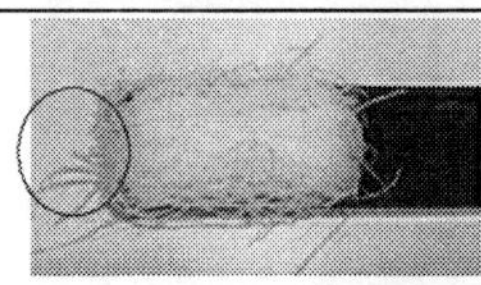	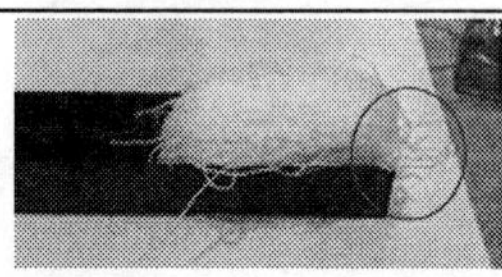
3	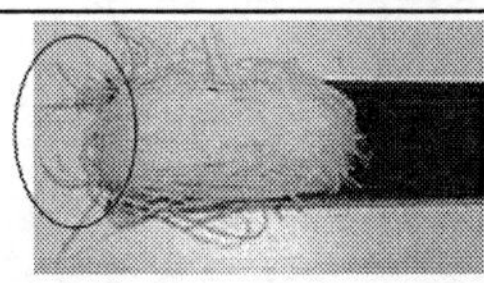	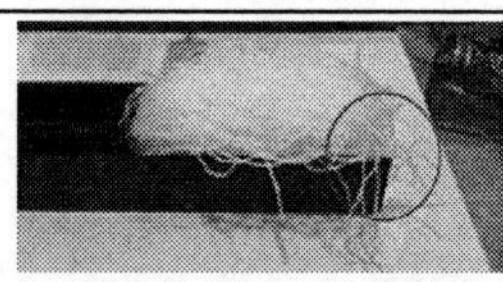
4	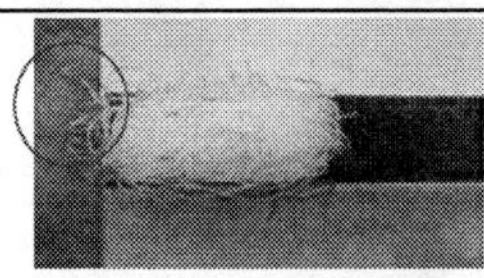	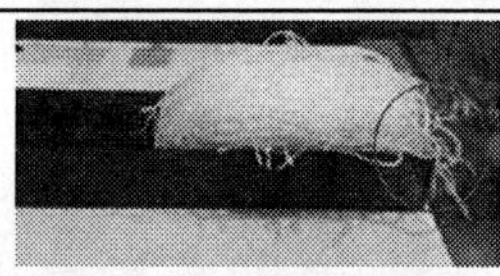
5		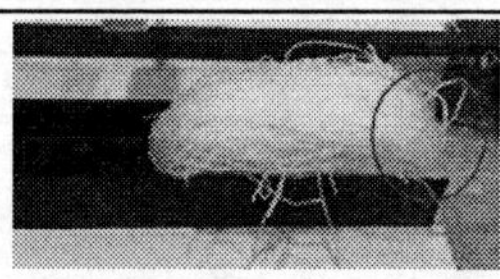

From the results of folding set, the averaged total time for only folding process is about 19 s as shown in Table 2. It is found that step 1 and step 3 that are fully automatic system, (in step 1 which is the process of folding vermicelli has an average time of 3.13 s and step 3 that is the process of vermicelli release to the tray has an average time of 2.71 s. However, the spending time in step 2 and step 4 that are majority of the spending time. These steps can be faster by improving to a fully automatic system.

Table 2. Spending time for first prototype of vermicelli folding set (unit: second)

	Step 1	Step 2	Step 3	Step 4	Total
Average time (sec)	3.13	7.25	2.71	5.78	18.87

3. The modifications and testing results

3.1. Folding and conveying set

After the folding process in the first prototype, it is observed that the shape of vermicelli has change because the vermicelli is soft, tangle, and messy. Therefore, folding and conveying set are redesigned into one combined set as shown in Figure. 4. and 5. By using a Cartesian structure, DC motor and belt transmission, and the pneumatic actuators, the modified design is compact. Unlike the previous model, this design can complete the entire task without the problem of moving the folded vermicelli into the aluminum boxes.

Figure 4. The modified prototype for folding machine with transporting system.

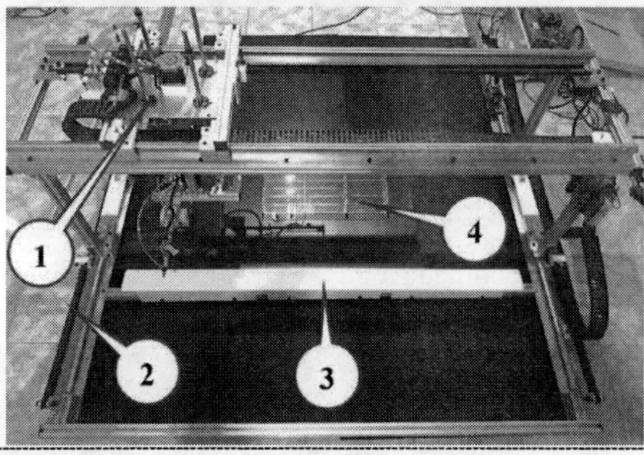

Figure 5. The components of folding and conveying set; (1) folding set, (2) conveyor set, (3) vermicelli base, and (4) testing box.

According to the vermicelli manufacturer, it is suggested that the shape should be changed. The reason for changing the folding technique is to reduce the overlapping parts that can cause the humidified spot after the drying process. These parts are shown in red circles in Figure 6. This is a crucial cause for unqualified products in vermicelli in production line.

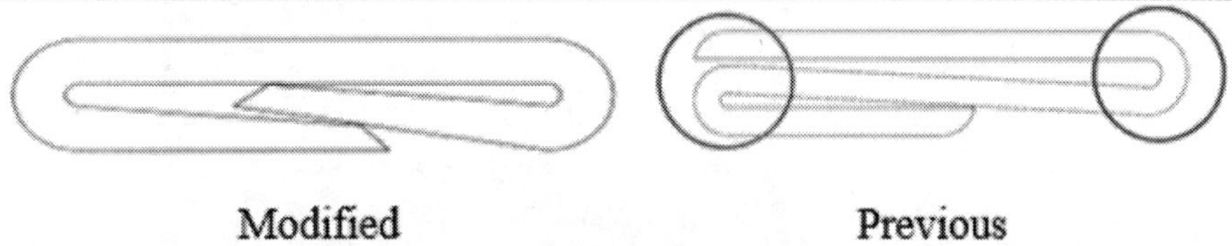

Figure 6. Overlapping parts in the modified folding technique, compared to the previous folding technique.

IOP Conf. Series: Materials Science and Engineering **639** (2019) 012015 doi:10.1088/1757-899X/639/1/012015

3.2. Testing of folding and conveying set (Transverse test)

The testing of folding and conveying set is examined by using dried vermicelli that is soaked in water for 5 minutes. Then, the vermicelli is arranged in a straight line with a length of 36 cm and weight of 110 grams. The vermicelli is place on base station, the procedure for folding are described in these following steps;

- The folding set moves from the HOME position to the vermicelli base.
- The folding set grips and folds the vermicelli, then moves into the testing box as shown in Figure 7. The spending time is recorded, and the experiments are repeated for 30 times.

After folding vermicelli with the fabricated folding and conveying machine, the data has been collected by taking pictures from the top view and side view. The examples of results are showed in Table 3.

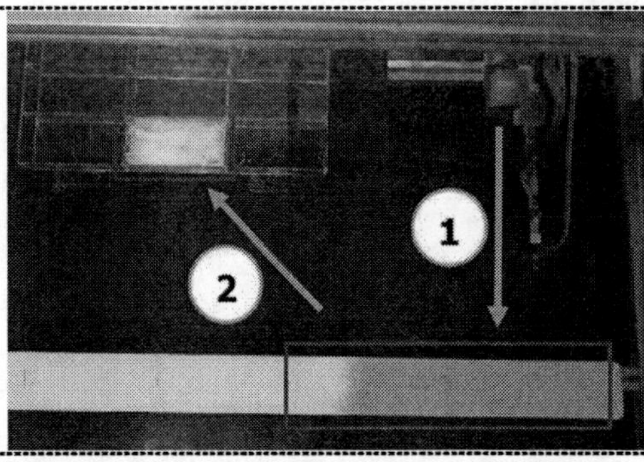

Figure 7. The steps for a folding and conveying set in transverse tests.

From the result of top and side view, it is noted that the shape of folding vermicelli is appropriate and neat. Unlike the previous method, the problem with "mustache" mostly disappeared, if the vermicelli is cut consistently as shown in Table 3.

Table 3. Testing results of the modified prototype for folding and conveying set in transverse tests

Trial No.	Top View	Side View
1	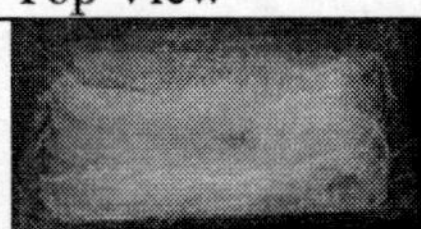	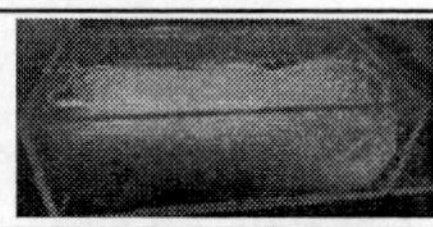
2	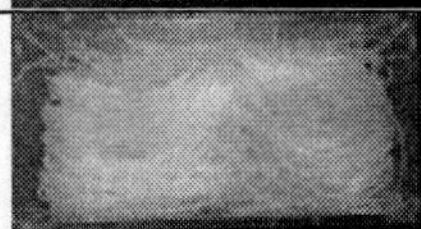	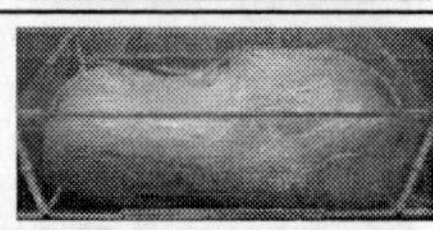
3	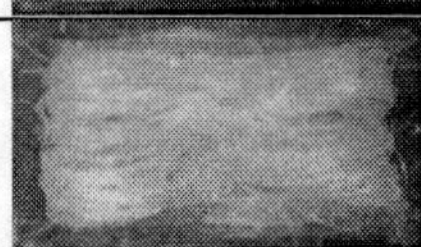	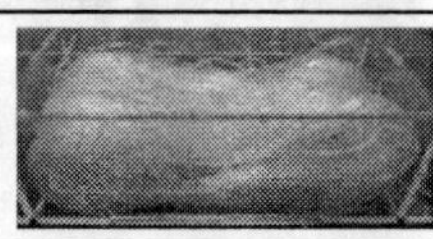
4	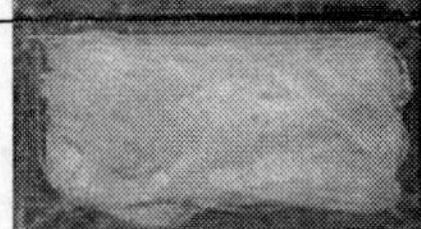	
5	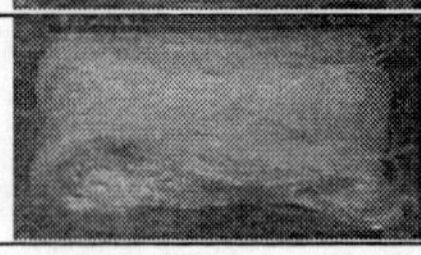	

IOP Conf. Series: Materials Science and Engineering **639** (2019) 012015 doi:10.1088/1757-899X/639/1/012015

From the spending time of the folding and conveying machine, it can be seen that all 30 trails have the same spending time of 23 s. as shown in Table 4. This proves the concept of automation that are driven from the same power supply and applied pressure. In step 1, the movement from HOME position to the base of vermicelli cutting station has an average time of 7 s. In step 2, folding vermicelli and transporting process have an average time of 16 s.

Table 4. Spending time for the modified prototype of vermicelli folding and conveying set transverse tests (unit: second)

	Step 1	Step 2	Total
	Moving	Folding	
Average time (sec)	7	16	23

4. Testing of folding and conveying set (Continuous test)

Another testing of folding and conveying set is continuous test by arranging the folded vermicelli in a linear pattern. The pattern begins from the Home position and the vermicelli is moved from position A to a, B to b, and C to c respectively. The procedures for folding are described in these following steps;

- Arrange vermicelli in a straight line with a length of 36 cm and weight of 110 grams and place on vermicelli base on position A, B, and C.
- The folding set moves from the HOME position to grip the vermicelli on position A base and places into testing box a, then moves from testing box a to position B base and place into testing box b. Then, the folding set moves from testing box b to position C base and place into testing box c as shown in Fig. 8. The spending time are recorded from the beginning until the arrangement in each box is complete. These experiments are repeated for 3 times.

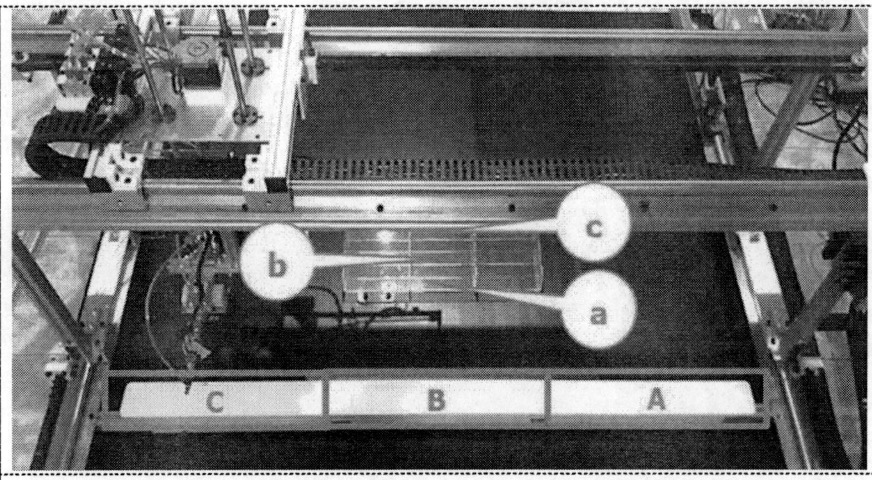

Figure 8. The steps for a folding and conveying set in Continuous tests.

The testing results are shown in Table 5. From the result, the shapes are in appropriate pattern, similarly to the linear pattern experiment. From the spending time of the folding and conveying machine in continuous test, it can be seen that there is some difference in time in each point due to the difference in moving distances. The average time for the entire process is 28 s as shown in Table 6.

Table 5. Testing results of the modified prototype of vermicelli folding and conveying set in continuous tests

Test no.	1	2	3
Top View	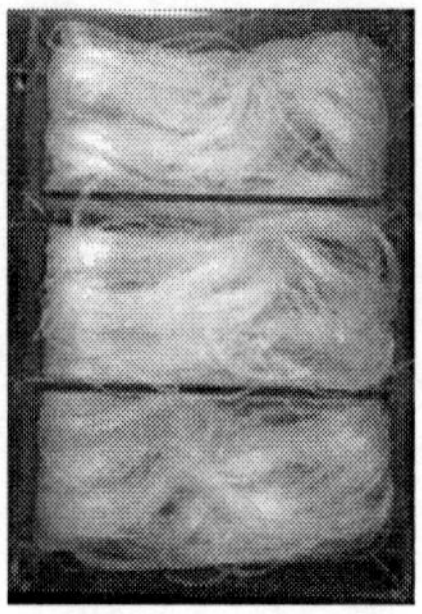	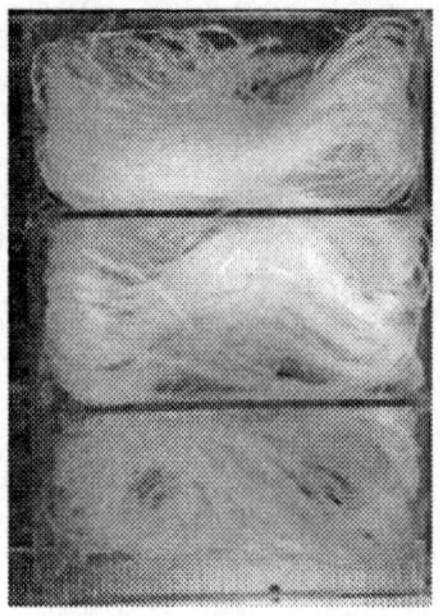	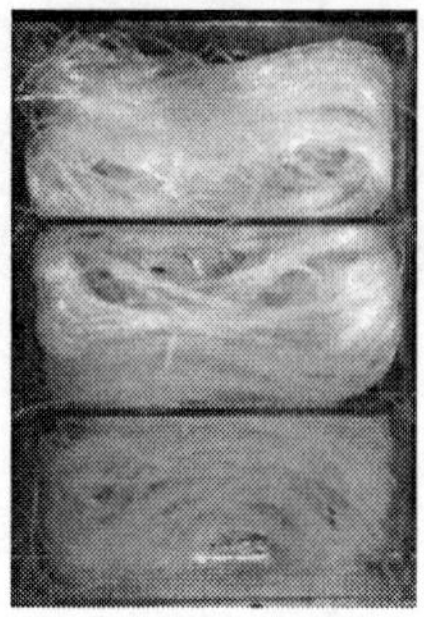

Table 6. Spending time for the modified prototype of vermicelli folding and conveying set in continuous tests (unit: second)

	A → a			B → b			C → c			Avg. Total
	Step 1 Moving	Step 2 Folding	Total	Step 1 Moving	Step 2 Folding	Total	Step 1 Moving	Step 2 Folding	Total	
Avg.	7	16	23	11	15	26	16	19	35	28

5. Conclusion

In the fabricated prototype, the folding and conveying set can pick all vermicelli from the base of the cutting station. No vermicelli falls during the folding and conveying process in the machine. The time-consuming for the transverse test was 23 seconds. For the continuous test; the average time for the complete arrangement was 28 seconds, due to the difference in travel distance. The problem in this design were the placement of vermicelli in the drying box that are not fit inside the testing box because the vermicelli's tip touched the box during conveying. This was resulted from the direction that the vermicelli was placed from the side view of the testing box. Alternatively, the vermicelli should be dropped from a higher position without touching the drying box. Another issue was the spending time for the folding and conveying set that is 28 seconds, compared to the spending time of 10-12 seconds in conventional process by human. However, the average spending time can be reduced if the folding and conveying set move in parallel. Moreover, if we added three more folding set in the same structure, the spending time can be reduced to 9.3 seconds per piece, which is approximately to the original process.

6. References

[1] Office of the Royal Development Projects Board (ORDPB), "Handbook 6: Vermicelli production from green beans 2nd Edition", Movement Gen 3 co.,ltd, 2012, pp. 5-12.
[2] S. Thanomsub, "Research and Development on Mungbean and Blackgram", Department of Agriculture, 2015, pp. 3-5.
[3] Ministry of Labour, Strategic Framework and Action Plan for Human Resource Development 20 years (BE 2560-2579), 2016, pp 11-15.
[4] Thai Industrial Standards Institute, The standard of vermicelli (TIS. 444-2549), 2007, pp 1-3.
[5] Sitthinan Co., Ltd., 2006, SITTHINAN Thai glass noodle manufacture , Available: http://www.sitthinan.co.th/aboutus/aboutus_th.htm [Feb 2nd, 2019].

Acknowledgments
The authors are grateful to the Thailand Research Fund (TRF) Grant no. RDG6250008 for providing a research grant to support this work.

IOP Conf. Series: Materials Science and Engineering **639** (2019) 012016 doi:10.1088/1757-899X/639/1/012016

Properties of Self-Compacting Concrete with Different Sources of Coarse Aggregates

Khamhou Saphouvong[1], Khampaseuth Thepvongsa[2], and Singvixay Simixay[3]

[1, 2, 3] Department of Civil Engineering, Faculty of Engineering, National University of Laos, Sokpaluang Campus, Lao-Thai Friendship Rd., Vientiane Capital, Lao PDR.
E-mail : k.saphouvong@nuol.edu.la

Abstract. Aggregate size and its source are crucial for concrete properties. This paper presents the evaluation of the properties of self-compacting concrete (SCC) which mixed with coarse aggregate from different sources. Coarse aggregate from three sources which are commonly available in Vientiane Capital, Mekong River (MK), Ngum River (NG) and Crushed Mountain Stone (MT) were used for this study. Six different mixed proportions with two maximum aggregate size 19mm and 12.5mm for each aggregate source were prepared. Ordinary Portland Cement Type I and Fly ash class C were used as binders. The replacement ratio of fly ash to cement was fixed at 20% and w/p ratio was 0.36. Sika Viscocrete-3180MR was used as water reducing admixture with 0.95% by weight of binders. The results show that, for the fresh concrete properties, the coarse aggregate with maximum size of 12.5mm from Mekong River is the most suitable aggregate comparing to other sources. For the hardened concrete with 28-day curing, the highest compressive strength is the mixed proportion MK12 for aggregate size 12.5mm. As seen, the coarse aggregate with 12.5mm of Mekong River is the most appropriate aggregate among three sources which has a flow-ability and higher compressive strength.

Keywords: self-compacting, aggregates, binders, flow-ability, compressive strength.

1. Introduction

Self-Compacting Concrete (SCC) is a flowing concrete mixture that is able to consolidate and completely filling in formwork under its own weight without the vibration during the placing of concrete, even in the presence of congested reinforcement. The hardened concrete is dense, homogeneous and has the same engineering properties and durability as traditional vibrated concrete. SCC was first introduced in the late 1980's by Japanese researchers. SCC is highly workable concrete that can flow under its own weight through restricted sections without segregation and bleeding [1]. Self-compacting concrete offers a rapid rate of concrete placement, with faster construction times and ease of flow around congested reinforcement. The fluidity and segregation resistance of SCC ensures a high level of homogeneity, minimal concrete voids and uniform concrete strength, providing the potential for a superior level of finish and durability to the structure. SCC is often produced with low water-cement ratio providing the potential for high early strength, earlier demolding and faster use of elements and structures. The elimination of vibrating equipment improves the environment on and near construction and precast sites where concrete is being placed, reducing the exposure of workers to noise and vibration. The improved construction practice and performance, combined with the health and safety benefits make. SCC is a very attractive solution for both precast concrete and civil engineering construction. EFNARC has published their "Specification & Guidelines for Self-Compacting concrete" which, at that time, provided state of the art information for producers and users [2]. Researchers have set some guidelines for mixture proportioning of SCC, which include i. reducing the volume ratio of aggregate to cementitious material [3], [4], ii. increasing the paste volume and water-cement ratio (w/c), iii. carefully controlling the maximum coarse aggregate particle size and total volume; For SCC, it is generally necessary to use super-plasticizer in order to obtain high mobility. Natural aggregates is the major incidents in concrete and its properties affected to concrete properties. Some researchers have proposed a mix proportioning system for SCC [5]. The maximum coarse aggregate particle size and total volume are essential to the rheological characteristics of SCC mixes .This study is being carried out to evaluate of the properties of self-compacting concrete (SCC) which mixed with coarse aggregate from different sources in Vientiane, Lao PDR.

2. Experimental details

2.1 Materials and mix proportions

Ordinary Portland cement (OPC) having a specific gravity of 3.13 is used for all concrete mixtures. Fly ash, class F, having a specific gravity of 2.0 is used as mineral admixture. High performance super plasticizer (SP), Sika Viscocrete-3180MR is used as water reducing admixture. Table 1 shows the physical properties of aggregates for this experiment. Three different sources of coarse aggregates are used in this study, Mekong River, Ngum River, and Crushed Mountain and the fine aggregate is river sand. All mixed proportion use the same fine aggregate. These aggregates conform to ASTM C33, standard specification concrete aggregates.

Table 1. Physical properties of standard aggregates

Aggregate type	Unit weight (g/cm^3)		Absorption (%)	Specific gravity	Los Angeles	Flakiness Index	F.M
	Loose	Rod					
Mekong River stone (MK)	1.63	1.65	0.81	2.56	26.8	36.61	-
Ngum River stone (NG)	1.53	1.51	1.25	2.55	36.5	14.96	-
Crushed Mountain Stone (MT)	1.36	1.53	0.6	2.66	22.35	41.43	-
River sand	1.55	-	1.42	2.38	-	-	2.6

Table 2. Mixed proportions for different combination

Designation of mixed proportion (Mixed ID)	Total binder (kg/m^3)	Cement C (kg/m^3)	Fly ash F.A (kg/m^3)	W (kg/m^3)	W/P ratio	SP %	SP (l/m^3)	Coarse Aggregate "G" (kg/m^3)	Sand "S" (kg/m^3)	G/S ratio
MK19	458	382	76	147	0.36	0.95	4.12	661	879	40/50
MK12	458	382	76	147	0.36	0.95	4.12	661	879	40/50
NG19	458	382	76	145	0.36	0.95	4.12	606	904	40/50
NG12	458	382	76	145	0.36	0:95	4.12	606	904	40/50
MT19	473	394	79	154	0.36	0.95	4.25	614	915	40/50
MT12	473	394	79	154	0.36	0.95	4.25	633	907	40/50

The mixed proportions in this study are shown in Table 2. Six series of mixed proportion of concrete were prepared with three different sources of coarse aggregate. The maximum of coarse aggregate size for each source is 19mm and 12.5mm. MK19, MK12, NG19, NG12, MT19, and MT12 are the samples that used the maximum size of coarse aggregate 19mm and 12.5mm for the coarse aggregate from Mekong River (MK), Ngum River (NG) and Crushed Mountain Stone (MT) respectively. Following guidelines of EFNARC for rheological properties of concrete in fresh state and using ACI 273R-07 method of mix design as reference, initial mix design was carried. The replacement ratio of fly ash to cement is 15% by weight of binders for all mixed proportions. Dosage of super plasticizer (SP) is 0.95% by weight of binders. As shown in the table, the ratio of coarse aggregate to fine aggregate is fixed at 40/50 and water powder ratio is 0.36.

2.2 Sample preparation and Testing Method

After testing the properties of mixing materials which shown in table 1, all designation mixed proportions as shown in table 2 were mixed one by one respectively. After mixing, the investigation of fresh concrete properties of self-compacting concrete were carried. The slump flow test is used to assess the horizontal free flow of SCC in the absence of obstructions. On lifting the slump cone, filled with concrete, the concrete flows. The average diameter of the concrete circle is a measure for the filling ability of the concrete. The time T500 is a secondary indication of flow. It measures the time taken in seconds from the instant the cone is lifted to the instant when horizontal flow reaches diameter of 500mm. The flow ability of the fresh concrete can be tested with the V-funnel test, whereby the flow time is measured. The funnel is filled with about 12 liters of concrete and the time taken for it to flow through the apparatus is measured, the flow time will increase significantly. According to guidance of EFNARC [2], a funnel test flow time less than 6s is recommended for a concrete to qualify for an SCC. The passing ability is determined using the L- box test10 .The vertical section of the L-Box is filled with concrete, and then the gate lifted to let the concrete flow into the horizontal section. The height of the concrete at the end of the horizontal section indication of passing ability. The specified requisite is the ratio between the heights of the concrete at each end or blocking ratio to be 0.8-1.0. Self-compacting concrete is characterized by filling ability, passing ability and resistance to segregation. Many different methods have been developed to characterize the properties of SCC. No single method has been found until date, which characterizes all the relevant workability aspects, and hence, each mix has been tested by more than one test method for the different workability parameters. Table 3 gives the recommended values for different tests given by different researchers [2], [6], for mix to be characterized as SCC mix. For hardened concrete, six cylindrical specimens with dimension of 15x30 cm for compressive strength were cast for 28-day curing age and six specimens with the same dimension were cast for tensile strength test.

Table 3. Recommended limits for different properties [2]

Property	Range
Slump flow Diameter	650-800mm
T_{500}	2-5 sec
V-funnel	6-12 sec
L-Box H_2/H_1	0.8-1.0

3. Results and Discussion

3.1 Properties of fresh concrete

The fresh properties of each mix were evaluated. The tests were carried out to determine the effect of maximum size and sources of coarse aggregate on the filling ability, passing ability, and segregation resistance of SCC. The rheological characteristics of mixes MK19, MK12, NG19, NG12, MT19, and MT12 are discussed as below. Figure 1 shows the results of slump flow tests. The value of slump flow represent the distance of horizontal free flow. As shown in the figure, almost mixed proportions for 19 and 12.5mm particle size aggregate, slump flow characteristics are between 650-750 mm, which satisfy the EFNARC requirement, 650-800mm; except, MT12, which is lower than 650mm. Thus, all the mixes are assumed to have good consistency and workability from the filling ability point of view. MK12 is

4% free flow better than NG12. However, the mix with crushed mountain stone show the wide range of variation. This variation illustrates the effect of the shape of coarse aggregate variables on filling ability of SCC mixes.

The value of T500 represent the time required for the concrete flow to reach a circle with a 500mm diameter as shown the results in Figure 2. The results of slump flow in figure 1 almost consistent. However, the results of T500 in figure 2 is vary. If the passing time is longer means the flow rate is slower. As seen in the figure 2, only two mixes, MT19 and NG12 are satisfy EFNARC requirement which ranges 2-5 sec. Some mixes for both, 12.5mm and 19mm maximum coarse aggregate size mixed, the flow rate is longer than requirement. This vary range of flow rate may affected by combination of size and shape of the aggregate. However, if time passing of T500 is greater than 5 seconds, those mixtures generally considered a high viscosity SCC which is required for resistance of segregation. This variation illustrates the effect of coarse aggregate variables on the filling ability of SCC mixtures. This agrees with the study carried out by Krishna1 et al [7] and Guru Jawahar et al [8].

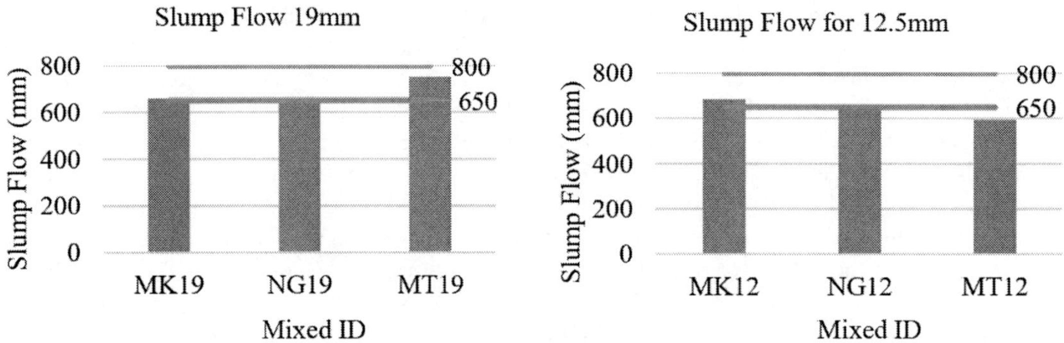

Figure 1. Slump Flow for NM19 and NM12.5

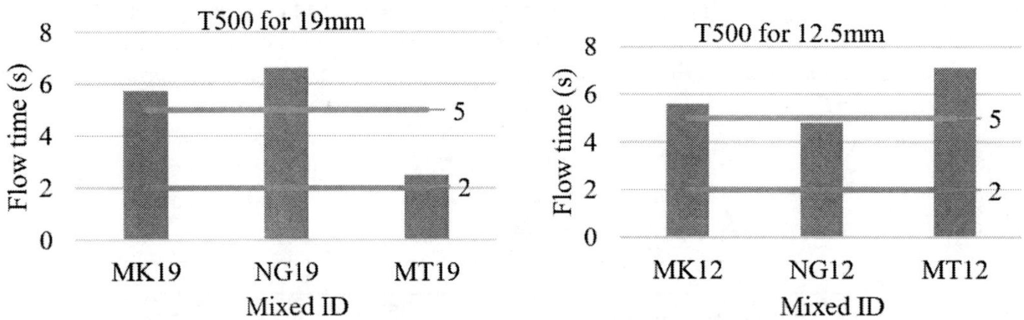

Figure 2. Time required passing 500mm Dia. Circle (T500) for NM19 and NM12.5

The L-Box is used to measure the passing and filling ability of SCC mixes. The value of H2/H1 represent the blocking ratio (BR). The L-Box test results are shown in Figure 3. The results show that the mixes with 12.5mm maximum size of coarse aggregate give values of (BR) lower than the mixes with 19mm maximum size of coarse aggregate. The smaller size presents the lower blocking ratio. It is corresponded to the study by Khaleef et al [9]. This is due to the tendency of the mixes with larger maximum size of aggregate to jam the flowing, while the mixes with the smaller maximum size of coarse aggregate will flow freely without stopping. Almost of mixed proportions are satisfy EFNARC requirement which ranges the (BR) from 0.8-1.0; except, MT12.

The V-Funnel is used to measure filling ability by measuring the flow rate of SCC mixtures. The test results of V-funnel are shown in Figure 4. The results reveal the influence of the maximum size of coarse aggregate on the flowing time. It can be seen from the figure, the larger maximum size of coarse aggregate leads to an increase in the flow time. From the results, the flow time of the mixes with 12.5mm maximum size of coarse aggregate is, 20% by average, faster than the mixes with 19mm maximum size of coarse aggregate. Almost of the mixtures are satisfy EFNARC requirement which ranges from 6-12 sec; except, NG12 which is the natural river aggregate with maximum coarse aggregate size of 12.5mm. The flow rate of NG12 is fastest comparing to the other mixes. Due to the shape of aggregate from Ngum River (NG) is round and smooth.

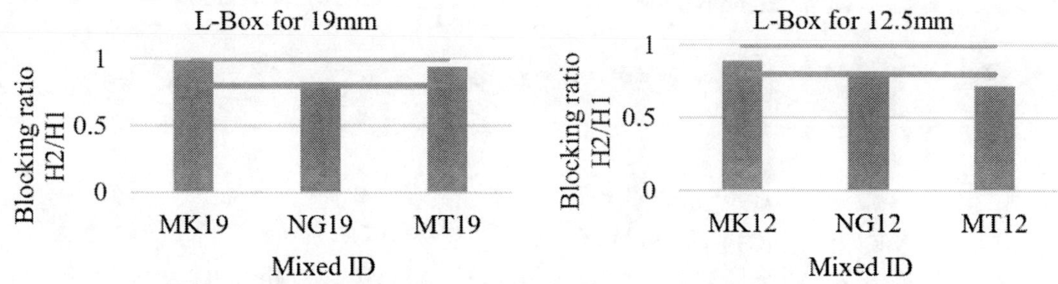

Figure 3. Blocking ratio (BR) of H2/H1 of L-Box test for NM19 and NM12.5

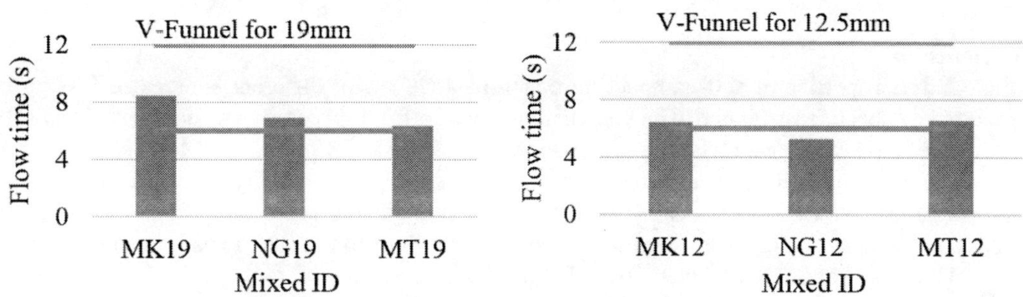

Figure 4. Flow time of V-funnel test for NM19 and NM12.5

3.2 Properties of harden concrete

From the test results presented in Figure 5 and Figure 6, it can be noticed that the compressive strength and tensile strength of the mixes with 12.5mm maximum size of coarse aggregate and the mixes with 19mm maximum size of coarse aggregate give the same tendency. The average compressive strength of the mixture with 12.5mm maximum size of coarse aggregate is almost the same with the mixture with 19mm maximum size of coarse aggregate as shown in Figure 5. Similarly, the tensile strength of the mixture with 12.5mm maximum size of coarse aggregate is almost the same with the mixture with 19mm maximum size of coarse aggregate as shown in Figure 6.

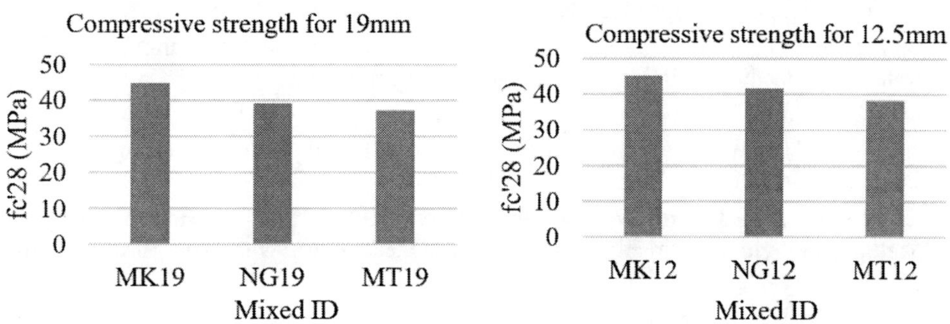

Figure 5. Compressive strength (f'_c) at 28 days for NM19 and NM12.5

Due to the Self-compacting concrete (SCC) is able to flow under its own weight and does not require any external vibration for compaction. Due to this highly workable concrete that can flow under its own weight through restricted sections without segregation and bleeding. It is sufficiently fill voids without tending to segregation resulted in a higher bonding strength in the interface transition zone (ITZ) around the aggregate particles. This is in agreement with Gencel et al [10]. The tensile strength test results of SCC in this study, by average, obtain 10% of compressive strength. The results show that tensile strength of SCC is commonly same as normal concrete by obtaining 10% of compressive strength.

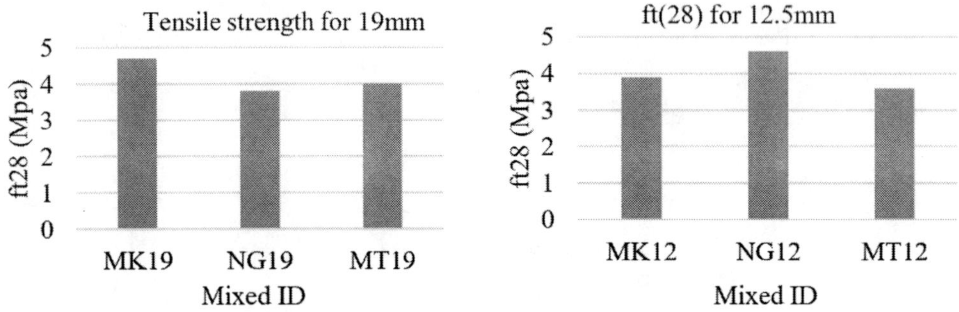

Figure 6. Tensile strength (ft) at 28 days.

4. Conclusion

In this study, properties of self-compacting concrete (SCC) with different sources of mixing coarse aggregate have been investigated. The conclusions derived from this study are summarized as follows:

1. The coarse aggregate with maximum size of 12.5mm and 19mm from Mekong River show the better rheological characteristics for the filling ability, passing ability. The 12.5mm of maximum aggregate size from Mekong River is the most suitable aggregate comparing to other sources.
2. The highest compressive strength is the mixed proportion MK12 for aggregate size 12.5mm. The tensile strength obtained 10% of compressive strength.
3. The coarse aggregate with 12.5mm of Mekong River is the most appropriate aggregate among three sources which has a flow-ability and higher compressive strength.

References

[1] Nagamoto N. and Ozawa K., (1997), "Mixture properties of Self-Compacting", High Performance Concrete, Proceedings, Third Canmet/ACI International Conferences on Design and Materials and Recent Advances in Concrete Technology, SP-172, V.M. Malhotra, American Concrete Institute, Farmington Hills, Mich., 623-637.

[2] EFNARC (European Federation of national trade associations representing producers and applicators of specialist building products), (2002), "Specification and Guidelines for self-compacting concrete", Hampshire, UK.

[3] Khayat K.H. and Ghezal A., (1999), "Utility of Statistical models in Proportioning Self-Compacting Concrete", Proceedings, RILEM International symposium on Self-Compacting Concrete, Stockholm, 345-359.

[4] Okamura H. and Ozawa K., (1995), "Mix Design for Self-Compacting Concrete", Concrete Library of Japanese Society of Civil Engineers, 107-120.

[5] Gencel, O., Ozel, C., Brostow, W., Martinez-Barrera, G., (2011), "Mechanical properties of self-compacting concrete reinforced with polypropylene fibres", Materials Research Innovations, Vol.15, No.3, 216-225.

[6] Nagataki S. and Fujiwara H., (1995), “Self-Compacting property of Highly-Flowable concrete”, Second Conference on advances in Concrete Technology, ACI SP-154,V.M. Malhotra, American Concrete Institute, 301-304.

[7] Krishna1, A.V., Krishna, B., Rajagopal Rao2, A., (2010), “Effect of different sizes of coarse aggregate on the properties of NCC and SCC”, International Journal of Engineering Science and Technology, Vol. 2(10), 5959-5965.

[8] Guru Jawahar, J., Premchand, M.M., (2012), “Effect of Coarse Aggregate Blending on Fresh Properties of Self-Compacting Concrete”, International Journal of Advances in Engineering & Technology, Vol. 3, Issue 2, 456-466.

[9] Khaleef, O.R., Al-Mishhadani, S.A., Abdul Razak, H., (2011), “The Effect of Coarse Aggregate on Fresh and Hardened Properties of Self-Compacting Concrete (SCC)”, The Twelfth East Asia-Pacific Conference on Structural Engineering and Construction, Science Direct, Elsevier, Procedia Engineering 14, 805-813.

[10] Gencel, O., Ozel, C., Brostow, W., Martinez-Barrera, G., (2011), “Mechanical properties of self-compacting concrete reinforced with polypropylene fibres”, Materials Research Innovations, Vol.15, No.3, 216-225.

IOP Conf. Series: Materials Science and Engineering **639** (2019) 012017 doi:10.1088/1757-899X/639/1/012017

Resource Scheduling Problem in Distribution Center

C. Hamontree[1], S. Prompakdee[2] and J. Koiwanit[3]

Department of Industrial Engineering, Faculty of Engineering, King Mongkut's Institute of Technology Ladkrabang, Bangkok, Thailand

Corresponding author: Chaowalit.ha@kmitl.ac.th

Abstract. Due to the rapid growth of the retail shop trend recently, the Distribution Center (DC) or warehouse had become to be a vital part in the supply chain in order to store goods from multiple suppliers and deliver to multiple retailers. Cross docking is a relatively the logistics technique used in the DC in order to move materials from inbound trailers to outbound trailers as quickly as possible. From the following problem, an ineffectual DC management leads to the resources utilization problem such as machine, forklift, industrial trucks, etc. The main objective of this research is to study and solve the scheduling problem of the usage of the forklift in the Cross docking of DC and to manage the inbound and outbound docks in order to minimize Tardiness jobs, and also to maximize the efficiency of machine utilization. This study has provided the scheduling, comparison from genetic algorithm and heuristic method. As the result came out we found that the forklift scheduling by the genetic algorithm method is given the better management result better than heuristic method.

1. Introduction

Materials handling is a central concern in storage space decisions and is mostly a cost-consuming activity. The objectives for materials handling are therefore cost focused and attempt to reduce handling costs while increasing space utilization. Materials handling can be improved through good load utilization, space layout and equipment choices. Typically, storage and order picking operations take up the bulk of handling activity in a warehouse. These operations include stock locating, stock arrangement, product sequencing, order splitting and item batching, all of which are labour intensive and expensive. Further, warehouses need to be configured to handle equipment and storage facilities to accommodate these activities and an inventory management system has to be in place. Cross-docking attempts to lessen or eliminate these burdens [6].

2. Research Theories

2.1. Cross Docking Problem

There are four major functions in a traditional warehouse receiving, storage, order picking, and shipping. The middle two functions are typically the most costly storage because of inventory holding costs; and order picking because it is labour intensive. Cross-docking is a logistics technique to transfer shipments directly from incoming to outgoing locations without storage in between. In a traditional DC, the warehouse maintains stock until a customer orders, then the product is picked, packed, and shipped. When replenishments arrive at the DC, they are stored until a customer is identified. In a cross-docking model, the customer is known before the product gets to the warehouse and there is no need to move it to storage [6]. In this case, cross-docking is very much like Just-In-Time (JIT) manufacturing, which is most viable when demand has low variance and there is high enough volume to justify frequent setups.

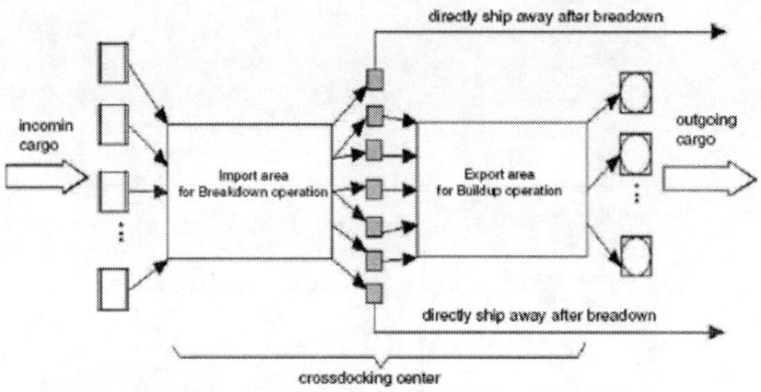

Figure 1. Typical Cross-Docking Flow

2.2. *Scheduling Problem*

It has been specified as "when labour, equipment and facilities are needed to produce a product or provide a service" [5]. Scheduling work on machines in a large manufacturing system is an extremely challenging task. A lot of research on this topic has illustrated poor performance in scheduling had brought numerous problems, such as failures to expedite jobs required by downstream work centers, long throughput time, high level of work in process inventory, excessive overtime and other operational costs and so on [3], [4]. Therefore, scheduling is one of the core activities in cross-docking environment.

2.2.1. Machine Scheduling Problem. The study of earliness and tardiness penalties in scheduling models is a relatively recent area of inquiry. For many years, scheduling research focused on single performance measures, referred to as regular measures, which are non-decreasing in job completion times. Most of the literature deals with such regular measures as mean flow time, mean lateness, the percentage of jobs tardy, and mean tardiness. The mean tardiness criterion, in particular, has been a standard way of measuring conformance to due dates, although it ignores the consequences of jobs completing early. However, this emphasis has changed with the current interest in Just-In-Time (JIT) production, which espouses the notion that earliness, as well as tardiness, should be discouraged. In a JIT scheduling environment, jobs that complete early must be held in finished goods inventory until their due date, while jobs that complete after their due dates may cause a customer to shut down operations. Therefore, an ideal schedule is one in which all jobs finish exactly on their assigned due dates. This can be translated to a scheduling objective in several ways. The most obvious objective is to minimize the deviation of job completion times around these due dates.

2.2.2. Genetic Algorithm. Genetic Algorithm (GA) was developed by John Holland in 1975. GA mimics the natural selection and natural genetic selection process concurrent with artificial system for optimising problems. This algorithm is applied from Darwin's principle of "Survival of fittest" [2]. Goldberg [1] stated that GA is different from other search methods, which are the superior pros of GA towards other methods. GA operates on a population of solutions rather than a single solution. GA works by encoding the parameter set of the problem into finite-length string. The strings, which represent the solution of the search problems are denoted as chromosomes. The alphabet is denoted as genes. The value of genes is referred as alleles. Implementing natural selection to scheduling problem, chromosomes represent the sequence of scheduling whilst genes are the position of the jobs in the schedule.

IOP Conf. Series: Materials Science and Engineering **639** (2019) 012017 doi:10.1088/1757-899X/639/1/012017

2.3. Simple Genetic Algorithm

The concept of GA starts by transforming potential solution of the problem into individual chromosomes. An initial set of chromosomes is called a population. Each individual chromosome has their own characteristics or fitness, which appraise through the GA operators: selection, reproduction (crossover), and mutation.

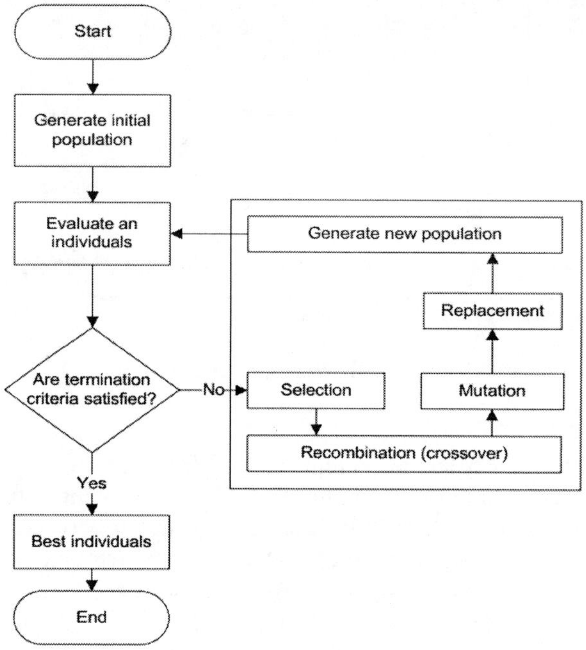

Figure 2. A Simple Genetic Algorithm cycle

The procedure starts from generating an initial population. Then, individuals are evaluated. If the individual is unsatisfied, the GA operation is preceded by starting from selection, recombination, and mutation respectively. The selection chooses a parent chromosome from the population. Secondly, the crossover produces new parent by recombining parts from parents, which is called offspring. Lastly, the mutation produces new parent by making changes within the offspring. The new population is chosen from better fitness, which is kept from each GA operator by operating through continuous iterations; called generations. Once having several generations, the algorithm converges to the best parent, which represents an optimal or suboptimal solution to the problem.

3. Research Theories

3.1. Problem Parameter Setting

The cross-docking scheduling problem described in the previous section can be modelled as a machine-scheduling problem. Each incoming container and each outgoing container is a job to be processed by teams that we think of as machines, where only a limited number are available. Further, machines handling incoming cargo can be thought to be parallel and likewise for machines handling outgoing cargo. This is because teams are able to operate simultaneously. Each incoming container can be thought of as a job that has a release time after which it can be processed, a due date and a processing time, assumed to be known beforehand. Each outgoing container has a due date, a processing time, and a number of source containers that feed it. Here, use 'container' as a generic packing form to include containers, pallets, and other packing used in DC. Use the terms 'team' and 'machine' and 'container' and 'job' interchangeably. The number of breakdown and build-up processing teams is also known together with penalties for earliness and lateness. An experiment is developed to study the suitability of the Genetic Algorithm (GA) solution for the cross-docking scheduling problem. The instance parameter to apply in this problem is shown in table 1.

T able 1. Setting up the optimization parameters

Parameter	Value
Number Generation	10,000 and 50,000
Experimental Population	500 and 1000
Crossover rate	0.4, 0.6 and 0.9
Mutation rate	0.05, 0.10 and 0.25

3.2. Setting Parameters

The trials number, population rate, crossover rate, and mutation rate are predefined automatically in the program. However, this value can be adjusted depending on the experiments. In this research, the trials number is 10,000 and 50000. The population sizes are 500 and 1000. The crossover rates are 0.4, 0.6 and 0.9. The mutation rates are 0.05, 0.1, and 0.25.

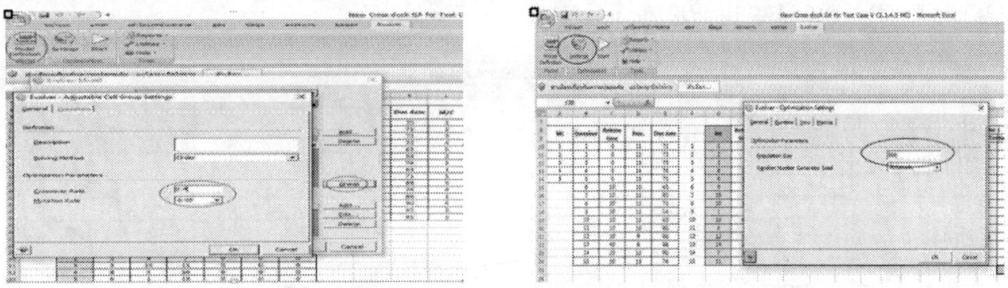

Figure 3. Model definition and setting parameters (Adjustable cell ranges)

4. Experimental Results

We have provided two methods for solving the cross-dock scheduling problem by *using Heuristic and GA method*. Both of these are appropriate for the problem. As cross-dock scheduling problem is NP-hard, so it will take a long time for an integer programming solver such as MPL (Not applied in this case) to reach an optimal solution. A part from heuristic methods, this problem has been solved by using Genetic Algorithm (GA). There are various parameters that should be considered in order to evaluate the performance of GA. The experiment is divided into two cases and three performance measurements are considered including of makespan, tardiness time, and machine utilization, which are described below.

Case I: One-Phase Single-Machine at inbound area
Case II: Two-Phase Parallel-Machine at inbound and outbound area

Parameter Notation Definition:

A: Release time for incoming container i
B: Processing time required to break down incoming container i
C: Due date for incoming container i
D: Number of source of outgoing container j
E: Processing time required to build up outgoing container j
F: Due date for outgoing container j

IOP Conf. Series: Materials Science and Engineering **639** (2019) 012017 doi:10.1088/1757-899X/639/1/012017

T a b l e 2. Instance Data Set Case II-IV

Case	Incoming Area		Outgoing Area	
	No. Container	No. Forklift	No. Container	No. Forklift
1	15	5	-	-
2	15	5	14	1

T a b l e 3. Data Set Case I-IV

	Number of Containers														
	1	2	3	4	5	6	7	8	9	10	11	12	13	14	15
A	0	0	0	0	0	10	10	20	30	10	10	30	40	20	30
B	15	15	10	15	15	10	10	10	10	10	10	10	10	15	15
C	70	70	70	70	80	60	90	70	50	60	90	90	90	90	70
D	2	4	3	1	5	6	8	3	9	12	13	12	6	12	
E	20	20	20	25	20	25	20	10	20	15	25	10	10	10	
F	80	80	80	90	90	80	110	110	70	80	110	110	110	110	

Case I: One-Phase Single-Machine Model at Inbound Area (Based Case)

In this case, we will investigate the one phase and single machine problem, which means only one machine at inbound area and only consider the breakdown operations and omit the build-up operations in outbound area, see in Figure 4.

list	Release time	Proc.	Total	Due date	M/C	Available for Next	Machine 1			Late	complete time
						M/C 1	Start Time	Comp Time	Idle Time		
1	0	15	15	70	1	15	0	15	0	0	15
2	0	15	15	70	1	30	15	30	0	0	30
10	10	10	20	60	1	50	30	50	0	0	50
4	0	15	15	70	1	65	50	65	0	0	65
3	0	10	10	70	1	75	65	75	0	5	75
5	0	15	15	80	1	90	75	90	0	10	90
6	10	10	20	60	1	110	90	110	0	50	110
11	10	10	20	90	1	130	110	130	0	40	130
7	10	10	20	90	1	150	130	150	0	60	150
8	20	10	30	70	1	180	150	180	0	110	180
14	20	15	35	90	1	215	180	215	0	125	215
9	30	10	40	50	1	255	215	255	0	205	255
12	30	10	40	90	1	295	255	295	0	205	295
15	30	15	45	70	1	340	295	340	0	270	340
13	40	10	50	90	1	390	340	390	0	300	390
									0	1380	

Figure 4. The experimental result of one-phase single-machine model at inbound area

M/C	Available for Next Task Assignment		Machine 1			Machine 2			Late	complete time
	M/C 1	M/C 2	Start Time	Comp Time	Idle Time	Start Time	Comp Time	Idle Time		
1	40	0	0	40	0	0	0	0	0	40
2	0	20	0	0	0	0	20	0	0	20
2	0	30	0	0	0	20	30	0	0	30
2	0	50	0	0	0	30	50	0	0	50
1	55	0	40	55	0	0	0	0	0	55
2	0	65	0	0	0	50	65	0	0	65
1	70	0	55	70	0	0	0	0	0	70
2	0	80	0	0	0	65	80	0	0	80
1	90	0	70	90	0	0	0	0	0	90
2	0	110	0	0	0	80	110	0	40	110
1	110	0	90	110	0	0	0	0	20	110
1	150	0	110	150	0	0	0	0	60	150
2	0	145	0	0	0	110	145	0	55	145
2	0	195	0	0	0	145	195	0	105	195
1	195	0	150	195	0	0	0	0	125	195
					0			0	405	

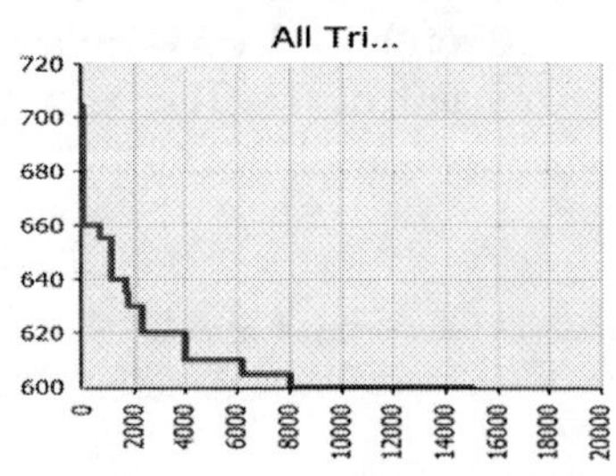

Figure 5. The experimental result of two-phase parallel-machine at inbound and outbound area

Case II: Two-Phase Parallel-Machine at Inbound and Outbound Area

This case is exactly the crossdocking scheduling model, which is two-phase parallel-machine scheduling problem. Two machines work at inbound area to consider the breakdown operations and the build-up operations in outbound area, see in Figure 5.

Inbound Area

Forklift	Complete time								
	1	2	3	4	5	6	7	8	9
1	40	55	70	90	110	150	195	-	-
2	20	30	50	65	80	110	145	195	-

Outbound Area

Forklift	Complete time								
	1	2	3	4	5	6	7	8	9
1	50	60	70	90	100	155	165	-	-
2	40	60	75	85	100	155	165	-	-

Figure 6. The experimental result of two-phase parallel-machine at inbound and outbound area

T a b l e 4. Comparison of the performance criteria

Method	Performance Criteria (Case 2)					
	Makespan (min)		Total Tardiness (min)		Utilization (%)	
	Inbound	Outbound	Inbound	Outbound	Inbound	Outbound
GA	182	155	100	175	100	42.4
Heuristics	195	165	115	180	90	36.1
Improved (%)	6.67	6.06	13.04	2.78	11.11	17.45

5. Conclusion

The experiment is finding the optimal solution for solving the forklift scheduling problem in cross-docking environment by using GA and heuristic method. There are five cases applied in this case study where the GA algorithm has shown that it is suitable method and given high performance better than heuristic method for solving the cross-dock scheduling problem. Moreover, it can be increased the machine utilization. This leads to improve the forklift scheduling and resource utilization.

References

[1] Goldberg, D.E, Genetic algorithms in search, optimization, and machine learning. Boston: Addison-Wesley Pub. Co., 2004.

[2] Holland, John H. Adaptation in Natural and Artificial Systems. Ann Arbor: University of Michigan Press, 1975.

[3] Jacobs, Larry W., and Lauer, Joachim . DSS for job shop machine scheduling. Industrial Management & Data Systems, 1994.

[4] Melnyk, Steven A., Vickery, Shawnee K., and Carter, Phillip L. Scheduling, sequencing, and dispatching: Alternative perspectives. Production and Inventory Management, 1986.

[5] Roberta S. Russell and Bernard W. Taylor- 4th ed., international ed. Upper Saddle River, N.J.: Prentice Hall, 2003.

[6] Y Li, A Lim and B Rodrigues, Crossdocking – JIT scheduling with time windows; Journal of the Operational Research Society, December, Volume 55, Number 12, 2004

IOP Conf. Series: Materials Science and Engineering **639** (2019) 012018 doi:10.1088/1757-899X/639/1/012018

Fast large-scale current oscillations in SOFCs with intermediate fuel utilisation

J D Sands

Department of Industrial Engineering, Faculty of Engineering, King Mongkut's Institute of Technology Ladkrabang, Bangkok 10520 , Thailand

E-mail: `jonathan.sa@kmitl.ac.th`

Abstract. In the push towards renewable energy technology fuel cells are set to replace many conventional means of power generation. Their efficiency and cleanliness compared to internal combustion engines is undisputed, however widespread adoption has yet to occur on a global scale. Two of the key issues affecting their mass adoption are their cost and their long-term stability. Whilst long-term testing has proven fuel cells a worthy competitor in the energy market, there are still reports in the literature of degradation caused by a number of different phenomena. In particular, most long-term testing focusses on intermediate fuel utilisation whereas many applications require high fuel utilisations. It is known that at high fuel utilisations several degradation mechanisms may come into play, sometimes resulting in agglomeration, cracking, decreased power output, and electrical oscillations. In this paper, current oscillations are reported in a solid oxide fuel cell under intermediate fuel utilisation which places it firmly in the applicable region for standard operation. Thus it is incumbent on fuel cell researchers to understand and control this unwanted behaviour in order to ensure long-term stability of a fuel cell stack. A prototype model is used here to explore potential causes using first principles.

Keywords - solid oxide fuel cells, fuel utilisation, instabilities, oscillations, nonlinear dynamics

1. Introduction

The continued industrialisation of an ever-increasing number of societies around the globe brings both opportunity, as well as new existential threats. Through the choice of technology, coupled with a deep scientific understanding, we aim to enhance the opportunities and mitigate the threats that new technology brings. In particular, it is noted that the expansion of industry, economic growth, energy production, and road transport is strongly correlated with increasing air pollution [1]. There exist many strategies to mitigate air pollution caused by industry through the removal of contaminants or traffic management (see for example [2, 3]), however it is noted that the most beneficial impact comes from the introduction of cleaner, more efficient technologies [4].

Solid oxide fuel cells (SOFC) are not only capable of operation that produces water as the only emission, but have been demonstrated to give high efficiencies in terms of energy production and storage [5, 6]. Furthermore, stacks on the order of kilowatts have been tested for long-term stability, one of which was able to reach 40,000 h of continuous operation, far exceeding the target for mobile applications of 10,000 h, and also meeting the requirements for decentralised energy conversion [7–9].

The correct management of fuel utilisation is critical to the long-term durability of a fuel cell stack, and it is well known that under high fuel utilisation, degradation leading to multiple steady states, autonomous oscillations in electrical output, and cell failure may occur [9–19].

IOP Conf. Series: Materials Science and Engineering **639** (2019) 012018 doi:10.1088/1757-899X/639/1/012018

Resolving these issues will help to lower the barrier to market and allow fuel cells to contribute towards technological development whilst simultaneously improving the quality of our air.

In [9] it is recommended to keep fuel utilisation below 85% using their particular SOFC configuration in order to avoid detrimental nickel oxidation at the anode. Additionally, small fluctuations and full oscillations can be seen in their galvanostatic tests at 80% and 90% fuel utilisations respectively. In order to operate with such high fuel utilisations we must fully understand the mechanisms occurring within the fuel cell under such conditions to enable better design of SOFCs that can tolerate such a load. High fuel utilisations were simulated in [13] for a single cell and a basic mathematical model was used to explore the potential origins of the observed steady state multiplicity. A more detailed analysis of the results presented in [13] that also explores the origins of autonomous oscillations will be presented in a forthcoming paper.

The scope of this paper is to present experimental results on fast, large-scale, autonomous current oscillations observed in an SOFC that was not operated under high fuel utilisation, placing it firmly in the practical operating range, and to explore potential causes using a first principles approach. Usually this kind of undesirable behaviour is observed under high fuel utilisations and attributed to a number of mechanisms in the literature. Observing unstable output at intermediate fuel utilisation is significant as it indicates that there is more to consider than simply reactant starvation. Oscillations at intermediate fuel utilisations have been observed before in [7] and [8], however the exact mechanism requires further clarification. A post-analysis of the cells in the stack showed several degradation mechanisms including mechanical damage, sealant issues, and nickel oxidation. Clearly a detailed understanding of the causes of the degradation will allow for more effective SOFC configurations and operating conditions to be designed.

The following section will provide details on experiments using tubular SOFCs under intermediate fuel utilisation. The occurrence of autonomous oscillations will be discussed in conjunction with steady state multiplicity in the results section, followed by a modelling approach used in [13] and [20] which allows for the exploration of potential mechanisms giving rise to this behaviour.

2. Experimental

An anode-supported tubular cell with the standard Ni/YSZ/LSCF configuration (dimensions are given in [13]), was ramped up to the operating temperature of 973 K at a rate of 10°/min in pure hydrogen in order to reverse any oxidation that may have occurred in the anode during storage. After maintaining a stable open circuit voltage (OCV) fuel mixtures of hydrogen and nitrogen were fed to the cell, simulating intermediate fuel utilisations. The cathode was exposed to ambient air. The cell voltage was held at a constant value using a potentiostat for at least 20 minutes while the current was recorded, after which, the cell voltage was reduced in steps of 0.1 V and the process repeated. Plots of the current against time were generated and the data were used to calculate the average current at each fixed voltage. These values were plotted in the IV-plane for each fuel composition and fitted with cubic splines in order to generate the characteristic curves. All fuel compositions are given as H_2/N_2 ml/min.

3. Results & Discussion

Using a 60/80 fuel composition provided a steady current output at 0.7 V, as can be seen in Fig. 1a. Upon dropping the cell voltage to 0.6 V some minor fluctuations began to appear alongside the expected increase in current density. The fluctuations continued to grow until distinct, large-scale oscillations were observed at 0.25 V. The limitations of the potentiostat meant that the current could not be recorded at voltages below 0.25 V since the changes in the cell output were too abrupt for the Chroma 6312 Modular DC Electronic Load to handle. A key feature of Fig. 1a is that the average current at each voltage level does not monotonically increase as

IOP Conf. Series: Materials Science and Engineering **639** (2019) 012018 doi:10.1088/1757-899X/639/1/012018

the voltage decreases, which is in contrast to the usual behaviour seen in polarisation curves for fuel cells. Considering the average current over a given region as a fixed point in a dynamical system, this behaviour corresponds with steady state multiplicity similar to that observed in [13].

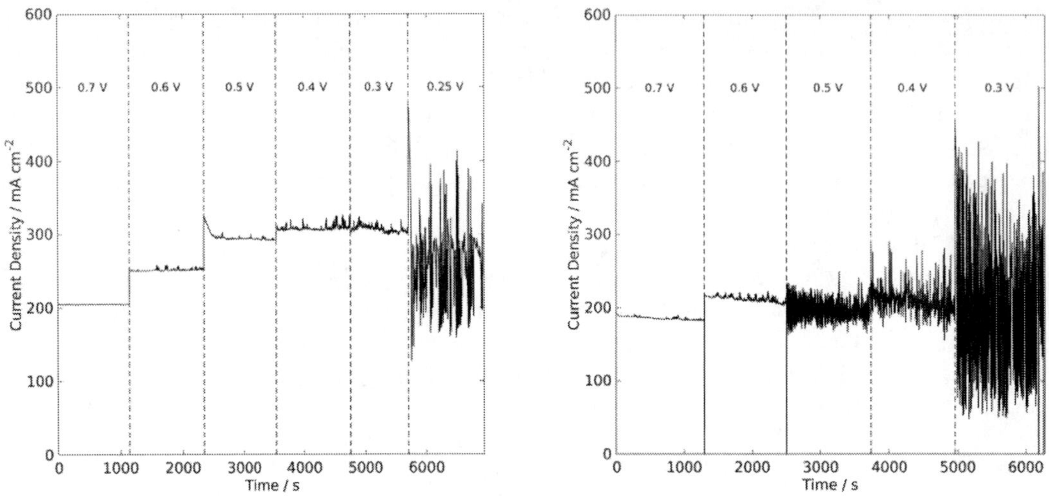

(a) (b)

Figure 1: Potentiostatic tests performed at simulated intermediate fuel utilisations: (a) Voltage range 0.25 V - 0.7 V fuel composition 60/80 ml/min (b) Voltage range 0.3 V - 0.7 V fuel composition 50/90 ml/min.

In Fig. 1b the current output of the cell was measured over the range 0.3 V - 0.7 V for a fuel composition of 50/90. Fluctuations were seen immediately at 0.7 V which steadily grew to large-scale oscillations as the cell voltage was decreased. Compared with the 60/80 composition, the amplitude of the oscillations at each voltage was larger for the 50/90 composition. Again, due to the potentiostat limitations, measurements were unobtainable below 0.3 V. Similarly to the other fuel composition, steady state multiplicity can be seen when taking the average current densities for each fixed voltage. Fig. 2 plots these values for both compositions along with tentative cubic spline interpolants. In fact, the exact shape of the polarisation curves under these conditions remains to be seen as the number of possible steady states has been hypothesised to be between one and five [16]. More detailed measurements are needed to fully determine the exact nature of the polarisation curves, however, first a study is required to determine the appropriate scan rates for various operating conditions since the dynamic response of a fuel cell can occur on both extremely long and extremely short time scales. For example, oscillations have been observed experimentally in SOFCs with periods on the order of seconds [14], minutes [12] and even hours [11]. It's certainly possible for oscillations due to different phenomena to occur simultaneously, resulting in mixed-mode oscillations. Therefore it is important to determine scan rates that truly reveal the character of the cells dynamic response.

Origins of Oscillatory Behaviour
Most oscillatory behaviour (particularly short time scale oscillations) in SOFCs has been observed under reaction depletion. In other words, when the fuel cell is operated under high fuel utilisation, local nickel oxidation takes place which changes the anode microstructure and causes electrical oscillations by various mechanisms. Due to the operating conditions imposed on the

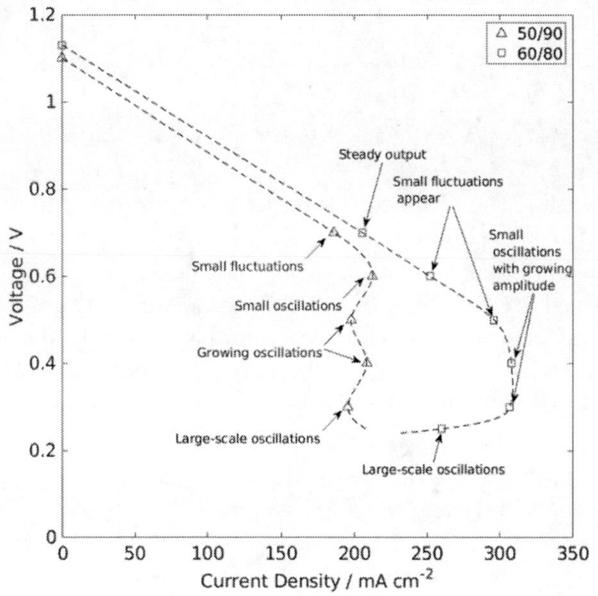

Figure 2: Interpolated polarisation curves using the average current at each applied voltage.

cell to induce these oscillations, quite often the average current output is quite low. However in the present experiments the short time scale oscillations occurred for intermediate fuel utilisation compositions and resulted in large oscillations around relatively high current densities that lie in the range of most practical applications.

It is therefore important to try to understand why these oscillations occurred in these experiments but not in others. It is repeatably demonstrable that high fuel utilisation inevitably leads to oscillations, although the full mechanisms are yet to be elucidated. On the other hand, such large, short time scale oscillations as observed here are more rare and appear to many fuel cell engineers to occur randomly. Of course the perception that they occur randomly stems from a lack of understanding of their cause. It is the purpose of this section to explore potential mechanisms and eliminate possibilities in the hope that future work in this area will be able to further clarify the situation.

In what follows, similar notation and generic equations to those in [13] are used, with the signs of the differential terms for various operating conditions given in Table 2 of [13].

Chemical Origins Given 2 chemical species in a fuel cell system, Z_1 and Z_2, a planar dynamical system can describe the chemical kinetics as,

$$\frac{dZ_1}{dt} = f_1(Z_1, Z_2), \qquad \frac{dZ_2}{dt} = f_2(Z_1, Z_2), \tag{1}$$

where the functions on the right hand sides represent reaction-diffusion equations. For oscillations to occur we examine the conditions under which a Hopf bifurcation may take place. According to the Hopf bifurcation theorem [21], we can expect to see oscillations when the eigenvalues of the linearised system about an equilibrium point form a complex conjugate pair with zero real part along with some other quite generic conditions. The eigenvalues of the above system are given by,

$$\frac{1}{2}\left\{\left(\frac{\partial f_2}{\partial Z_2}+\frac{\partial f_1}{\partial Z_1}\right)\pm \sqrt{\left(\frac{\partial f_1}{\partial Z_1}\right)^2-2\frac{\partial f_1}{\partial Z_1}\frac{\partial f_2}{\partial Z_2}+4\frac{\partial f_1}{\partial Z_2}\frac{\partial f_2}{\partial Z_1}+\left(\frac{\partial f_2}{\partial Z_2}\right)}\right\}, \tag{2}$$

which requires that the 2 terms in the first line (non-radical terms) be of opposite sign in order for a Hopf bifurcation to take place. According to [13] for a purely chemical system the reactants would have both $\frac{\partial f_1}{\partial Z_1} < 0$ and $\frac{\partial f_2}{\partial Z_2} < 0$ meaning that oscillations would not be expected to occur. However if the chemical system is autocatalytic then one of the reactants can be the product of the other one meaning that the sign of one of the derivatives would reverse. Under such conditions we would expect a Hopf bifurcation and therefore the onset of oscillations arising from nonlinear chemical dynamics. Such a behaviour was observed for a methane reforming SOFC in [14].

Electrochemical Origins Given a chemical species in a fuel cell system, Z, and an overpotential, η, a planar dynamical system can model the cell dynamics as follows:

$$\frac{dZ}{dt}=f\left(Z,\eta\right), \qquad \frac{d\eta}{dt}=\frac{1}{C_{dl}}\left(i-j_Z(Z,\eta)\right). \tag{3}$$

The first equation represents the chemical kinetics and the second equation represents charge conservation (i is cell current density, j_Z is the Faradaic current). Examining the eigenvalues of the above system and with $\frac{\partial f}{\partial Z} < 0$ we then require $\frac{\partial i}{\partial \eta} - \frac{\partial j_Z}{\partial \eta} > 0$. However, Kirchoff's law requires that $U = U_0 - \eta - ir_e$ where U is the cell voltage, U_0 is the OCV, and r_e is the electrolyte resistance. Also Faraday's law states that $j_Z = nFR_Z$ where n is the valency number, F is Faraday's constant and R_Z represents the reaction terms determined by mass action. This gives $\frac{\partial i}{\partial \eta} < 0$ and $\frac{\partial j_Z}{\partial \eta} > 0$ meaning that a Hopf bifurcation will not arise under these conditions. However, if the first equation represents a surface reaction then it's possible that $\frac{\partial f}{\partial Z} > 0$, and if the discriminant of the eigenvalues is negative, a Hopf bifurcation can occur. It can be shown using the conditions in [13] that the discriminant can never be negative and we thus rule out the possibility of oscillations occurring as a result of the interactions between basic mass action kinetics and charge conservation dynamics.

Temperature-Related Origins Given a chemical species in a fuel cell system, Z, and variable temperature, T, a planar dynamical system can model the cell dynamics as follows:

$$\frac{dZ}{dt}=f\left(Z,T\right), \qquad \frac{dT}{dt}=h\left(Z,T\right). \tag{4}$$

The second equation now represents heat transfer within the fuel cell and to the surroundings. Following the same method of analysis, the eigenvalues reveal that a Hopf bifurcation is indeed possible if either the reaction rate constant or the electrolyte resistance has a strong dependence on the temperature (enough to cancel out the negative term in the non-radical part of the eigenvalues). Consistent with these findings, the occurrence of nonlinear dynamics related to temperature fluctuations has been theoretically investigated before and found to occur depending on the electrolyte properties [16]. Furthermore, if the first equation represented a surface reaction rather than the usual chemical reaction, then a Hopf bifurcation becomes possible under reaction depletion.

Origins from Nickel Redox Cycling Given a chemical species in a fuel cell system, Z, and changing nickel content of the anode, X, a planar dynamical system can model the cell dynamics as follows:

$$\frac{dZ}{dt} = f(Z, X), \qquad \frac{dX}{dt} = g(Z, X). \tag{5}$$

The second equation now represents the rate of change of the nickel content of the anode. Similarly to the other 3 cases the eigenvalues of the above system can easily be determined. The non-radical part of the eigenvalues requires that $\frac{\partial f}{\partial Z}$ and $\frac{\partial g}{\partial X}$ be of opposite signs. From the typical forms of the rate equations associated with the changing nickel content it can be shown that $\frac{\partial g}{\partial X} < 0$. This means that unless we consider surface reactions a Hopf bifurcation will not result from the above system. Considering a surface reaction, the condition that the discriminant be negative is only satisfied if the surface coverage of the reactant is very low. Under these conditions nickel oxidation will occur and oscillations become possible via a Hopf bifurcation.

4. Conclusions

Fast, large-scale current oscillations were observed in a tubular SOFC where reaction depletion was not an issue. A common explanation for instabilities is that reaction depletion causes nickel oxidation and via one mechanism or another gives rise to the oscillations. For these experiments this was clearly not the case and so a brief analysis was conducted to investigate other potential causes. From the cases considered above, it seems most likely that the observed oscillations could be attributed to either an autocatalytic surface reaction scheme involving some intermediate species (since the only reactant supplied to the anode was hydrogen), or to nonlinearities arising from a strong temperature dependence of the reaction rate constant or electrolyte conductivity. Further work is required to determine the unique attributes of oscillations caused by each of these phenomena, some of which will be presented in a forthcoming paper.

References

[1] Roy R 2014 *The Cost of Air Pollution: Health Impacts of Road Transport* (OECD Publishing) http://dx.doi.org/10.1787/9789264210448-en
[2] Schnelle K B, Dunn R F and Ternes M E 2016 *Air Pollution Control Technology Handbook (Second Edition)* (Boca Raton: CRC Press)
[3] Keuken M, Sanderson E, van Aalst R, Borken J and Schneider J 2005 *Contribution of traffic to levels of ambient air pollution in Europe Health Effects of Transport-Related Air Pollution* ed Krzyzanowski M *et al* (Denmark: World Health Organization) chapter 2 pp 53-84
[4] Fenger J 1999 *Atmos. Environ.* **33** 4877-4900
[5] Jensen J H, Larsen P H and Mogensen M 2007 *Int. J. Hydrog. Energy* **32** 3253–57
[6] Braun R J, Klein S A and Reindl D T 2006 *J. Power Sources* **158** 1290–1305
[7] Blum L, Packbier U, Vinke I C and de Haart L G J 2013 *Fuel Cells* **13** 646–653
[8] Groß-Barsnick S M, Fang Q, Batfalsky P, Niewolak L, Blum L and Quadakkers W J 2019 *Fuel Cells* **19** 84-95
[9] Fang Q, Blum L, Peters R, Peksen M, Batfalsky P and Stolten D 2015 *Int. J. Hydrog. Energy* **40** 1128–36
[10] Yoshizumi T, Taniguchi S, Shiratori Y and Sasaki K 2012 *J Electrochem. Soc.* **159** F693-F701
[11] Marina O A, Pederson L R, Coyle C A, Thomsen E C and Edwards D J 2011 *J. Electrochem. Soc.* **158** B36-B43
[12] Wang Z, Lü Z, Wei B, Chen K, Huang X, Pan W and Su W 2011 *Electrochim. Acta* **56** 6688-95
[13] Sands J D and Wongyao N 2018 *J. Electroanal.Chem.* **812** 59-67
[14] Sands J D, Uddin J and Needham D J 2016 *J. Electrochem. Soc.* **163** F856–62
[15] Park K, Lee S, Bae G and Bae J 2015 *Renew. Energy* **83** 483-490
[16] Mangold M, Krasnyk M and Sundmacher K 2006 *J. Appl. Electrochem.* **36** 265-275
[17] Huang T J, Huang M C, Chen W J and Chou C L 2009 *Chem. Eng. J.* **153** 164-9

[18] Parackal B 2017 *An investigation of low temperature direct propane fuel cells* (University of Ottawa: Master's thesis)

[19] Kawasaki T, Matsuda J, Tachikawa Y, Lyth SM, Shiratori Y, Taniguchi S and Sasaki K 2019 *Int. J. Hydrog. Energy* **44** 9386-99

[20] Hanke-Rauschenbach R, Mangold M and Sundmacher K 2011 *Rev. Chem. Eng.* **27** 23-52

[21] Wu J and Zhou K 1995 *New numerical methods for high dimensional hopf bifurcation problems Bifurcation and Chaos: Theory and Applications* ed Awrejcewicz J (Berlin: Springer-Verlag) pp 47-69

Characterization and fabrication of bio-composite filaments for fused deposition modeling 3D printing

P Phengchan, A Chaijaruwanich*, W Nakkiew and S Pitjamit
Advanced Manufacturing Technology Research Center (AMTech), Department of Industrial Engineering, Faculty of Engineering, Chiang Mai University, Chiang Mai, 50200, Thailand.

Email address: anirut@eng.cmu.ac.th

Abstract. The metal bone screws are generally used in bone fractures. The bio-composite bone screws will be fabricated soon by using hydroxyapatite (HA) to reinforce polylactic acid (PLA) and polycaprolactone (PCL) filament for 3D printing. In this study aimed to characterize morphological and physical properties of bio-composite filaments. HA was synthesized from bovine bone. X-ray diffraction analysis indicated that the obtained HA and HA standard showed a similar chemical composition. The morphological observation of bio-composite filaments by scanning electron microscope (SEM) revealed the distribution of HA in bio-composite filaments. The elemental composition of the filament by energy dispersive spectrometer indicated the consisting of calcium, carbon and oxygen elements. The diameter measurement of all bio-composite filaments indicated the values ranged from 1.66 to 1.86 mm which were non-significantly different value in each composite filament and similar with a filament standard.

Keywords: Poly-lactic acid, Polycaprolactone, Hydroxyapatite, Bio-composites filament, Fused deposition modeling

1. Introduction

Bone fractures are common small animal orthopedic injuries which can be treated by using bone screw. The bone screws development has been performed to increase the treatment abilities especially in canine humerus, femur, and tibia fractures. Nevertheless, the bone screws removal is required due to the complications in short term and long-term implantation. Titanium and stainless steel did not provide the living cells with the biological environment according to the bone screw materials [1]. This research is looking for biomaterials that can eliminate metal bone screws limitations. At present, the use of 3D printing in biomedical technology is rising rapidly. There have been several studies of bone screws applications using 3D printing for orthopedic surgery, according to the rapid process of 3D printing [2]. The most satisfied 3D Printing technique is fused deposition modeling (FDM). Thermoplastic filament will be feed through the heating nozzle. The melted filament will be printed layer by layer until the end. The advantage of FDM technique is production speed, material costs and design flexibility [3]. Various types of biomaterials can be printed using the FDM technique. Previous studies have been focused on a biodegradable polymer and bio-ceramics that activate cell creation includes polylactic acid (PLA) that can be degraded by microorganism and turn into carbon dioxide (CO_2) and vapor (H_2O) and sourced by corns, tapiocas and sugarcanes which has linear thermoplastic fiber and widely used in commercial term [4], also with Polycaprolactone (PCL) is thermoplastic and semi-crystalline polymer that has ester group in its component which synthesized by ring opening polymerization of ε-caprolactone with catalysts and heat transfer [5]. Hydroxyapatite (HA) has compound from calcium and Phosphorus combination which can be found in mammals' bones and synthetic [6]. To increase biological and mechanical properties, it can be fabicated into a structure or combined with other biomaterials. Thus, HA is used to create bio-composite filament for bone screws 3D printing as a reinforcing material for PLA/PCL.

IOP Conf. Series: Materials Science and Engineering **639** (2019) 012019 doi:10.1088/1757-899X/639/1/012019

In this study, the filaments were extruded, and their morphological and physical characteristics of the obtained filaments will be investigated before printing. The factors of filaments extrusion consisted of HA ratio, extrusion temperature, and extrusion speed. Then, the bio-composite filament will be tested diameter stability property. Finally, appropriate bio-composite filament will be printed and tested with mechanical and biological properties also with simulated the actual treatment in Finite element analysis in the future study.

2. Materials and Methodology

2.1 Synthesis of Hydroxyapatite (HA)
First, bovine bone was boiled to remove tissue completely. Then, the bones were soaked in oxidizing solution for 72 hrs. Next, bones were calcined by incubation at 950 °C in oven for 12 hrs. After cooling, bones were milled and sieved to a powder. [7]

2.2 Fabrication composite filaments
First, PLA granules (4043D NatureWorks, LCC) was incubated at 60 °C in oven (PROTRONICS INTERTRADE CO.,LTD. BPG—9070A oven) for 4 hrs, then PLA and PCL were mixed with 70:30 ratio with 0% and 5% wt HA using alumina balls with shaking for 15 min in the cylindrical Teflon mold to homogeneous material [8]. Filaments were fabricated in different factor values of the extrusion process (Table 1). A total 4 conditions consisted of 0%wt HA, 180 °C, 10 rpm, 0% wt HA, 190 °C, 50 rpm, 5% wt HA,180 °C, 10 rpm and 0% wt HA, 190 °C, 50 rpm. by Wellzoom extrusion machine

Table 1. Factor values of filament extrusion

Factors	Value	
	Low	High
% wt HA	0	5
Extrusion temperature (°C)	180	190
Extrusion speed (rpm)	10	50

3. Testing

3.1 Scanning electron microscope and energy dispersive spectrometer
The top, middle and bottom sites of each sample were cut into 5 mm lengths. HA powder distribution was observed on 1000x magnification by Scanning Electron Microscope (SEM) (EFI Quanta 200 3D) as shown in Figure 2. Energy Dispersive Spectrometer (EDS) analysis of elements in the composite filaments was determined by JSM-5910LV.

3.2 Diameter measurement
Filament prototypes were sampled and measured the diameter stability by analog micrometer (Mitutoyo 0.25 mm – 0.01 mm). A total 10 points of each filament was measured. The obtained diameter values were analyzed on Minitab version 16 program and T-test analysis was performed.

4. Result

4.1 X-ray Diffraction

X-ray Diffraction: XRD of HA powder referred by the recent research [9] for checking the component phase with X-ray diffraction techniques. The XRD result of the synthesized material corresponded to hydroxyapatite (HA) when comparing with HA standard component phase $Ca_{10}(PO_4)_6(OH)_2$ as shown in Figure 1.

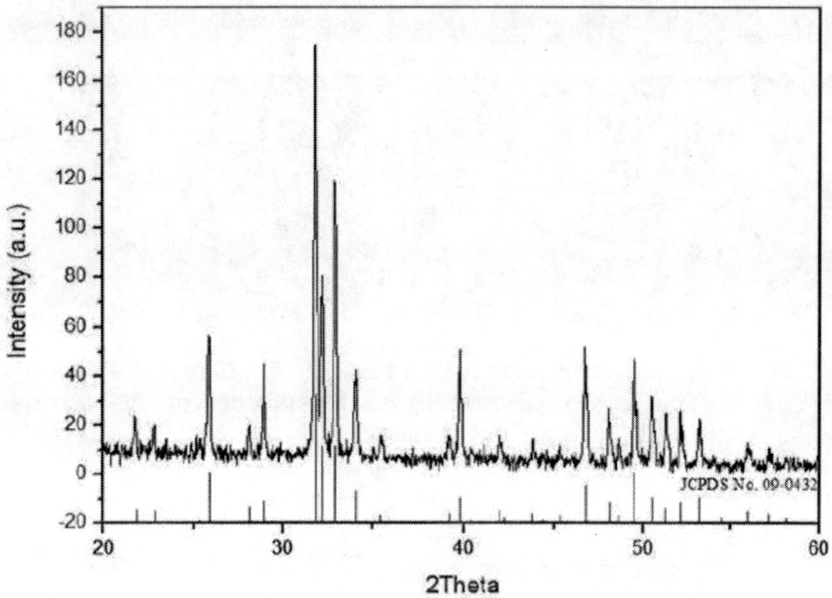

Figure 1. XRD pattern of synthesized HA powder.

4.2 Scanning Electron Microscope and Energy Dispersive Spectrometer

In morphological testing, use SEM with condition that specific contains HA powder. In some conditions with low temperature (180 °C) non-melting plastic pellets, as shown in Figure 2. And the condition at 190 °C was a suitable condition for PLA melting. Moreover, HA powder was found to be successfully integrated into a composite matrix. The elemental composition of the composite filament has been given in Figure 3. According to the EDS spectrum, the sample mainly consisted of C, O and Ca components. C, O and Ca are primarily derived from the carbonated HA [10] which at high extrusion temperature and speed found that weight percent and atomic of Calcium is higher.

IOP Conf. Series: Materials Science and Engineering **639** (2019) 012019 doi:10.1088/1757-899X/639/1/012019

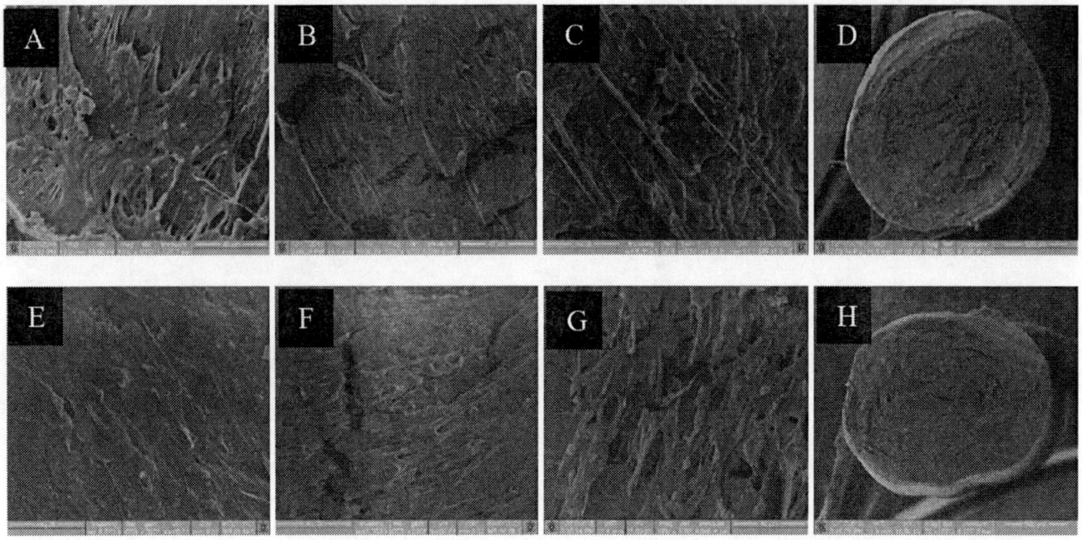

Figure 2. SEM images composite filament at 5% wt HA, 180 °C, 10 rpm (A-D) and 5% wt HA, 190 °C, 50 rpm (E-H). Each images showed top (A,E), middle (B,F), bottom (C,G) and overall (D,H) of filament.

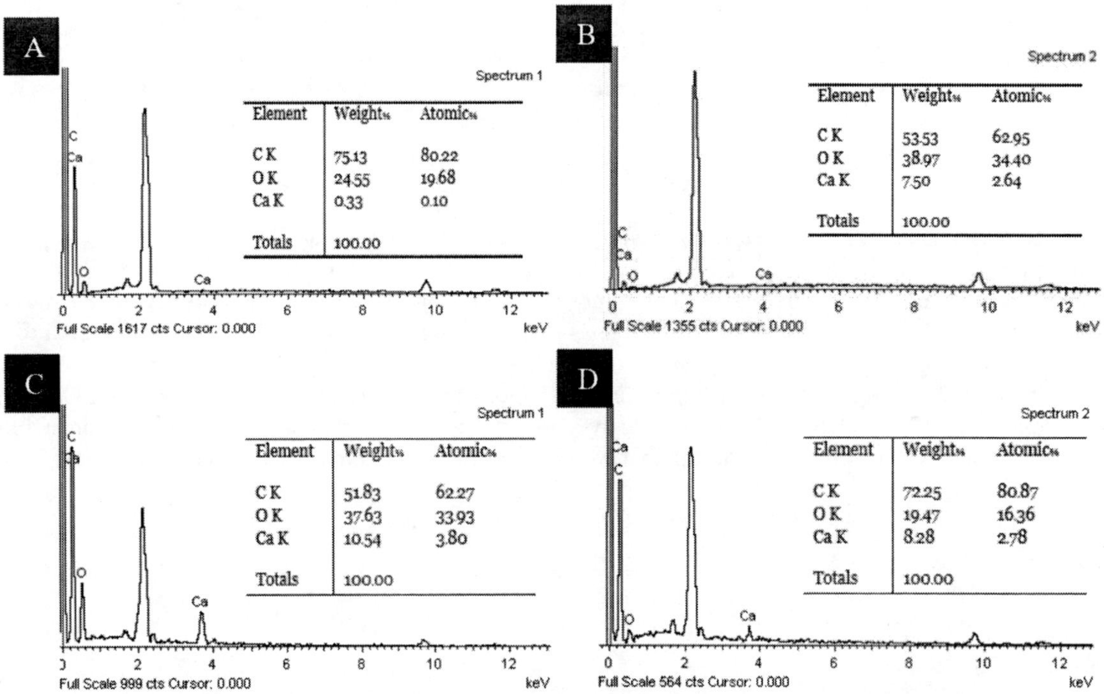

Figure 3. EDS spectrum (A) and (B) of the composite filament at 5% wt HA, 180 °C, 10 rpm and EDS spectrum (C) and (D) of the composite filament at 5% wt HA, 190 °C, 50 rpm.

4.3 Diameter measurement analysis

The diameter values of each filament measurement are shown in Table 2. The obtained diameter values of all composite filaments ranged from 1.66 to 1.86 mm which is non-significantly different by T-test (Table 3)

Table 2. Measurement of filament diameter (in mm) for various conditions

Sample conditions	Replicates									
	1	2	3	4	5	6	7	8	9	10
0%HA,180°C,10rpm	1.73	1.66	1.68	1.71	1.68	1.77	1.77	1.76	1.81	1.76
5%HA,180°C,10rpm	1.86	1.81	1.7	1.76	1.86	1.74	1.77	1.75	1.73	1.71
0%HA,190°C,50rpm	1.78	1.81	1.83	1.81	1.75	1.76	1.75	1.73	1.74	1.76
5%HA,190°C,50rpm	1.74	1.78	1.81	1.71	1.64	1.79	1.82	1.74	1.79	1.7

Table 3. Results of 1sample t-test.

variable	Mean(mm)	StDev	SE Mean	95% CI	T	P
0%wtHA,180°C,10rpm	1.73	0.0490	0.0155	(1.6979, 1.7681)	-1.10	0.301
5%wtHA,180°C,10rpm	1.77	0.0570	0.0180	(1.7282, 1.8098)	1.05	0.320
0%wtHA,190°C,50rpm	1.77	0.0339	0.0107	(1.7477, 1.7963)	2.05	0.071
5%wtHA,190°C,50rpm	1.75	0.0567	0.0179	(1.7114, 1.7926)	0.11	0.914

5. Conclusion

According to the purpose of the end study, HA will be used to reinforce PLA/PCL bone screws which will be fabricated by 3D printing process. HA should provide the fracture area with a good biological environment to speed up the bone healing process by providing a good surface for proliferation, cell adhesion, and migration. In addition, biological enzymes can degrade HA and PLA/PCL. Thus, the 3D printing HA bone screws reinforcing PLA/PCL eliminates the removal process and provides a good biological environment for bone regeneration. HA synthesized in this study could use for a composition material. The bio-composite filament from 5% wt of HA, extrusion 190°C temperature and 50 rpm extrusion speed condition shows the highest HA distribution after observation by SEM. The bio-composite filaments consisted of calcium, carbon and oxygen elements. The diameter value of bio-composite filament obtained in this study was similar to the filament standard for fused deposition modeling 3D printing. Bio-composite filament for 3D printing will be tested for future work and other properties such as mechanical properties, cytotoxicity, biodegradable and finite element analysis will be observed.

6. Acknowledgment

Authors would like to acknowledge Advanced Manufacturing Technology Research Center (AMTech), Department of Industrial Engineering, Faculty of Engineering, Chiang Mai University for their facility support.

7. References

[1] Navarro, M., Michiardi, A., Castano, O., & Planell, J. A. (2008). Biomaterials in orthopaedics. *Journal of the Royal Society Interface*, *5*(27), 1137-1158.

IOP Conf. Series: Materials Science and Engineering **639** (2019) 012019 doi:10.1088/1757-899X/639/1/012019

[2] Rengier, F., Mehndiratta, A., Von Tengg-Kobligk, H., Zechmann, C. M., Unterhinninghofen, R., Kauczor, H. U., & Giesel, F. L. (2010). 3D printing based on imaging data: review of medical applications. International journal of computer assisted radiology and surgery, 5(4), 335-341.

[3] Mohamed, O. A., Masood, S. H., & Bhowmik, J. L. (2016). Mathematical modeling and FDM process parameters optimization using response surface methodology based on Q-optimal design. Applied Mathematical Modelling, 40(23), 10052-10073

[4] Oksman, K., Skrifvars, M., & Selin, J. F. (2003). Natural fibres as reinforcement in polylactic acid (PLA) composites. *Composites science and technology*, *63*(9), 1317-1324.

[5] Wright, D. D. (2004). Degradable polymer composites. In *Encyclopedia of Biomaterials and Biomedical Engineering* (pp. 423-432).

[6] Elkayar, A., Elshazly, Y., & Assaad, M. (2009). Properties of hydroxyapatite from bovine teeth. *Bone and Tissue Regeneration Insights*, *2*, 31.

[7] Pitjamit, S., Sriprapha, P., & Nakkiew, W. (2016). Suitable forming condition of hydroxyapatite and bioactive glass composites for a bone fixation plate using Taguchi experimental design. *Engineering and Applied Science Research*, *43*, 466-469.

[8] Russias, J., Saiz, E., Nalla, R. K., Gryn, K., Ritchie, R. O., & Tomsia, A. P. (2006). Fabrication and mechanical properties of PLA/HA composites: a study of in vitro degradation. *Materials Science and Engineering: C*, *26*(8), 1289-1295.

[9] Watcharaprapapong, P., Nakkiew, W., Wattanuchariya, W., & Pitjamit, S. (2018). Effect of forming conditions of poly-lactic acid/hydroxyapatite to tensile strength of canine bone fixation plate using full factorial experimental design. In *MATEC Web of Conferences* (Vol. 192, p. 01049). EDP Sciences.

[10] Huang, Y., Song, G., Chang, X., Wang, Z., Zhang, X., Han, S., ... & Zhang, X. (2018). Nanostructured Ag+- substituted fluorhydroxyapatite-TiO2 coatings for enhanced bactericidal effects and osteoinductivity of Ti for biomedical applications. *International journal of nanomedicine*, *13*, 2665.

IOP Conf. Series: Materials Science and Engineering **639** (2019) 012020 doi:10.1088/1757-899X/639/1/012020

Preliminary Study of Epoxy Coating Process on Basketry Inlay with 3D Printed Model

Pholchai Chotiprayanakul

Department of Industrial Engineering, Faculty of Engineering, King Mongkut's Institute of Technology Ladkrabang, Bangkok, Thailand

Atsushi NODA

National Institute of Technology, Anan College

Corresponding author: pholchai.ch@kmitl.ac.th

Abstract. In Japan and South-East Asia Countries, Basketry is a heritage handcraft made with natural material such as bamboo strip and wheat straw. Shape and size of basketry products are depended on the basketry pattern and now FDM (Fused Deposition Modelling) 3D printer is used to a print the basketry pattern. This paper will present a bit further in epoxy coating on basketry products to make it more rigid and better shining appearance either while the coated basketry still have the 3D printing model inside as a structure or it will be taken off after coating epoxy. There are a lot of makers who try to use the 3D printed model for resin casting mold but many are fail to get the casted out of the 3D printed mold. Thus we do this preliminary study to investigate the adhesive strength of the casting epoxy resin holding on the 3D FDM printed model. An experiment was designed to find the effect of the curing time and the bamboo fiber adding for resin reinforce. In experiment, a flat dog-bone shape specimen for tensile testing is redesigned by separating into 2 identical parts and they are printed with ABS plastic filament. We glue epoxy resin on 12.5 X 10 mm. contact area in middle of the specimen. We left them cure and dry for 6, 12, and 24 hours before tensile testing then repeat the experiment again with bamboo strip added in between the contact area. The result of the experiment shows that bamboo strip reinforce and curing time affect the adhesive strength.

1. Introduction
In Thailand, basketry is a household industry mostly handcrafted which comes all range from low cost product until premium luxury product. To evaluate this industry with 3D printing method, 3D printed model is used as a pattern or an inlay structure of the basketry. By the other process, epoxy coating is one of effective methods to smoothen the surface of 3D printed model. Commercial as XTC-3D® [1] provide better smooth surface and strengthen the printed model. Shaker Alkaraki and et al. uses metalize spray coating a 3D printed antenna [2] make the model response to electromagnetic signal. Some researchers as Joseph T. Belter and Aaron M. Dollar [3] inject epoxy resin into the model's lacunas to strengthen the 3D model for manufacturing. This paper intends to combine those two processes: basketry forming and epoxy coating together.
There are two possible combinations: 1) use 3D printed model as pattern when finish forming basketry and coated with epoxy, the pattern will be taken of the product 2) use 3D printed model as an inlay structure which is left and stay inside the basket as a rigid structure. Epoxy coating will effect on the

product if we want to get the 3D printed model off, or epoxy coat should be bonded on the 3D pinted model strong enough for proper using.

This paper presents the study of adhesive strength of the epoxy coating on 3D ABS printed model by setting a mechanical tensile experiment. First, we design test specimen and testing procedure. Finally, the test result will give the adhesive shear strength.

2. Flat Dog-Bone Specimen

In tensile test, specimens are designed to control the broken position. We select ASTM D 638 Type II for our dog-bone specimen with 6 mm. thickness and print it with XYZ Davinci for 0.3 mm. layer and 80% infill. Fig XXX shows the dog-bone specimen's dimension and tensile tested. Result of pre-tensile test from 3 specimens gives the average tensile strength around 38 MPa where maximum force is at 2.8 kN.

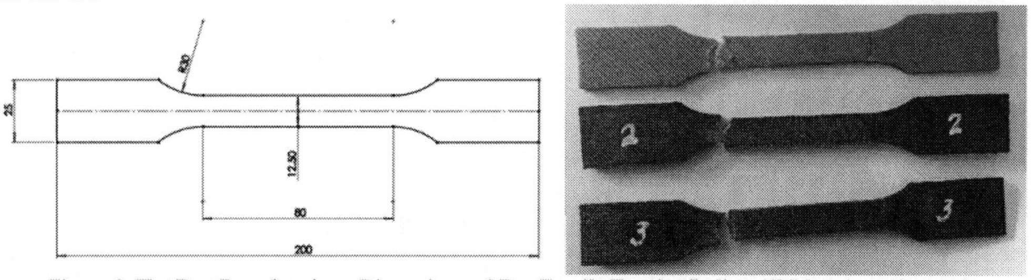

Figure 1. Flat Dog-Bone Specimen Dimension and Pre-Tensile Test for finding ABS Specimen Strength

2.1. Design of Flat Dog Bone Specimen

This maximum force is used for re-designing to the epoxy fused dog-bone specimen. The epoxy fused dog-bone specimen is divided to 2 identical parts which can be re-assembled back to dog-bone shape with epoxy fused at the joining area. Adhesives shear strength of epoxy is about 17 MPa (2500 psi) thus the joining area must smaller than167 mm.^{2} that gives the length of the joining area must less than 13.36 mm.. The length of the joining area is selected to be 10mm. as shown in Fig XXX.

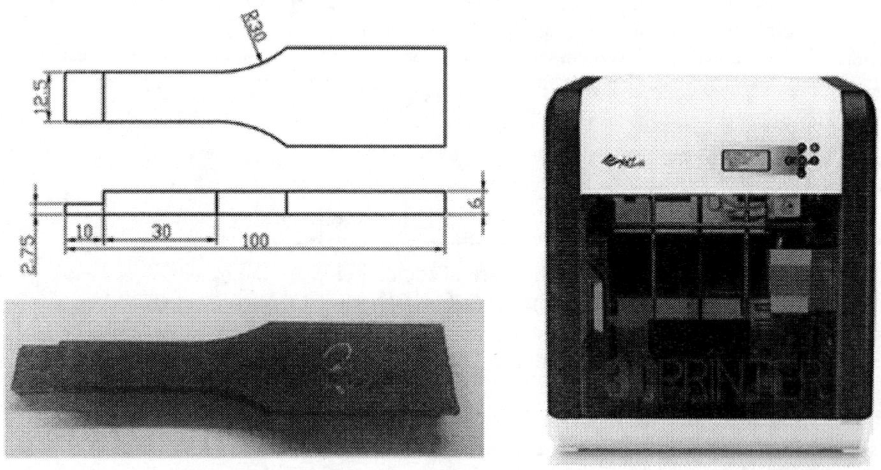

Figure 2. Re-design of Flat Dog-Bone and XYZ Davinci 3D Printer

2.2. Preparation of Flat Dog Bone Specimen

All specimens were printed with ABS plastic filament on XYZ Davinci 1.0 for 0.3 mm. layer and 80% infill. After specimens were printed, joining area of every specimen was inspected and polished until its surface flat, smooth, and dimension. Then A-B epoxy resin were mixed by weight proportion on weight scale and small amount of epoxy was fully filled on 12.5 X 10 mm. contact area in middle of the specimen and left them cure and dry for 6 hours, 12 hours, and 24 hours . After epoxy was set, the specimens were brought into tensile test to find the maximum force that epoxy can hold on the ABS specimen. Second test set, we added bamboo strip in between the contact area and cover with epoxy again also left the epoxy to set for 6, 12, and 24 hours and then repeat the tensile test.

3. Tensile Test

After specimens were prepared and left for epoxy to set, the specimens were pulled in universal tensile testing machine Testomatic model AX M500-100kN. For setting parameters, a pulling rate was set at 2.5 mm. per second and the force limit was set at 200N. In the test, maximum force was considered as the first priority parameter in order to calculate the adhesives shear strength of epoxy join the ABS plastic. The maximum force would be divided by the joining area which is 12.5×10 mm.^{2} that is given the adhesives shear strength of adhesives shear strength of epoxy join the ABS specimen.

Figure 3. Universal Tensile Testing Machine Testomatic AX M500-100kN

4. Result

For tensile testing, two control variables are investigated as curing time and bamboo strip reinforce. For curing time, there are 3 sets of curing time different and each set contains 3 samples. The test were repeated with adding bamboo strip reinforce in between joining area in middle of the specimen. The bamboo strip thickness is thinned to 0.3 mm. by sandpapering and measuring it with vernier calliper. From tensile tester, the machine gives report as graph of stress-strain and value of maximum force that specimen can take before break apart. Figure 1Figure 4 and Figure 5 show the some testing results which are test set of 12 hours curing time and with and without bamboo strip reinforce.

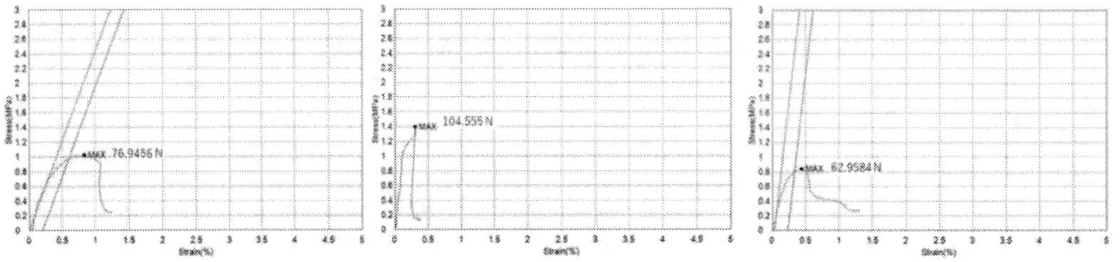

Figure 4. A tensile test result of curing time 12 hours and without bamboo strip reinforce, average maximum force at 81.4863 N.

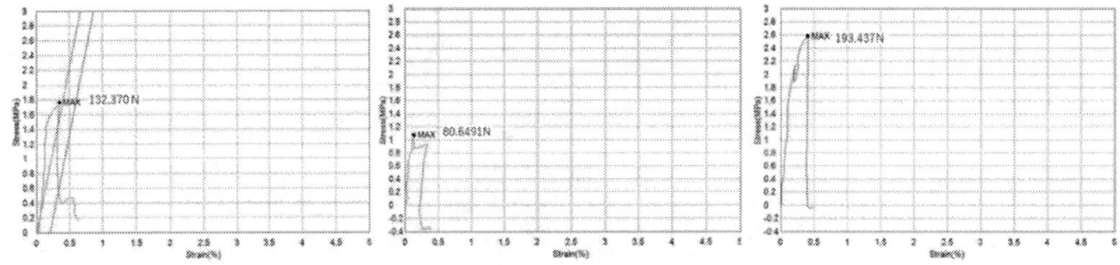

Figure 5. A tensile test result of curing time 12 hours and with bamboo strip reinforce, average maximum force at 135.485 N.

Table 1. Result of Tensile Test shows the Maximum Tension Force in Newton

	Maximum Tension Force (N)			
Curing Time	without Bamboo Strip Reinforce		with Bamboo Strip Reinforce	
6 Hrs	52.2611	55.56	93.611	85.74
	64.8249		57.9597	
	49.6134		105.667	
12 Hrs	76.9456	81.48	132.37	135.48
	104.555		80.6491	
	62.9584		193.437	
24 Hrs	80.2432	87.35	126.529	140.58
	97.435		152.533	
	84.3792		142.695	

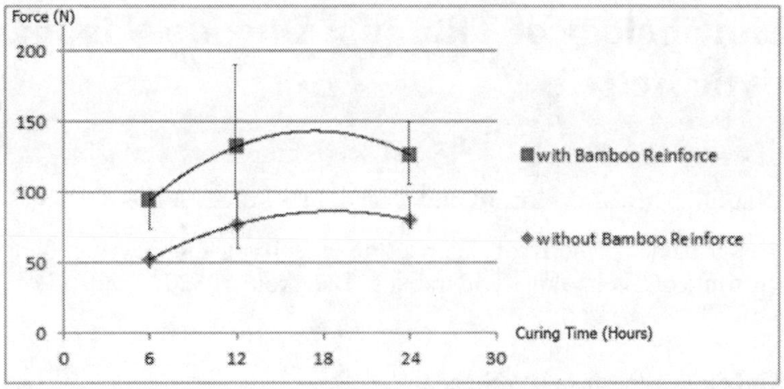

Figure 6. Result of the Tensile Test with 2 Variables Bamboo Reinforce and Curing Time

Table 1 and graph in Figure 6 show as expect for curing time by the recommend of 200AB epoxy that it will fully cure by 24 Hours. Also data illustrations that the effect of the bamboo strip reinforces will increase the adhesive shear strength because the bamboo strip makes epoxy more rigid than only epoxy coat. When specimen with only epoxy can be easy deform and wobble losing it contact in some area (not sudden completely tear apart), this will make longer strain as seen in graph of Figure 4 and Figure 5.

In this study, the ultimate adhesive shear strength for epoxy and ABS plastic can be calculated by the maximum force of fully cure epoxy by experiment divided by the contact area (12.5×10 mm.^{2}) that given adhesive shear strength for epoxy-ABS equal to 0.65N/mm.^{2} and epoxy-ABS reinforce with bamboo strip equal to 1.12N/mm.^{2}

5. Conclusion

This preliminary study aims to find effect of epoxy curing time and bamboo strip reinforce on adhesive shear strength. When we put epoxy resin coating on the surface of 3D FDM printed model or on the basketry which has 3D FDM printed pattern inside. An experiment was designed and implemented. Result of the experiment shows the magnitude of adhesive shear strength of AB epoxy and ABS plastic model. The bamboo strip is used to mimic the basketry is tested. The testing result shows the epoxy with bamboo strip reinforce is slightly stronger sticky on the ABS plastic than put on epoxy only. This means, we will use more effort to take epoxy-coated basketry of the ABS pattern, or the epoxy-coated basketry with ABS structure inside is stronger than coating only epoxy.

References

[1] Smooth-On XTC-3D, https://www.smooth-on.com/product-line/xtc-3d/, accress on March 2019
[2] Shaker Alkaraki and et al (2018), Compact and Low-Cost 3-D Printed Antennas Metalized Using Spray-Coating Technology for 5G mm-Wave Communication Systems, IEEE Antennas And Wireless Propagation Letters, Vol. 17, No. 11, November 2018
[3] Joseph T. Belter and Aaron M. Dollar (2015), Strengthening of 3D Printed Fused Deposition Manufactured Parts Using the Fill Compositing Technique, PLoS ONE 10(4): e0122915. doi:10.1371/journal.pone.0122915, Nicola Pugno, Università di Trento, ITALY.
[4] Stephen Burke Driscoll (1998), The Basics of Testing Plastics: Mechanical Properties, Flame Exposure, and General Guidelines, Astm Intl, June 1,page 20-24

IOP Conf. Series: Materials Science and Engineering **639** (2019) 012021 doi:10.1088/1757-899X/639/1/012021

Effect of Morphology of Titanium Dioxide Nanoparticles on Photocatalytic Activity

Naphat Thiwakornkitkul and Teeraporn Suteewong*

Department of Chemical Engineering, Faculty of Engineering, King Mongkut's Institute of Technology Ladkrabang, Bangkok, 10520, Thailand

Email: teeraporn.su@kmitl.ac.th

Abstract. Dispersing nanoparticles into the matrix is a simple but effective method to improve properties of coating layer. Titanium dioxide (TiO_2), known for its photocatalytic activity (PCA) to degrade organic substances and to protect microbial infection, is considered to be potential candidate. However, intense light is required to achieve high PCA. Scattering layer composing of hollow TiO_2 nanoparticles, known as light harvester in solar cell devices, may be suitable for photocatalysis at lower light intensity. In this work, we studied the effect of hollow morphology to the PCA of TiO_2 nanoparticles. Hollow TiO_2 nanoparticles were successfully synthesized using hard-template-assisted sol-gel method. Transmission Electron Microscope (TEM) images show hollow TiO_2 nanoparticles possessing 125-nm hollow core and 50-nm TiO_2 shell. X-ray Diffraction (XRD) results revealed no crystalline peak for both calcined and non-calcined samples, but difference for each sample was reported by Ultraviolet-Visible Diffuse Reflectance Spectroscopy (UV-DRS). UV-DRS spectrum showed the higher diffuse reflectance for both hollow TiO_2 samples than the dense counterpart, suggesting multiple light scattering and high reflection. However, energy bandgaps of amorphous and calcined hollow TiO_2 samples are higher than dense counterparts.

1. Introduction

Incorporating dispersed nanoparticles into the polymer matrix have been used to improve or to introduce new properties to coating layer. [1] Focusing on the dispersed nanoparticles possessing multifunctionalities is considered to be an interesting choice. Titanium dioxide (TiO_2) is one of widely-used additives in coatings and photocatalyst industries due to its expedient advantages e.g. low-toxicity, photoactive as well as chemical and thermal stability. [2] Extensive studies have been focused on its photocatalytic activity (PCA), such as dyes-sensitized solar cell (DSSCs), hydrogen production, self-cleaning surface, medical devices, and antimicrobial properties. [2-6] However, this property is photo-intensive. Its PCA might be limited to only environment that possesses high intensity of light. Hence, the improvement of light intensifying properties of TiO_2 is necessary.

Since the functions of coating stem from integrated components, its efficiency could be enhanced through modification of TiO_2 particles by manipulating particle size, surface chemistry, crystallinity and morphology. [7-9] With the concept of light intensification, hollow morphology is of great interest due to its unique optical properties. Hollow cavity inside the shell can perform internal multiple light scattering and reflection, which results in the enhancement of light harvesting ability comparing with their dense counterpart. In addition, the hollow particle possesses large number of active sites due to high specific surface area and short electron diffusion pathway. This high surface area to volume ratio also increases the possibility of contact between reactants and active sites. [9-13] There are reports on the improvement of mechanical, electrical and chemical properties of hollow nanoparticles of silica (SiO_2), polystyrene and zinc oxide as well as their applications. [10,11]

Similarly, studies of hollow TiO_2 has been focused on green energy aspects. Only a few have been examined on their advantages in terms of PCA for photodegradation or antimicrobial applications. Moreover, most of the available literatures involved composite hollow particles. [12,13] Yamaguchi *et al.* reported the comparative studies of inactivating chemical and biological substances using hollow and solid TiO_2 nanoparticles. However, the representative of hollow morphologies in their work was actually an opened-hole, cup-liked structure. [14] More precise understanding in structural effects could allow us to improve the photocatalytic activity characteristic of TiO_2.

Herein, we studied the effects of hollow structure of TiO_2 particles on photocatalytic activity in term of bandgap energy, with respect to dense particles. Facile sol-gel approach was employed for syntheses of hollow TiO_2 particles with the aid of SiO_2 nanoparticle template. After the formation of TiO_2 shell, silica core was removed using chemical etching. [15] Resulting particles were then calcined to form anatase phase. Particle size and morphology were investigated using transmitting electron microscope (TEM). X-ray diffraction spectroscopy (XRD) was used for characterization of crystallinity. Finally, Ultraviolet-Visible Diffuse Reflectance Spectroscopy (UV-DRS) was employed to evaluate optical properties of the samples as diffuse reflectance and bandgap. Comparison of photocatalytic activity was done according to the bandgap values.

2. Materials and Methods

2.1. Chemicals and Materials

Tetraethyl orthosilicate (TEOS, 99%, Aldrich), Titanium tetraisopropoxide (TTIP, 99%, Aldrich), Titanium tetrabutoxide (TBT, Aldrich), Methanol (MeOH, 95%, Fluka), Ethanol (EtOH, 95%, Aldrich), Ammonium hydroxide (28%, Acros) Sulfuric acid (H_2SO_4, 18 M, Aldrich), Sodium hydroxide (NaOH, 99%, Fluka) and Calcium chloride ($CaCl_2$, Fluka) were used as received. Deionized (DI) water was used throughout the work.

2.2. Preparation of SiO_2 Nanoparticle Template

Ammonium hydroxide, DI water and EtOH were first mixed in a round-bottom flask for 10 minutes. TEOS was slowly added into the prepared mixture and stir for 4 hours. Unreacted chemicals were removed by centrifugation. Re-disperse the resulting particles in ethanol and repeat the washing step twice.

2.3. Preparation of Hollow TiO_2 Nanoparticles [15]

Known weight of prepared SiO_2 particle was dispersed in the mixture of EtOH and DI water and refluxed at 85 °C. Prepare the solution of TBT in EtOH at volume ratio of 1:4.5 (TBT: EtOH) before slowly adding titania precursor into the refluxed solution. After 12 h, remove unreacted chemicals by centrifugation and re-disperse the resulting particles in DI water. Add NaOH solution (2.5 M) into the particle mixture and stir for 6 hours before cleaning with DI water. Dry the sample at 80 °C overnight before eliminating the remaining Na atoms on the particle surface by dispersing the prepared particles in diluted H_2SO_4 solution for 30 minutes. Centrifuge to remove undesired chemicals. Calcine dried samples at the desired temperature and time.

2.4. Preparation of Dense TiO_2 Nanoparticles [16]

Prepare 50 ml of MeOH in round-bottom flask. Add $CaCl_2$ solution with the desired concentration as a size regulator and stir for 10 minutes. Gradually add 850 μl TTIP. Age resulting solution for 24 hours while gently stirring and then remove the unreacted chemicals using centrifugation. Re-disperse the particles in EtOH and repeat the centrifugation twice. Dry particles at 80 °C. Calcine dried sample at the desired temperature and time.

2.5. Characterization

Transmission Electron Microscopy (TEM) images were investigated with TEM (Philips Tecnai 12) microscope. X-ray Diffraction (XRD) patterns of dense and hollow TiO_2 particles before and after calcination were obtained on XRD Rigaku Smartlab. Optical properties were measured by Ultraviolet-Visible Diffuse Reflectance Spectroscopy (UV- DRS) (Shimadzu UV-2600).

3. Results and Discussion

Figure 1 shows TEM images of dense and hollow TiO_2 particles before (a, c) and after calcination (b, d), respectively. Before calcination, particle size distribution of dense TiO_2 particles was approx. 150 nm with relatively uniform contrast throughout particle. As-made hollow TiO_2 particles in Figure 1c showed well-defined hollow structure with 50-nm thick shell and approx. 150 nm hollow core. The shells were the assemblage of much smaller TiO_2 nanoparticles. The inner shell surrounded the hollow core was denser than the outer part. To transform the as-made amorphous TiO_2 to crystalline anatase structure, the samples were calcined at different temperatures (400-950 °C) and times. It was found that, regardless of morphologies, TiO_2 particles were severely aggregated when calcination temperatures were higher than 500 °C and longer than 5 h (data not shown). Calcination temperature at 400 °C for 3 h was then selected. After calcination, slightly shrinkage in size were observed in both calcined dense and hollow samples. The intensity contrast of calcined dense TiO_2 particles were less uniform. In the case of calcined hollow TiO_2, the hollow morphology still remained, but particle size became smaller (approx. 140 nm) with denser shell due to further condensation of titania. In addition, their particle shape and hollow core became slightly irregular.

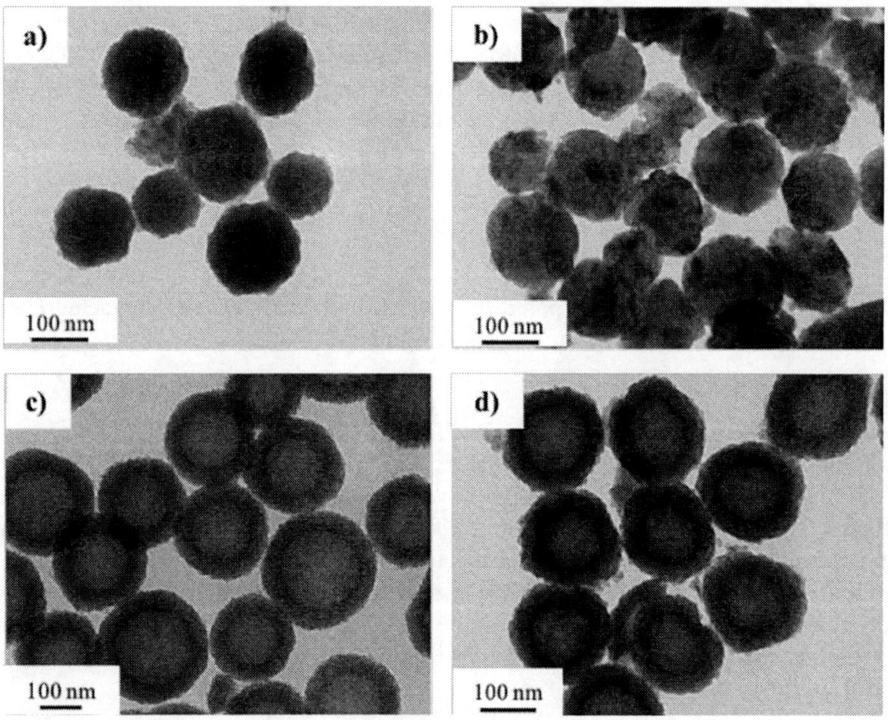

Figure 1. TEM images of dense TiO_2 before (a) and after (b) calcination and hollow TiO_2 before (c) and after (d) calcination, respectively.

Figure 2 shows the XRD patterns of dense and hollow TiO_2 before and after calcination at 400 °C. Compared to the reference anatase, only the diffraction pattern of calcined dense TiO_2 (DTC) showed peaks that corresponded to anatase phase. Diffraction pattern of the calcined hollow TiO_2 exhibited non-resolved peaks, suggesting incomplete phase transformation. This might be due to the loosely assemblage of TiO_2 nanoparticles in the shell layer around the hollow core. However, the difference in optical properties between calcined and amorphous hollow TiO_2 samples was observed in UV-DRS analysis.

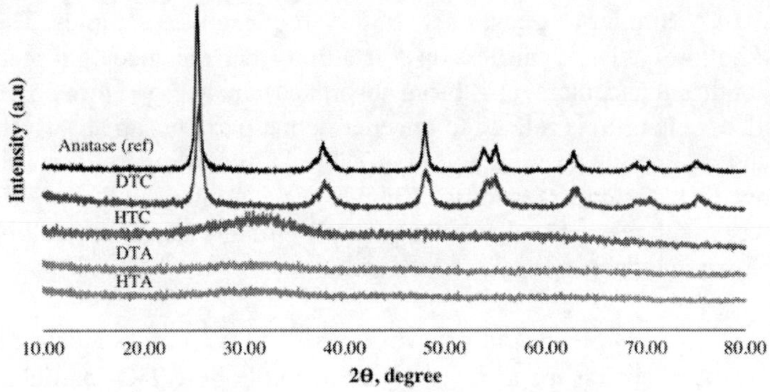

Figure 2. XRD patterns of anatase (reference) as well as dense and hollow TiO_2 particles before and after calcination (denoted as amorphous dense TiO_2 (DTA), amorphous hollow TiO_2 (HTA), calcined dense TiO_2 (DTC) and calcined hollow TiO_2 (HTC), respectively).

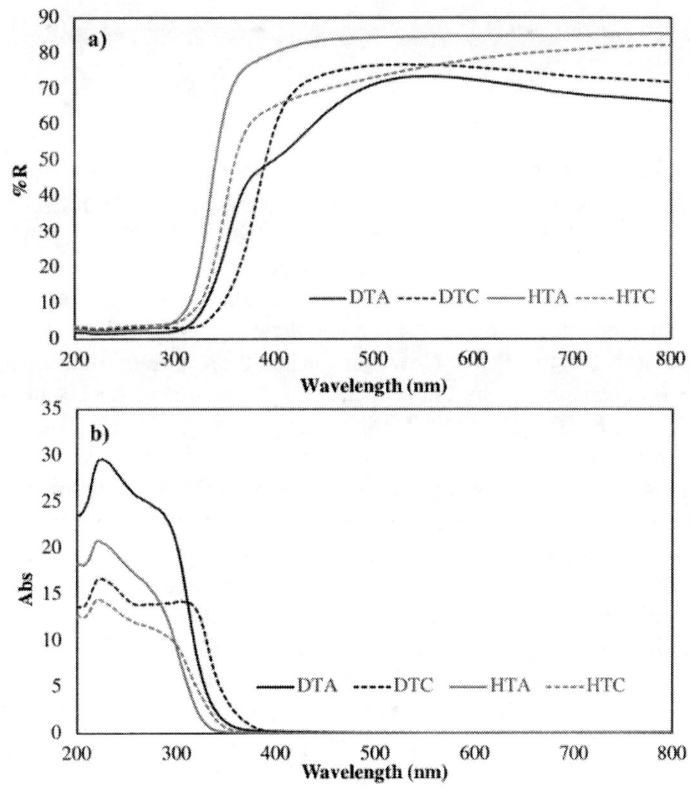

Figure 3. UV-DRS spectra of (a) %reflectance (%R) and (b) absorbance (Abs) of amorphous dense TiO_2 (DTA), amorphous hollow TiO_2 (HTA), calcined dense TiO_2 (DTC) and calcined hollow TiO_2 (HTC), respectively.

Optical properties were characterized by UV-DRS and the resulting spectra are illustrated in Figure 3(a-b). The diffuse reflectance results (%R) in Figure 3a were directly measured as raw data with the instrument while the results shown in Figure 3b were obtained using Kubelka-Munk function. Amorphous hollow TiO_2 nanoparticles exhibited higher %R and boarder range of wavenumber compared to dense TiO_2. Similar trend was also observed in calcined samples. These results suggested that the presence of hollow cavity, regardless on crystallinity, can enhance light scattering and reflection, as reflected in reflection enhancement. [14] From absorbance analysis (Figure 3b), the first wavenumber of response (cut-off wavelength) is related to the energy that particle can absorb (bandgap). Using data from Figure 3b, bandgap energies were reported in Table 1. According to the results, both calcined and non-calcined hollow TiO_2 started to response at higher wavenumber, indicating that they involved the higher bandgap energy with respect to dense TiO_2 samples. In the case of crystalline phase, the calcined samples exhibited lower bandgap energy.

Table 1. Cut-off wavelengths ($\lambda_{cut\ off}$) and calculated bandgaps at $\lambda_{cut\ off}$ of dense and hollow TiO_2 particles before and after calcination.

Sample	$\lambda_{cut\ off}$ (nm)	Bandgap (eV)
DTA	352	3.53
DTC	370	3.36
HTA	333	3.73
HTC	347	3.58

From Table 1, both hollow TiO_2 samples possessed higher bandgap energy compared with dense TiO_2 counterparts. Note that the bandgap energy of calcined dense TiO_2 sample is close to those reported generally in bulk anatase TiO_2 (3.2 eV). [2-6, 17] While the bandgap of hollow TiO_2 became smaller after thermal treatment, implying that there was partial polymorph transition. Based on absorbance spectra and calculated band gap energy, it could be said that in term of PCA, dense TiO_2 samples required lower energy to activate the photocatalysis. This made them response to lower energy spectrum of light, which provide good trait for the PCA. However, as mention earlier, the concept of multiple light scattering and reflection was proved by the reflectance spectra, consistence with studies in the field of DSSC applications. [12] Moreover, the results reported by Yamaguchi and co-workers also support that PCA can be enhanced through multiple light scattering. [14]

To further confirm the effect of morphology on PCA, photodegradation test was performed by coating a thin layer of palmitic acid (organic model) on TiO_2 particle-incorporated polyurethane films prior to UV irradiation at 365 nm for 10 h. FTIR spectra of coated films before and after irradiation are presented in Figure 4(a-b). Before irradiation, intensities of peaks attributed to C-H stretching vibration at 2828-2916 cm^{-1} and C=O stretching vibration at around 1700 cm^{-1} of palmitic acid were sharp and strong. After irradiation, intensities of these characteristic peaks at 2828-2916 and 1700 cm^{-1} became lower, which indicated the photodegradation of palmitic acid on the surface of both TiO_2 particle-incorporated polyurethane films upon UV irradiation. Higher reduction of C-H and C=O vibrations was observed in the irradiated polyurethane films incorporated with crystalline dense TiO_2, which could be attributed to crystallinity-dependent PCA. [17] Despite being amorphous, FTIR spectra of hollow TiO_2 particle-incorporated polyurethane film (Figure 4b) also showed significant depletion of C-H and C=O stretching of palmitic acid after irradiation. This result supported the UV-DRS analysis and suggested that hollow structure took part in the enhancement of light scattering and hence photodegradation of palmitic acid.

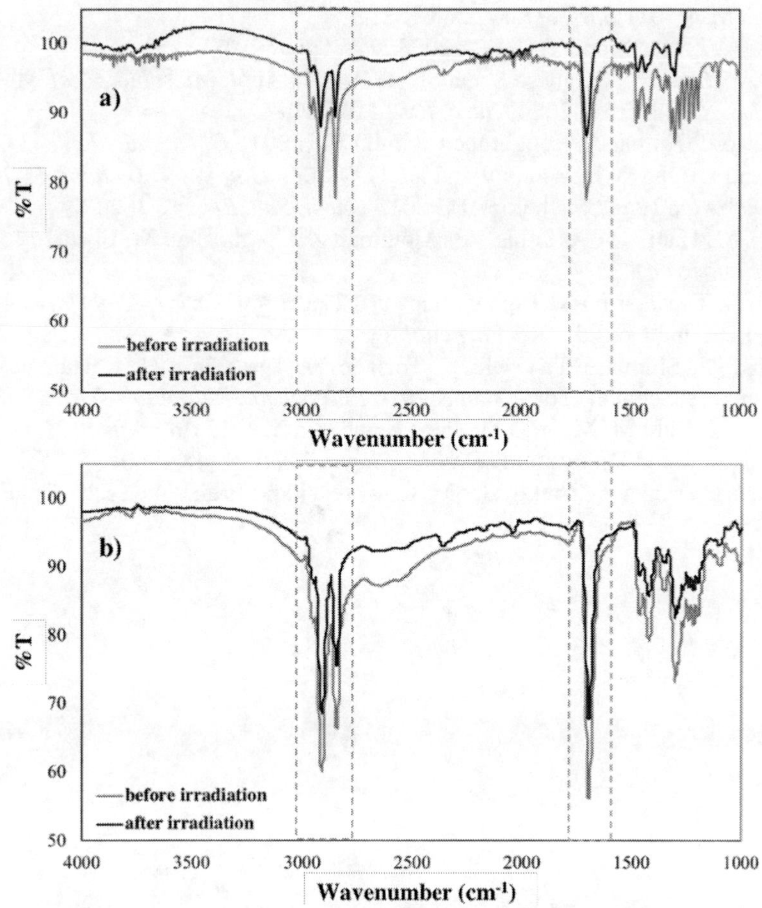

Figure 4. FTIR spectra of palmitic acid-coated polyurethane films incorporated with (a) crystalline dense and (b) hollow TiO_2 particles before and after UV irradiation at 365 nm for 10 h.

4. Conclusions

Dense and hollow TiO_2 particles were successfully synthesized using sol-gel method. After calcined at 400 °C for 3 h, particle shapes of both dense and hollow TiO_2 particles remained but their particle sizes became smaller. Although there was no crystalline peak observed in the XRD patterns, %R of calcined hollow TiO_2 particles was higher than that of non-calcined hollow TiO_2. All hollow TiO_2 exhibited higher reflectance than dense counterparts, indicating that hollow cavity can enhance the light scattering. However, band gap energy of all hollow TiO_2 was slightly higher than dense samples. Photodegradation results indicated the degradation of palmitic acid on the surface of amorphous hollow TiO_2-incorporated polyurethane films.

Acknowledgements

This work is supported by Faculty of Engineering, King Mongkut's Institute of Technology Ladkrabang (2558-01-01042).

References

[1] Balazs A C, Emrick T and Russell T P 2006 *Science* **314** 1107
[2] Cargnello M, Gordon T R and Murray C B 2014 *Chem. Rev.* **114** 9319
[3] Pelaez M, Nolan N T, Pillai S C, Seery M K, Falaras P, Kontos A G, Dunlop P S M, Hamilton J W J, Byrne J A, O'Shea K, Entezari M H and Dionysiou D D 2012 *Appl. Catal. B-Environ.*

IOP Conf. Series: Materials Science and Engineering **639** (2019) 012021 doi:10.1088/1757-899X/639/1/012021

125 331
[4] Henderson M A 2011 *Surf. Sci. Rep.* **66** 185
[5] Fujishima A, Zhang X and Tryk D A 2008 *Surf. Sci. Rep.* **63** 515
[6] Foster H A, Ditta I B, Varghese S and Steele A 2011 *Appl. Microbiol. Biot.* **90** 1847
[7] Dahl M, Liu Y and Yin Y 2014 *Chem. Rev.* **114** 9853
[8] Tian J, Zhao Z, Kumar A, Boughtonc I R and Liu H 2014 *Chem. Soc. Rev.* **43** 6920
[9] Gao M, Zhu L, Ong W L, Wanga J and Ho G W 2015 *Catal. Sci. Technol.* **5** 4703
[10] Li G L, Mohwald H and Shchukin D G 2013 *Chem. Soc. Rev.* **42** 3628
[11] El-Toni A M, Habila M A, Labis J P, Alothman Z A, Alhoshan M, Elzatahry A A and Zhang F 2016 *Nanoscale* **8** 2510
[12] Dadgostar S, Tajabadi F and Taghavinia N 2012 *ACS Appl. Mater. Interfaces* **4** 2964
[13] Zhang P, Li A and Gong J 2015 *Particuology* **22** 13
[14] Yamaguchi Y, Shimodo T, Usuki S, Torigoe K, Terashima C, Katsumata K I, Ikekita M, Fujishima A, Sakai H and Nakata K 2016 *Photochem. Photobiol. Sci.* **15** 988
[15] Joo J B, Lee I, Dahl M, Moon G D, Zaera F and Yin Y 2013 *Adv. Funct. Mater.* **23** 4246
[16] Han C, Luque R and Dionysiou D D 2012 *Chem. Commun.* **48** 1860
[17] Smirnova N, Fesenko T, Zhukovsky M, Goworek J and Eremenko A 2015 *Nanoscale Res. Lett.* **10** 1

Development and Control of Rotary Inverse Pendulum by LQR with Integral Action

Surachat Chantarachit

Department of Mechatronics Engineering,
Rajamangala University of Technology Thanyaburi, Thailand
E-mail: surachat_c@rmutt.ac.th

Abstract. This research a basic well-known hardware, is introduced. The inverse pendulum is the focused in this research. In order to prove the control algorithm LQR with integral action. The dynamic model is obtained by Euler-Lagrange equation. Motor actuator is included in the model in order to get the correct model. The model is linearize by Taylor series expansion. The model parameters are identified by software simulation. The inversed pendulum prototype is manufactured and assembled. The Linear Quadratic Regulator with Integral (LQR+I) controller is presented and applied to control the inverse pendulum. The control simulation and experiment results are present in the performance of this controller

Keyword : Rotary Inverse Pendulum, Conventional LQR, LQR with Integral action

1. Introduction

Inverse Pendulum is the basic tool to prove the designing of control algorithm. By simple hardware and mechanism, the inverse pendulum is the good example in control system, because of dynamic model of inverse pendulum is non-linear model but it not complicate. The model type of inverse pendulum can be divide into two types. The first is the convention inverse pendulum or linear inverse pendulum [1] (LIP). The limitation of LIP is the boundary of the movement, it depend on the length of the guide of the hardware. In order to eliminate this limit hardware problem, the rotary inverse pendulum [2] (RIP) can increase working space. By this technique, the experiment results can be collected more. By simply of structure and dynamic model, many control algorithms are applied to prove the concept. Neural network is the one of non-linear algorithm. [3] are applied to control the inverse pendulum by present ability to adjust any parameter in the system. The linear controller also applied to control the inverse pendulum. [4] applied LQR controller to manipulate pendulum upright position. To increase the performance of the LQR, [5] put the fuzzy controller to adjust the controller gain of the LQR. The controller gain can change depend on error of the system. This paper present the conventional LQR with integral action. The performance of LQR with integral is evaluated in this paper.

2. Hardware Design

The rotary inverse pendulum is designed by increasing working space. The rotary inverse pendulum compose of two parts; the first part is moving part and the second part is stationary part. The rotary inverse pendulum model is presented in figure 1.

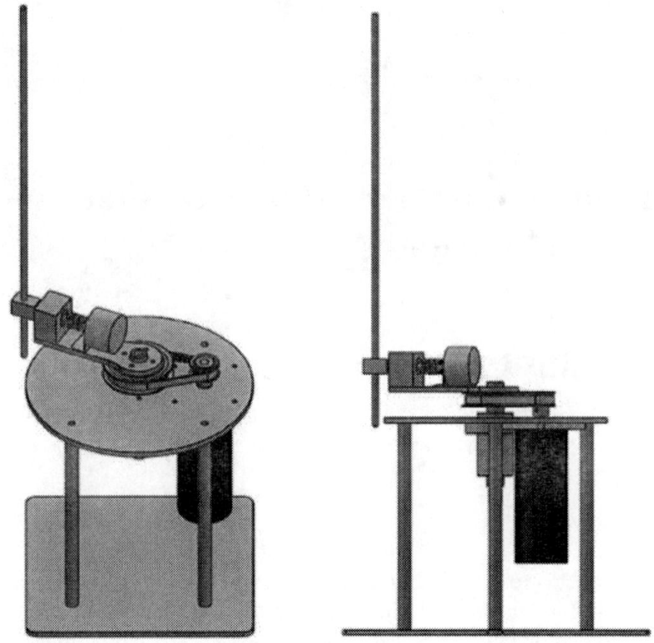

Figure 1 CAD modelling of Rotary Inverse pendulum

The moving part presents the pendulum rod and cart. To measure angle of the pendulum rod, rotary encoder is directly attached to the shaft which links to the pendulum rod. There is encoder electronic device on the moving part, then the cable of the encoder is the obstacle of full rotation of the moving cart. To solve the problem of the cable, Slip ring is the equipment which attached to divide moving part and stationary part. The moving part can be fully rotating. The RTD1024 is rotary encoder with resolution 1024 pulse per revolution (2.84 pulse per degree). And the significant power to rotate the pendulum is 100 watts of DC motor. Those power transmits through pulley and belt with ratio (18:7).

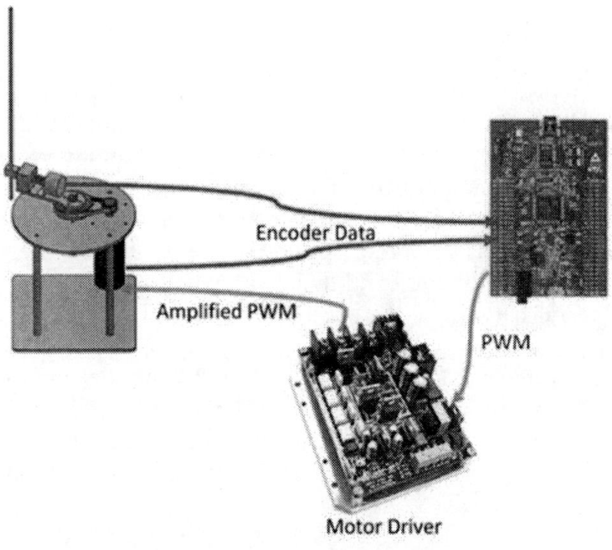

Figure 2 Electronic Diagram of Inverse pendulum

The 32 Bit of micro-controller is applied to compute encoder data and control rule. The electronic diagrams of the inverse pendulum shown in figure 2. And the structure of the inverse pendulum is built from aluminium and steel as shown in figure 3.

IOP Conf. Series: Materials Science and Engineering **639** (2019) 012022 doi:10.1088/1757-899X/639/1/012022

Figure 3 Inverse pendulum

3. Dynamic Model

The dynamic model of inverse pendulum is spilt into two parts the first part is inverse pendulum another part is pendulum arm. Both of them are the moving part. The 2D model is presented in figure 4.

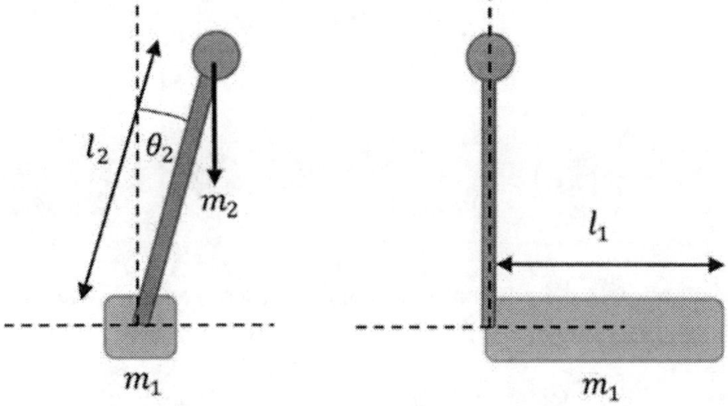

Figure 4 2D inverse pendulum model

The Euler-Lagrange equation (1) is solved the dynamic model of the inverse pendulum. Friction in the system is small and negligible.

$$\frac{d}{dt}\left(\frac{\partial L}{\partial \dot{q}}\right)-\frac{\partial L}{\partial q}=f \tag{1}$$

When - L is the difference between total kinetic energy and potential energy
- q is the state variable
- f is the external input

Then the Lagrange equation is

$$L=\frac{1}{2}J_{hub}\dot{\theta}_1^2+\frac{1}{2}m_2\left(l_1\left(\dot{\theta}_1\right)+l_2cos\theta_2\left(\dot{\theta}_2\right)\right)^2+\frac{1}{2}m_2\left(-l_2sin\theta_2\left(\dot{\theta}_2\right)\right)^2+\frac{1}{2}J_{pendulum}\dot{\theta}_2^2-m_2gl_2cos(\theta_2)\quad(2)$$

From the equation (1), the non-linear model is rearranged to equation (3)

$$M(q)\ddot{q}+C(q,\dot{q})+G(q)+D=F\quad(3)$$

when

$$M=\begin{bmatrix}M_{11}&M_{12}\\M_{21}&M_{22}\end{bmatrix}\ C=\begin{bmatrix}C_1\\C_2\end{bmatrix}\ G=\begin{bmatrix}G_1\\G_2\end{bmatrix}\ D=[0]\ F=\begin{bmatrix}F_1\\F_2\end{bmatrix}$$

$$M_{11}=J_{hub}+l_1^2m_2$$
$$M_{12}=l_1l_2m_2\mathrm{Cos}\theta_2$$
$$M_{21}=M_{12}$$
$$M_{22}=J_{pen}+l_2^2m_2$$
$$C_1=-l_1l_2m_2\mathrm{Sin}\theta_2\dot{\theta}_2^2+B_{eq}\dot{\theta}_1+\frac{K_t^2K_e^2}{G_{ratio}R}\dot{\theta}_1$$
$$C_2=0$$
$$G_1=0$$
$$G_2=-l_2m_2g\mathrm{Sin}\theta_2$$
$$F_1=-\frac{K_tK_e}{G_{ratio}R}\mathrm{Vol}$$
$$F_2=0$$

And the linear model is solved from equation (3) by Taylor series expansion. The linear model of the inverse pendulum is shown in equation (4) and rearranged to state space form.

$$\dot{x}=Ax+Bu\quad(4)$$

or:

$$\begin{bmatrix}\dot{\theta}_1\\\dot{\theta}_2\\\ddot{\theta}_1\\\ddot{\theta}_2\end{bmatrix}=\begin{bmatrix}&0\ 0\ 1\ 0&\\&0\ 0\ 0\ 1&\\M^{-1}*\begin{bmatrix}0&0&-B_{eq}-\frac{K_e^2K_t^2}{RG_{ratio}}&0\\0&gl_2m_2&0&0\end{bmatrix}\end{bmatrix}\begin{bmatrix}\theta_1\\\theta_2\\\dot{\theta}_1\\\dot{\theta}_2\end{bmatrix}+\begin{bmatrix}0\\0\\M^{-1}*\begin{bmatrix}-\frac{K_eK_t}{RG_{ratio}}\\0\end{bmatrix}\end{bmatrix}Vin$$

M. is the symmetrical positive definite matrix of the robot inertia and evaluated at the upright position.

4. Control Algorithm

This section is presented the implemented control algorithm. By adding integral term to conventional LQR to improve the performance.

4.1 Conventional LQR Controller

The conventional LQR controller is the reference controller refer to [5] that optimally determines the gains by compromising the state and control-input cost. The LQR cost function (J) is expressed by

$$J=\int_0^{\infty}\left(x^TQx+u^TRu\right)dt\quad(5)$$

The Q and R matrices are the state and control weighting matrices. The control signal follows equation (5)

IOP Conf. Series: Materials Science and Engineering **639** (2019) 012022 doi:10.1088/1757-899X/639/1/012022

$$u = -Kx \tag{6}$$

The optimal gain matrix is solved by the algebraic Riccati equation (6-7).

$$A^T P + PA + Q - PBR^{-1}B^T P = 0 \tag{7}$$

$$K = R^{-1}B^T P \tag{8}$$

The metrics Q and R are the weighting metrics which define depend on significant state.

4.2 LQR with Integral Action Controller

The LQR with Integral (LQR+I) is improvement of controller form the conventional LQR. The LQR+I increase integral state to control the system by modify state of the LQR as show in equation (9)

$$\dot{x} = \begin{bmatrix} \dot{x}_e \\ \dot{z}_e \end{bmatrix} = \begin{bmatrix} \dot{x} - \dot{x}_r \\ x - x_r \end{bmatrix} = \begin{bmatrix} A & 0 \\ C & 0 \end{bmatrix} \begin{bmatrix} x_e \\ z_e \end{bmatrix} + \begin{bmatrix} B \\ 0 \end{bmatrix} u = \tilde{A}x + \tilde{B}u \tag{9}$$

when - x_e is error state

- z_e is $\int edt$
- x_r is the reference input

And the control signal of the LQR+I is

$$u = -Kx = -\begin{bmatrix} K & K_i \end{bmatrix} \begin{bmatrix} x_e \\ z_e \end{bmatrix} = -K(x - x_r) - K_i C \int (x - x_r) dt \tag{10}$$

Increasing state of the integral term improves steady error of the system. This term can eliminate error and overshoot of the system caused by incomplete dynamic model, friction in the system. The block diagrams of the LQR+I shows in figure 5.

This technique can produce good dynamic and static performance.

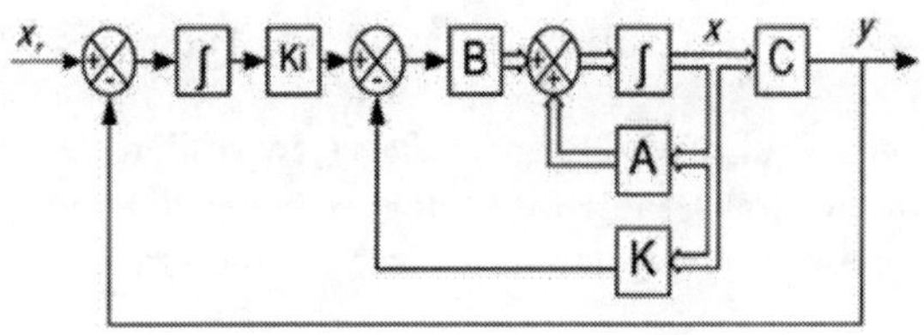

Figure 5 LQR+I Block diagrams

5. Simulation Results

In this section, the balancing performance of inverse pendulum are simulated and compared with convention LQR and LQR+I. The simulation results are simulated by MATLAB/Simulink. The initial condition of simulation is set to 5 degree of angle. The simulation results are presented in figure 6 and 7

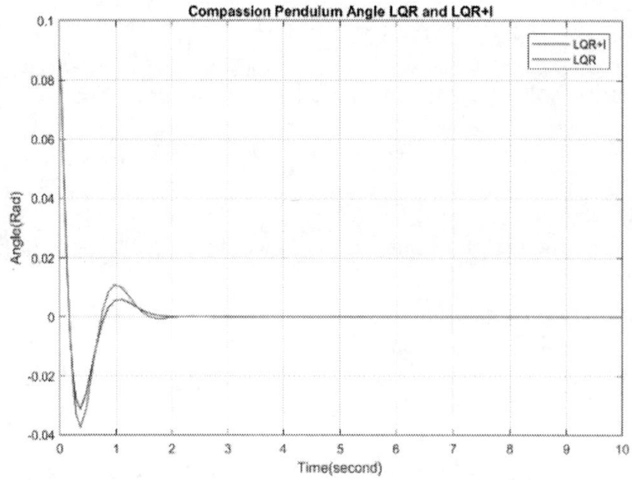

Figure 6 Comparison pendulum angle of LQR and LQR+I

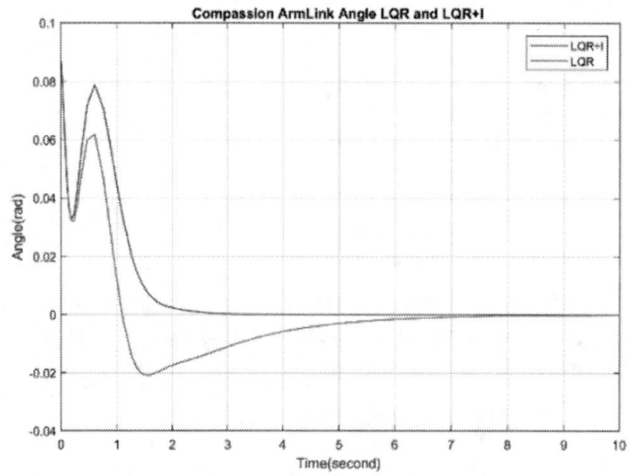

Figure 7 Comparison arm link angle of LQR and LQR+I

The inverse pendulum angle result of LQR and LQR+I controller can convert error response to set point. And the response of arm links angle by LQR+I is better than convention LQR. By adding the integral term, the first response of the system is slower but the system can eliminate error better than LQR.

6. Experimental result

After simulate the response results, the inverse pendulum system is test by the real hardware. The control rule of the LQR is implemented by velocity control and convert it to discrete control. The discrete control rule is implement to microcontroller as shown in equation (11).

$$u[n]=u[n-1]+[K \quad K_i]\begin{bmatrix} e_{\theta 1}[n]-e_{\theta 1}[n-1] \\ e_{\theta 2}[n]-e_{\theta 2}[n-1] \\ (e_{\theta 1}[n]-2e_{\theta 1}[n-1]+e_{\theta 1}[n-2])/\Delta t \\ (e_{\theta 2}[n]-2e_{\theta 2}[n-1]+e_{\theta 2}[n-2])/\Delta t \\ e_{\theta 1}[n]\Delta t \\ e_{\theta 1}[n]\Delta t \end{bmatrix} \quad (11)$$

The experimental results are presented in inverse pendulum angle and arm link angle as shown in figure 8 and 9.

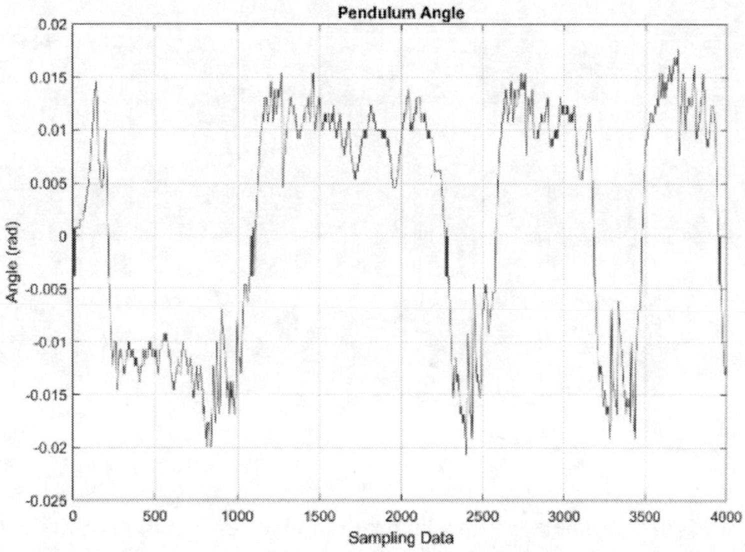

Figure 8 The experimental result of inverse pendulum of LQR

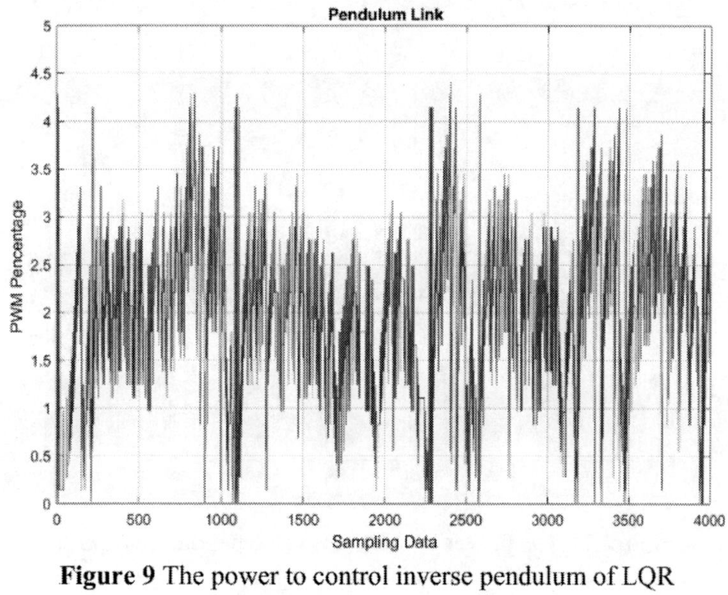

Figure 9 The power to control inverse pendulum of LQR

From the experiment results, both controller algorithm can balance pendulum to upright position. The amplitude of balancing angle of the LQR+I is smaller than LQR by 40 percentage. Increasing the performance of the inverse pendulum, LQR+I consume power than conventional LQR. By this controller, the inverse pendulum can balance longer time than conventional LQR. By adding the integral term to the controller, the performance is better in term of error because the behaviour of integral action. In term of response, adding integral action lead to slowly response compare with convention LQR.

7. Conclusion

The inverse pendulum is simulated and test by LQR and LQR+I controller. Both control algorithms are compared. The Performance of the LQR+I can reduces overshoot and eliminates steady state error. By this algorithm, the performance is improved but high energy consume also.

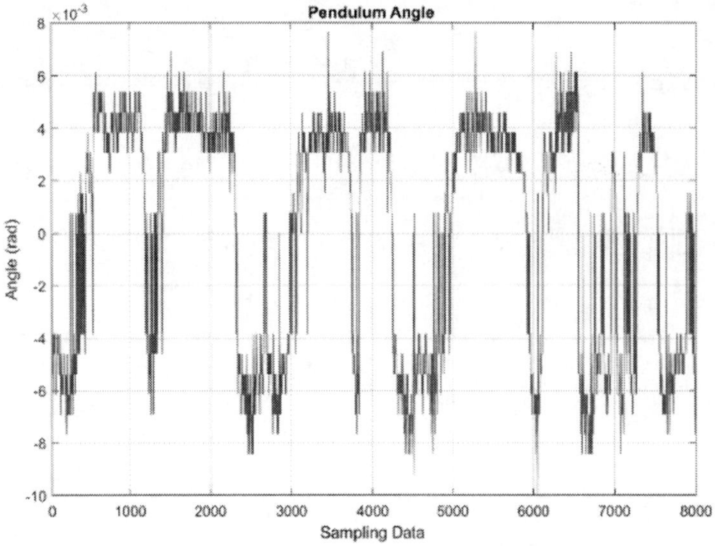

Figure 10 The experimental result of inverse pendulum of LQR+I

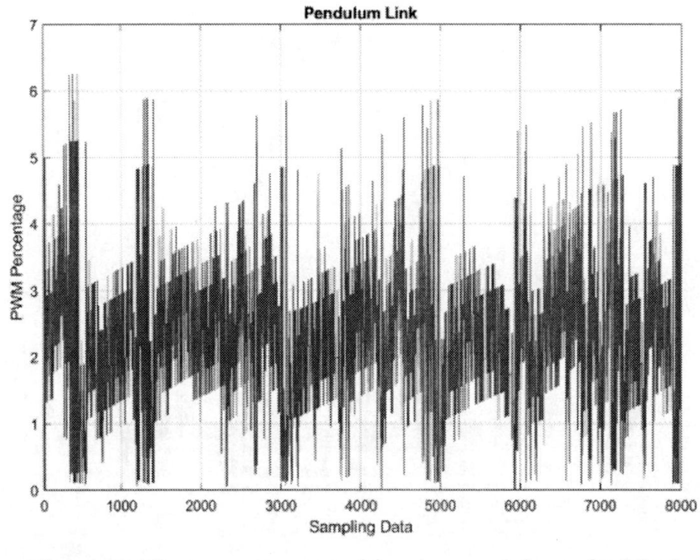

Figure 11 The power to control inverse pendulum of LQR+I

References

[1] Muskinja, N., & Tovornik, B. (2006). Swinging up and stabilization of a real inverted pendulum. IEEE transactions on industrial electronics, 53(2), 631-639.

[2] Akhtaruzzaman, M., & Shafie, A. A. (2010, August). Modeling and control of a rotary inverted pendulum using various methods, comparative assessment and result analysis. In Mechatronics and Automation (ICMA), 2010 International Conference on (pp. 1342-1347). IEEE.

[3] Quyen, N. D., Van Thuyen, N., Hoc, N. Q., & Hien, N. D. (2012). Rotary inverted pendulum and control of rotary inverted pendulum by artificial neural network. In Proc. Natl. Conf. Theor. Phys (Vol. 37, pp. 243-249).

[4] Kumar, K. P., & Rao, G. S. K. Modeling and controller designing of rotary inverted pendulum (RIP)-comparison by using various design methods. In Proceedings.

[5] Chantarachit, S. (2018). Design and simulate of LQR-Fuzzy controller for unicycle robot with double flywheels. In MATEC Web of Conferences (Vol. 192, p. 02001). EDP Sciences.

IOP Conf. Series: Materials Science and Engineering **639** (2019) 012023 doi:10.1088/1757-899X/639/1/012023

Strain-based Evaluation of Bridge Monitoring using Numerical Model Analysis

N Sinsamutpadung[1] **and E Sasaki**[2]

[1]Department of Civil Engineering, Faculty of Engineering, King Mongkut's Institute of Technology Ladkrabang, 1 Chalongkrung Road, Ladkrabang , Bangkok, 10520, Thailand

[2]Department of Civil Engineering, Tokyo Institute of Technology, 2-12-1-M1-23 Ookayama, Meguro-ku, Tokyo, 152-8552, Japan

Email: natdanai.si@kmitl.ac.th

Abstract. According to visual inspection on the bridge site, the target bridge has major cracks in RC deck in the center span. This certain type of damage could affect structural performance and should be monitored. The structural health monitoring system was set up on the target bridge and the strain-based evaluation could be used to quantitatively determine bridge condition. The FE model of the bridge was modeled with various types of RC cracking pattern to represent the severity of bridge damage. The relationship between damage level and strain measurement is generated, As a result, Bridge monitoring curve based on strain index has been established.

Keywords: Structural Health Monitoring, RC deck bridge, Finite element model

1. Introduction

In the past three decades, there number of constructed bridges to serve the expansion of civilization. In present, these bridges have aged and deteriorated along with time [1]. The structural performance of these bridges has become lower and could affect the safety of people. The bridge maintenance is required. However, budget and manpower are limited. In order to conduct efficient management, the severity of bridge damage should be evaluated and bridge prioritizing should be performed.

In general, visual inspection is used to determine the severity of damage in bridges. However, the process required an experienced engineer to perform the inspection at the bridge site. Since there are a limited number of engineers, the inspection is usually performed every 2–5 years per bridge [1, 2]. According to the time gap between inspections, the damage could be progressive too far before detection which led to high repairing cost. Hence, early damage detection can improve the efficiency of budget allocation. In addition, there are certain types of damage that couldn't detect by visual inspection alone (e.g., cracking depth in RC deck, corrosion inside the steel girder). Therefore, an alternative evaluation method can be useful.

Structure health monitoring system is developed [3, 4]. The bridge condition could be remotely assessing via sensor (e.g., strain gauge, accelerometer, displacement transducer). Thus, the unusual behavior indicating bridge damage could be real-time detected. So far, the system is used as an event detector. For example, when the large displacement or strain occurred, the system will alert engineer to perform the inspection at the bridge to ensure the structure safety. Recently, there are studies on

IOP Conf. Series: Materials Science and Engineering **639** (2019) 012023 doi:10.1088/1757-899X/639/1/012023

bridge evaluation based on acceleration data [5, 6]. Bridge conditions can be remotely evaluated by post-processing of acceleration data. The method could save engineer trips to the bridge site for inspection. Besides acceleration, it is well known that strain and deflection can reflect structural behavior and might improve the resolution of bridge evaluation from previous studies. Therefore, there is a possibility that strain gauge and displacement transducer can be used to evaluate the bridge condition.

In this study, the steel box girder bridge with reinforced concrete deck (RC deck) is chosen as the target bridge. Visual inspection is performed to identify damage types. Crack patterns in RC deck are observed and categorized. Then, FE model of the target bridge is modeled with various cracking patterns to represent the severity of bridge damage. Deflection and strain distribution at the lower surface of RC deck can be determined. Quantitative evaluation of bridge condition can be performed by matching the severity of bridge damage and measurement index. As a result, bridge monitoring curve based on strain index is established.

2. Target Bridge

The steel box girder bridge with the RC deck designed as a non-composited girder has been chosen as the target bridge as shown in figure 1. The bridge has three continuous spans which side span length is 71.3 meters and the middle span is 88 meters. Visual inspection is performed on the bridge. It is found that the bridge has major cracks in the center span of the RC floorboard. The influence of the crack and its progression should be monitored. Thus, sensor measurements (strain gauge and displacement transducer) are attached at the lower surface of the center span (Figure 2). However, sensor measurement value also depends on locations of the sensor (e.g., large strain near crack tip, deflection at the middle of span). Thus, sensitivity analysis to determine an appropriate location for each sensor should be conducted.

3. Numerical Model

3.1. Bridge details

Based on the drawing of components (i.e., main girders, transverse girders, RC deck, and rubber bearing), the finite element model (FE Model) of the target bridge has been created by using ABAQUS 6.14 [7] as shown in figure 3. The RC deck and rubber bearing were modeled by using solid element while steel girder was modeled by using shell element. The material properties of concrete and steel were assigned based on the value indicated in the specification and the rubber bearing stiffness was calculated by using the estimate equation. Moreover, abutment and pier bearing have difference dimensions. Therefore, the rubber bearing stiffness calculated individually at each location. The material properties used in the model are shown in table 1.

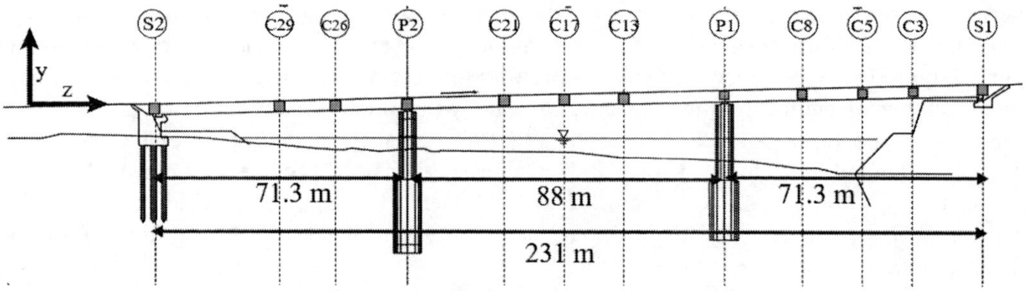

Figure 1. Side view of Target Bridge

IOP Conf. Series: Materials Science and Engineering **639** (2019) 012023 doi:10.1088/1757-899X/639/1/012023

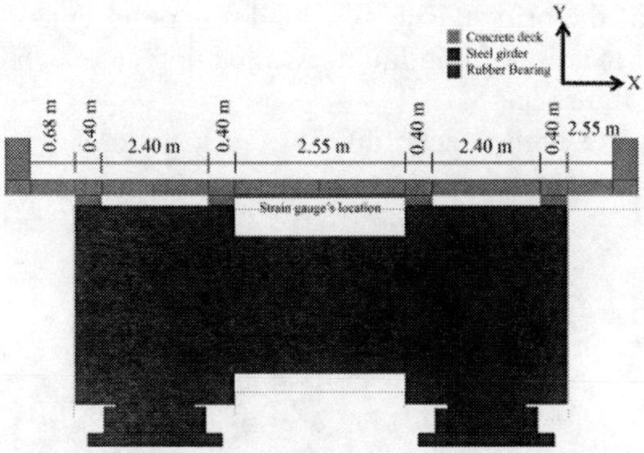

Figure 2. Cross sectional of Target Bridge

Figure 3. FE model of Target Bridge

Table 1. Material properties

Index	*Material*	*Young Modulus*
1	Concrete	28 GPa
2	Steel	205 GPa
3	Abutment bearing rubber	54.14 MPa
4	Pier bearing rubber	60.14 MPa

3.2. Crack modelling

Based on the visual inspection, crack patterns found in the bridge are transverse crack (T-crack), longitudinal crack (L-crack), and combined crack (TL-crack). In order to observe the influence of damage, crack patterns were modeled with a series of transverse crack and combined crack direction. Moreover, various crack intervals are considered for representing the severity of bridge damage. For example, short crack interval represents that there are a number of cracks which means the bridge condition is severely damaged.

In order to introduce the RC deck cracking into the FE model, the stiffness reduction by using orthotropic elasticity properties [7] was assigned on elements along the crack to represent the damage. The orthotropic property in terms of stress (σ), nominal strain (ε), shear strain (γ), young modulus (E), shear modulus (G), and Poisson's ratio (ν) can be expressed as shown in equation (1) and equation (2).

Three types of an element were defined as T-element, L-element, and TL-element. The crack patterns are shown in figure 4. The orthotropic properties of each type of element are shown in table 2.

$$\begin{Bmatrix} \varepsilon_{11} \\ \varepsilon_{22} \\ \varepsilon_{33} \\ \gamma_{12} \\ \gamma_{13} \\ \gamma_{23} \end{Bmatrix} = \begin{bmatrix} 1/E_1 & -\nu_{21}/E_2 & -\nu_{31}/E_3 & 0 & 0 & 0 \\ -\nu_{12}/E_1 & 1/E_2 & -\nu_{32}/E_3 & 0 & 0 & 0 \\ -\nu_{13}/E_1 & -\nu_{23}/E_2 & 1/E_3 & 0 & 0 & 0 \\ 0 & 0 & 0 & 1/G_{12} & 0 & 0 \\ 0 & 0 & 0 & 0 & 1/G_{13} & 0 \\ 0 & 0 & 0 & 0 & 0 & 1/G_{23} \end{bmatrix} \begin{Bmatrix} \sigma_{11} \\ \sigma_{22} \\ \sigma_{33} \\ \sigma_{12} \\ \sigma_{13} \\ \sigma_{23} \end{Bmatrix} \quad (1)$$

$$G_{ij} = \frac{\sqrt{E_i E_j}}{2(1+\nu)} \quad (2)$$

3.3. Loading conditions

The FE model is subjected to load at the center span of the bridge. The axle load was represented by two nodal load with 50 kilonewtons (kN) at the wheel location. In addition, two loading cases which are load on center and load on lane is used as shown in figure 5.

For each analytical configuration, analysis cases were named according to crack patterns, crack intervals, and loading locations. The analytical configuration can be list as shown in table 3.

a) Transverse direction (T-crack) **b)** Combined direction (TL-crack)

Figure 4. Crack patterns

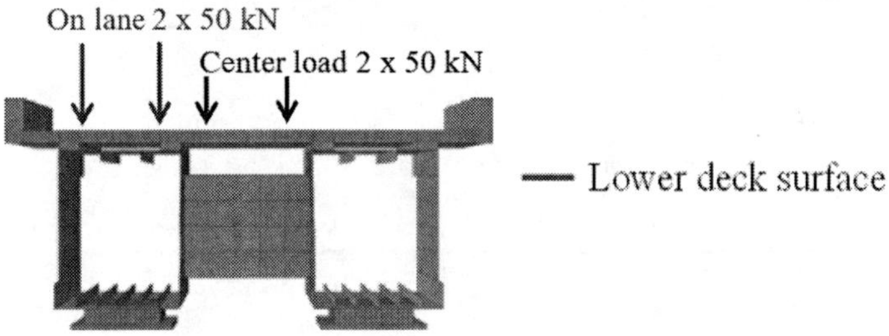

Figure 5. Loading conditions

Table 2. Orthotropic properties for cracks

Index	Name	*Young modulus* E_{xx}	E_{yy}	E_{zz}
1	■ T-element	5 GPa	1 GPa	1 GPa
2	■ L-eleement	1 GPa	5 GPa	1 GPa
3	■ TL-element	1 GPa	1 GPa	1 GPa

Table 3. Analytical configurations

Index	*Name*	*Crack patterns*	*Crack interval (m)*	*Loading location*
1	No-crack-C	No crack	-	Center
2	No-crack-OL	No crack	-	On lane
3	1.0T-C	Transverse crack	1	Center
4	1.0T-OL	Transverse crack	1	On lane
5	0.5T-C	Transverse crack	0.5	Center
6	0.5T-OL	Transverse crack	0.5	On lane
7	0.25T-C	Transverse crack	0.25	Center
8	0.25T-OL	Transverse crack	0.25	On lane
9	1.0TL-C	Combined direction	1	Center
10	1.0TL-OL	Combined direction	1	On lane
11	0.5TL-C	Combined direction	0.5	Center
12	0.5TL-OL	Combined direction	0.5	On lane
13	0.25TL-C	Combined direction	0.25	Center

IOP Conf. Series: Materials Science and Engineering **639** (2019) 012023 doi:10.1088/1757-899X/639/1/012023

4. Bridge monitoring curve

The analysis results were obtained from the lower surface of the deck (X-axis) as indicated in figure 3.

4.1. Deflection

Deflection is calculated from vertical displacement along the lower deck and shown in figure 5. The influence of cracking in the RC deck on deflection can be observed. The bridge with more severe damage will have larger deflection. The deterioration level of the bridge can be distinguished. Especially, in the case of load at center, the difference of each damage level can be clearly observed. To represent the severity of damage using deflection consider as a straightforward index for evaluating the bridge condition.

However, deflection measurement is not so convenience in practical monitoring because displacement transducer is required to set an initial position before start measurement. There is a possibility that the transducer slips out from the original position (e.g., strong wind, earthquake, animal). On the other hand, a strain gauge can be attached with a protector that can withstand the environment circumstance. Therefore, the bridge assessment based on strain index is considered to be more practical.

4.2. Transverse strain

Transverse strain along the lower deck surface is shown in figure 4. As in agreement with deflection index, it is found that transverse crack has an influence on strain along the lower deck surface. The severity of damage can be evaluated using strain measurement at lower deck surface. For example, 0.25T-C, 0.25TL-C which considered as severe damage can be observed. Hence, the bridge monitoring curve based on strain index has been established.

Comparison between Figure 5a and Figure 5b, the loading condition has a significant influence on the order of measured strain. Loading at the center of the deck caused a larger strain level than loading on lane. However, the severity of damage also can be observed in a smaller range of strain (figure 5c). Therefore, there should be noted that the strain measurement range is required to have enough resolution to capture the damage.

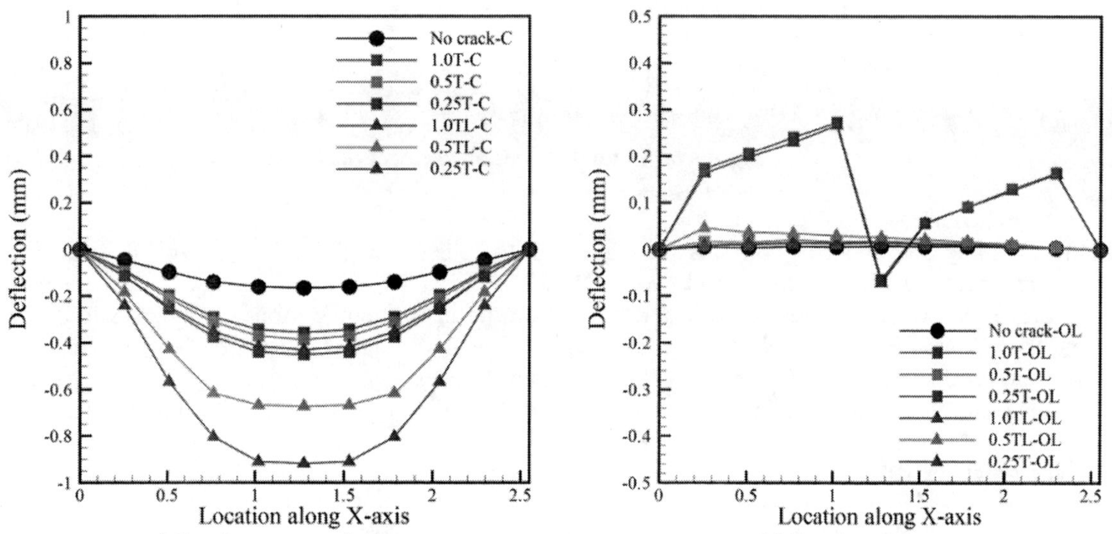

a) Load at center deck b) Load on lane

Figure 5. Deflection at lower deck

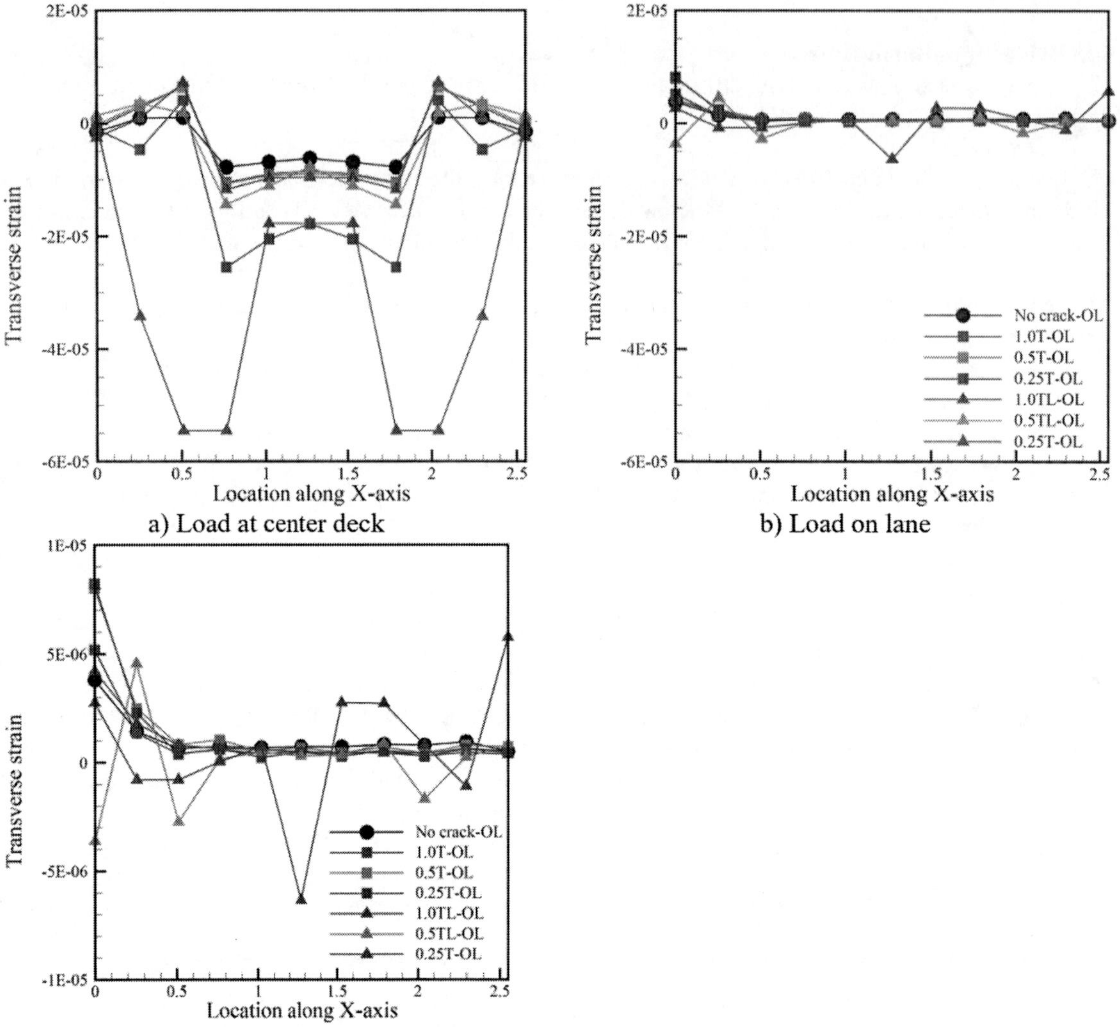

a) Load at center deck

b) Load on lane

c) Load on lane ($\pm 10^{-5}$ strain range)

Figure 6. Strain distribution at lower deck

5. Conclusions

In this study, visual inspection was performed on the steel box girder bridge with the reinforced concrete deck (RC deck). Major cracks were found in the center span of the RC deck. Then, FE model of the target bridge is modeled with crack based on the inspection. Various cracking patterns were introduced into FE model to represent the severity of bridge damage. The relationship between bridge damage level and measurement index (deflection and strain) is generated and can be used to evaluate bridge condition. As a result, bridge monitoring curve based on strain index has been established.

Acknowledgement

The research work shown in this paper was supported by Council for Science, Technology and Innovation, Cross-ministerial Strategic Innovation Promotion Program (SIP), Infrastructure Maintenance, Renovation, and Management" (funding agency: MLIT) and it has been conducted under the collaborative study with Omron Social Solutions, Co., Ltd.

IOP Conf. Series: Materials Science and Engineering **639** (2019) 012023 doi:10.1088/1757-899X/639/1/012023

References

[1] Ministry of Land, Infrastructure, Transport and Tourism, MLIT. 2014. *http://www.mlit.go.jp/road/sisaku/yobohozen/yobohozen.html*

[2] AASHTO, 1974 *Manual for Maintenance Inpsection of Bridges* American Association of State Highway and Transportation Officials

[3] J. Shinae, J. Hongki, M. Kirill, A. R. Jennifer, S. Sung-Han, J.m. Hyung-Jo, Y. Chung-Bang, F. S. Billie, and A. Gul, 2010 Structural health monitroing of a cable-stayed bridge using smart senson technology: deployment and evaluation, Smart Structures and Systems, vol. 6, no. 5-6, pp. 439-459.

[4] E. Sasaki, G. V. Minesawa, T. Shimozato, Y. Arizumi, and S. Nakamine, 2013 A wireless SHM system solutions for a long span interisland bridge in Oki-nawa, The 12th Japan-Korea Joint Symposium on Steel Bridges.

[5] E. Sasaki, P. Tuttipongsawat, N. Sinsamutpadung, H. Nishida, and K. Takase, 2018 Development of a remote monitoring system with wireless power-saving sensons for analyzing bridge conditions, 6th International Symposium on Life-Cycle Civil Engineering, October 2018, Ghent, Belgium.

[6] E. Sasaki, P. Tuttipongsawat, N. Sinsamutpadung, H. Nishida, K. Takase, 2018 Condition Evaluation of a Highway Bridge with RC Deck Using Monitoring Data Obtained by Wireless sensors, 1st International Conference on Concrete and Steel Technology, Engineering & Design (CASTED2018), May 2018, Phillipines.

[7] ABAQUS Inc. 2014. ABAQUS Analysis User's Manual, Version 6.14.

IOP Conf. Series: Materials Science and Engineering **639** (2019) 012024 doi:10.1088/1757-899X/639/1/012024

Measurement model of good governance in government procurement

T Chaitongrat[1*], N Leungbootnak[2], W Kusonkhum[2], W Deewong[3], S Liwthaisong[1], and K Srinavin[2]

[1] Faculty of Architecture, Urban Design and Creative Arts, Mahasarakham University, Mahasarakham, Thailand
[2] Department of civil engineering, Faculty of Engineering, Khon Kaen University, Khon Kaen, Thailand
[3] Civil Architecture Program, Faculty of Technology, Udon Thani Rajabhat University, Thailand

E-mail: tanayut.c@msu.ac.th

Abstract. This research aimed to develop competency factors for Good Governance in Government Procurement. The research adopted structural relationship models and is descriptive in nature. The research sample was drawn from 353 government sectors. The research tool was an expert questionnaire which had content validity and reliability of 0.986. The data was statistically analyzed by 1) exploratory factor analysis and 2) second-order confirmatory factor analysis. The research found that the model for the Good Governance in Government Procurement had high construct validity by both exploratory and confirmatory factor analysis. The model composed of 1 factor and 14 sub-factors from 15 selected variables.

1. Introduction

Good Governance was a universal principle recognized by international institutions that it was a crucial factor in making an organization be successful [1]. Also, it was recognized among international development organizations as the concept and key factor that helped to promote sustainable development. The "level of corporate governance" and the "level of political responsibility" are the main criteria used to determine the funding for any country [2]. with serious policy implementation by reducing poverty, developing democracy, and developing the management of transparency with responsibility. [3].

Good Governance was absolutely essential for Thailand. [4], After the economic crisis in 1997, Good governance had a negative impact on all sectors of the society with some partial causes from national administration and management as well as the organizations in public and private sectors which were defective and ineffective including wrongdoing, corruption, and the lack of ethics. This had a negative impact on all sectors of society with some partial causes from national administration and management and the organizations in public and private sectors which were defective and ineffective; including wrongdoing, corruption and the lack of ethics [1]. Under the Government Procurement and Supplies Management Act B.E. 2560, published in Government Gazette Volume 134, Section 24 Kor, dated on

February 24, 2017 [5], Good Governance was an important principle in the development of Electronic Government Procurement (e-GP) which was the central government procurement information center to provide public and private agencies with access to public procurement and government procurement resources in Thailand. From the study about a variety of perspectives on various aspects regarding the use of Electronic Government Procurement: e-GP, there were no effective electronic procurement policies and practices to achieve the stated objectives [6]; Vaidya et al., 2004), and most applications did not take into enough efficiency and effectiveness in terms of value and risk from the use of electronic government procurement [7]. Moreover, [8]. believed that the appropriate management of electronic government procurement could reduce corruption in government agencies, and public procurement practices often reflected tensions between public expectations about high governance standards, management requirements for operating performance. [9] [10]. Electronic government procurement had also proven itself to be a more effective and sufficient tool [11].

Currently, it was not known exactly what were the critical factors of good governance in public procurement; especially, electronic government procurement (E-Government Procurement: e-GP). Until now no study on Good Governance in electronic government procurement using component analysis has been found. [12]. Therefore, the researchers team is keen to develop the component of Good Governance in the government procurement in order to strengthen the administration and management of the government procurement system. Good governance in the government procurement will bring peace and benefit Thailand in terms of fast and sustainable development.

2. Conceptual framework of the study

Good governance originated as a condition of the World Bank's funding to developing countries [13] to prevent inefficient operation and government corruption of the borrower's government [1] [5]. Then, there were many organizations focusing on the Good Governance such as the United Nations Economic and Social Commission for Asia and the Pacific (UNESCAP), the United Nations Development Program (UNDP), and the Organization for Economic Co-operation and Development (OECD).

Good governance occurred and linked to societies in both Macro and Micro levels, and the Institute on Governance (IOG) [12] stated that good governance had covered areas into four levels: (1) Global Governance which referred to all the factors involved in the globalization process such as organizing international community governance rules to solve international problems or between states to unconstitutional states, including the rules and international regulation on commodity stability of economic, capital, and goods, (2) National Governance – covering all relevant political, economic and governance standards, and consisting of Administrative law, Political Cost-effectiveness, and New Institutionalism [14-16], (3) Organizational Governance consisted of guidelines or management paradigms. If it was a government organization, it would be two approaches to public management. The first one focused on management principles that paid attention to work or performance which was derived from the mandates and was subject to the constraints as they were, which was the concept of short-term public management principle. This principle was weighted for the permanent job within the organization [11] [17]. The second one emphasized the importance of the role of public managers as mandatory. Authorized Public managers had the political power within the organization to build networks of relationships and co-operation between different sectors of the society, which was the concept of long-term public management principle [18], and (4) Community Governance was the community operation by good governance factors.

The researcher created a model showing the assumptions of factors of good governance in government procurement, as shown in "Fig. 1", The research hypothesis consisted of (1) the studied variables which could be classified into the exploratory factors analysis of good governance in government procurement, and (2) confirmatory factor analysis of good governance in government procurement which was developed with integrity.

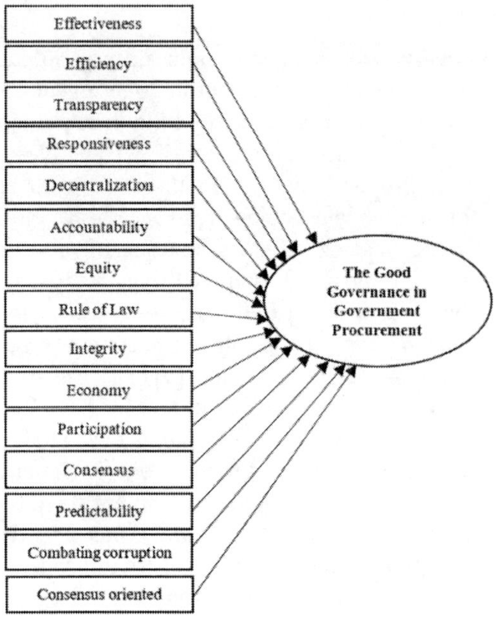

Figure 1. Conceptual framework of Study.

3. Methods

The study methodology of this study was the descriptive research by using structural relationship model and consisted of two steps.

3.1. Population and Samples

Consisted of 3,025 projects of government procurement (Information from January 2014 to January 2015). The samples used for collecting data were from 353 projects of government procurement project in accordance with the formula of Taro Yamane (1967: 886) at a confidence level of 95% from the statistical analysis condition of Factor Analysis; that is, there must be more than 300 samples [19] with Systematic Random Sampling [20].

3.2. Study instruments and their quality control

Consisted of a questionnaire divided into two sections: 15 items for the general information of respondents and 15 items for the good governance in government procurement. The questions on the questionnaire were the 5-level rating scale type, and quality of the study instruments was all content validity and reliability (α) of 0.986.

3.3. Data Collecting

Data Collecting was from Web Surveys [20] published on the website of Government Procurement system (www.gprocurement.go.th) between April 2016 and May 2016 with the samples, and 353 questionnaires were returned, representing 100 percent of the identified samples.

3.4. Data Analysis of Research

Data Analysis of Research was divided into two sections consisted of (1) Exploratory Factor Analysis from the data of qualified persons, using the statistical value of Exploratory Factor Analysis from Principal Component Factor Analysis and Orthogonal rotation by Varimax with consideration of factor loading focusing on weighting higher than 0.30. Then, the statistical value of each factor derived from the exploratory factor analysis was taken into comparison with the factors from the literature review; including the name of the factor with the analysis of Cronbach's alpha for each factor [21]. (2) Confirmation factor analysis from the data collection in Section 3.3 from the statistical value of confirmatory factor analysis for checking the harmony of each variable with empirical data by Pearson's product-moment correlation coefficient was used as the basic information for factor model analysis. Composite factor model test with consistency index included Chi-Square, Relative Chi-Square (CMIN / DF), Goodness of Fit Index (GFI), Root Mean Square Error of Approximation (RMSEA), and reliability was considered on the significance of Standardized Factor Loading, Standard Error (SE), t-test, and Observed Variance which were described by Squared Multiple Correlations (SMC).

4. Result

4.1. The results of exploratory factor analysis (EFA)

The results of suitability tests of the studied variables were analyzed by using the Kaiser-Meyer-Olkin Measure of Sampling Adequacy, and it was found that KMO was 0.956 which was greater than 0.80, indicating that the variable of this set was suitable for the factor analysis at an outstanding level according to the criteria of Kim and Mueller [22]. Bartlett's Test of Sphericity was also used, and it was found that there were statistically significant correlations at 0.001, which could be used for factor analysis.
Principal Component Analysis: The Communality of each variable was used for the factor analysis of good governance in government procurement with a total of 15 factors, ranging from .552 to .812, size of medium to large, and tended to be able to fit into one of the components.

The result of Orthogonal Rotation by the Varimax method was shown in Table 1. The number of factors, variance of variables, percent variance, and cumulative percent variance of good governance factors in government procurement based on the factor with the Eigenvalue as greater than 1 that there was only one factor and could explain the variance of all variables equivalent to 68.863 percent.

4.2. Results of confirmatory factor analysis

The results of the consistency analysis showed that the confirmatory factors analysis about good governance model in government procurement had statistical values that were unqualified. Therefore, it was necessary to adjust the model to be more harmonious. The researcher had adapted the model based on the recommendation from the Model Modification Indices (MI) by cutting off one unqualified variable, namely the Responsiveness (GG03), the modified model analysis, and the chi-square test without significance (P-value = 0.074). The chi-squared correlation was 1.334. The NFI index was 0.989. The GFI index was 0.978. The CFI index was 0.997. The standardized RMR was 0.006. Moreover, the RMSEA was 0.031. As a result, the whole statistic values became the qualified ones which showed that the model of good governance in government procurement was consistent with empirical data as shown in "Fig. 2" and Table 2.

Based on the results of Table 1, it was found that the good governance model in government procurement had factor loading at the range between 0.686 and 1.102, and the variance describing the good governance model in government procurement was 45.8 to 81.4 percent. The highest factor loading was the morality (β = 1.102); followed by transparency (β = 0.934), cost-effectiveness (β = 0.902), predictability (β = 0.899), equality (β = 0.892), and the lowest one was consistency (β = 0.073).

5. Discussion

The result of the exploratory factor analysis of good governance in government procurement was found that all 15 studied variables could be grouped into one factor. The factor had Eigen value greater than 1.00. Variability of all variables could be explained by the factor or 68.86%. The factor loading of each variable varied from 0.780 to 0.882. The analysed variables could be used for the study of good governance in government procurement. This was because these variables were documented, and research supported. The researcher had also studied widely relevant papers and research papers. Thus, the key variables we got were consistent and covered the larger structure of good governance that could be grouped into one factor already and the result of confirmatory factor analysis of good governance in government procurement was found that confirmatory factor model of good governance consisted of one main factor and 14 sub factors.

Considering the importance of morality in government procurement, it was the first important principle with similar factor loading. It also corresponded to the research of [10] [12] [14-15] [23] that adhered to goodness by promoting the state officials to abide by this government procurement in performing their duties as examples to society and encouraging the stakeholders to develop themselves. The transparency of government procurement was consistent with the Government Procurement and Supplies Management Act B.E. 2560 [5] [23], which was the disclosed action to give a chance to have a fair competition with the equal treatment for all entrepreneurs and to have a reasonable and enough time to submit proposals with the clear evidence of operation and disclosure of procurement and inventory management information at all steps.

Cost-effectiveness was consistent with the Government Procurement and Supplies Management Act B.E. 2560 [5] which required quality or features that served the intended purpose of use by the government agency with well-priced items and had an appropriate and clear facilities management plan.

Good governance in government procurement consisted of two main dimensions: (1) the process of building relationships between the government, society, private sectors, and the public to make the administration work effectively, transparently, and (2) good or fair administration and governance consisted of three factors namely transparency, accountability, and efficiency [1] [12] [23].

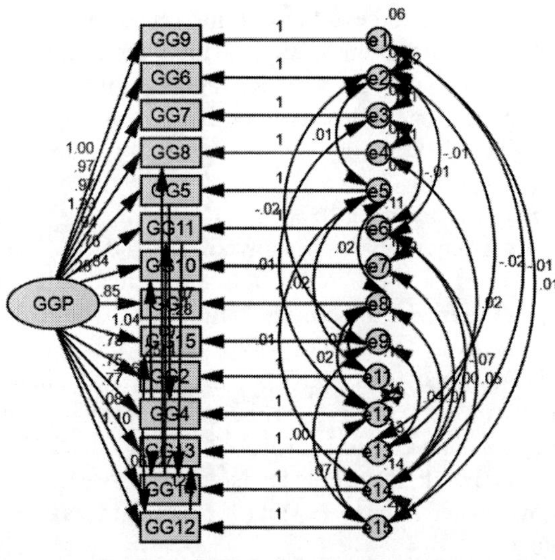

Chi-squre = 54.701, CMIN/DF = 1.334, df = 41, p = .074
NFI = .989, GFI = .978, CFI = .997, RAR = .006, RMSEA = .031

Figure 2. The second confirmatory factor model of good governance in government procurement.

6. Conclusion

The results of the exploratory factor analysis of good governance in government procurement showed that there was 1 factor of factor structure from 14 selected variables.

The results of confirmatory factor analysis of good governance in government procurement was found that the factor model had good structural integrity since the model was consistent with the empirical data with the qualified statistic of harmony measurement, and the chi-squared statistic did not have significance (P-value = 0.074). The Chi-square correlation coefficient was 1.334. The NFI index was

0.989. The GFI index was 0.978. The CFI index was 0.997. The standardized RMR was 0.006, and the RMSEA value was 0.031. All statistics met the required criteria and could be sorted according to the top 5 main factors from the highest to the lowest, consisting of morality, transparency, cost-effectiveness, predictability, and equality, respectively

Table 1. The first confirmatory factor model of good governance in government procurement.

No. Item	Good Governance	EFA	CFA			
		Factor Loading	*Standardized Factor Loading*	*SE*	*t-test*	*SMC*
1 GG08	Morality	0.870	1.102	0.118	10.380**	0.776
2 GG15	Transparency	0.806	0.934	0.061	17.087**	0.678
3 GG09	Economy	0.882	0.902	1	**	0.814
4 GG12	Predictability	0.780	0.899	0.249	4.409**	0.458
5 GG06	Equity	0.881	0.892	0.042	23.033**	0.796
6 GG07	Rule of Law	0.880	0.885	0.38	25.329**	0.784
7 GG05	Accountability	0.854	0.849	0.041	23.152**	0.720
8 GG01	Effectiveness	0.818	0.768	0.044	19.013**	0.589
9 GG02	Efficiency	0.799	0.753	0.042	18.408**	0.568
10 GG10	Participation	0.839	0.747	0.059	14.391**	0.667
11 GG04	Decentralization	0.794	0.688	0.080	9.436**	0.556
12 GG13	Combating corruption	0.793	0.686	0.060	12.703**	0.610
13 GG11	Consensus	0.843	0.641	0.173	4.396**	0.706
14 GG14	consensus-oriented	0.793	0.073	0.385	0.212**	0.582

Remark: ** referred to the 0.01 level of significance and * referred to the 0.05 level of significance

Table 2. Goodness of – fit measurement of factor model and empirical data regarding good governance model in government procurement.

Index	Criteria	Before adjusting model		After adjusting model	
		Statistic	*Result of Consideration*	*Statistic*	*Result of Consideration*
X^2/df	< 3.00	9.167	Unqualified	1.334	Qualified
p-value of X^2	> 0.05	<0.001	Unqualified	0.074	Qualified
NFI	> 0.90	0.851	Unqualified	0.989	Qualified
GFI	> 0.90	0.730	Unqualified	0.978	Qualified
CFI	> 0.95	0.864	Unqualified	0.997	Qualified
RMR	< 0.08	0.019	Unqualified	0.006	Qualified
RMSEA	< 0.08	0.152	Unqualified	0.031	Qualified
Reliability	>0.70	0.967	Qualified	0.966	Qualified

Note: the p-value Also Sensitively to sample size more than 200.

7. Acknowledge

We would like to thank you to the Mahasarakham University Development Fund for their fund support to this study.

References

[1] OPDC 2009 *A Handbook of Good Governance Rating* (Bangkok: Office of the Public Sector Development Commission, Thailand)
[2] Elkington J 2006 Governance for sustainability *Corp. Govern. Int. Rev.* **14(6)** 522–9
[3] Brand H 2007 Good governance for the public 's health *European Journal of Public Health* **17** 541
[4] Nayana R 2014 *Good Governance Concepts and Components* (Bangkok: Centre for Good

Governance, Thailand)

[5] CGD *2017 The Government Procurement and Supplies Management Act B.E. 2560* (Bangkok: The Comptroller General's Department, Thailnad)

[6] Concha G, Astudillo H, Porrúa M and Pimenta C 2012 E-Government procurement observatory, maturity model and early measurements *Government Information Quarterly* **29(1)** 43–50

[7] Trkman P and McCormack K 2010 Estimating the benefits and risks of implementing e-procurement *IEEE Trasactions on Engineering Management* **57(2)** 338-49

[8] Neupane A, Soar J, Vaidya K & Yong J 2012 The potential of e-procurement technology for reducing corruption *International Journal of Information Technology and Management* **11(4)** 273-87

[9] Curtin University of Technology 2017 International Survey of E-procurement Systems, *World Bank Publication*, Available online at http://idbdocs.iadb.org/wsdocs/getdocument.aspx?docnum =1064452

[10] Tanja A, Yasemin P and Andreas S 2008 One Size Fits All? How the European Union Promotes Good Governance in Its Near Abroad, *SFB 700 Working Paper 18,* Berlin: Germany

[11] Boudreau J A, Jouve B and Keil R 2006 Comparing metropolitan governance the cases of Montréal and Toronto *Progress in Planning* **66 (1)** 7-59

[12] Narong L, Tanayut C and Preenithi A 2018 An Exploratory Factor Analysis of Government Construction Procurement Problems *MATEC Web of Conferences* 192 02057(2018) https://doi.org/10.1051/matecconf/201819202057

[13] Maldonado N 2010 *The World Bank's Evolving Concept of Good Governance and its Impact on Human Rights* (Bonn: University of Bonn, Germany)

[14] Samnak N 2009 *Regulations of the Office of the Prime Minister on Good Governance B.E.2542* (Bangkok: Office of the Civil Service Commission, Thailand)

[15] Phiphat N 2013 Public and Private Governance Integrating Governance for Development *Romphruek Journal* **2(February-May)** 145-70

[16] Lachit C 2013 Good Governance Context of Thailand, *Romphruek Journal* **2(February-May)** 78-98

[17] Kumar S R 2015 Good Governance in Public Procurement An Evaluation of The role of an E-Procurement System *ProQuest Dissertations and Theses*

[18] Bussabong C, and Boonmee L 2007 *Research Report of Good Governance Indicator* 3rd ed (Bankok: King Prajadhipok's Institute, Thailand)

[19] Richard F, Fellows M, Anita M and Liu M 2008 *Research methods for Construction* (Oxford: Wiley-Blackwell)

[20] Neuman W 2011 *Social research methods qualitative and quantitative approaches* (Boston: MA Pearson)

[21] Pallant J, 2007 *SPSS Survival Manual A Step by-step Guide to Data Analysis using SPSS for Windows* (Buckingham: Open University Press)

[22] Andy F, 2013 *Discovering statistics using IBM SPSS statistics* 4th ed (London: Sage Publications Ltd.)

[23] Tanayut C, Narong L, Preenithi A and Patrick M 2017 Application of Confirmatory Factor Analysis in Government Construction Procurement Problems in Thailand, *International Transaction Journal of Engineering, Management, & Applied Sciences & Technologies* **8(3)** 22

IOP Conf. Series: Materials Science and Engineering **639** (2019) 012025 doi:10.1088/1757-899X/639/1/012025

Fundamental Analyses and Observations of Liquid Droplet on Aluminum Surface for Heat Exchangers

N. Charoensuk[1], C. Sae-kung[2], J. Charoensuk[1], S. Hirai[3], A. Treetong[4] and V. Lilavivat[2,*]

[1] Department of Mechanical Engineering, King mongkut's Institute of technology of lardkrabang, Lat Krabang, Khet Lat Krabang, Bangkok, Thailand
[2] National Metal and Materials Technology Center, National Science and Technology Development Agency 114 Thailand Science Park, Pathum Thani, Thailand
[3] Department of Mechanical Engineering, Tokyo Institute of technology, Tokyo, Japan
[4] National Nanotechnology Center,National Science and Technology Development Agency 111 Thailand Science Park, Phahonyothin Road, Khlong Nueng, Khlong Luang, Pathum Thani 12120 Thailand

E-mail: visarn.lil@mtec.or.th,

Abstract. Aluminum is one of the most popular materials in the industry because of lightweight Furthermore, the heat exchanger is using more aluminum rather than cumbersome and costly copper inside the system. Nowadays, all of the low-temperature heat exchanger fins are aluminum. However, when the heat exchanger is used for a long time the fin of the heat exchanger will accumulate contamination inside such as dust. The existence of contamination has strongly effect heat exchanger system performance. In this study, the physical of the water droplet with the different surface roughness (Ra) on the aluminum surfaces were studied. Hence, in this experiment, the liquid water droplet behaviors on the aluminum surfaces were observed by the contact angle. The interaction between liquid droplets and solid surface on the flat plate affect the contact angle of the water droplet. Liquid water droplet movements were analyzed by considering the change of the contact angle, were used to explain liquid droplet behaviors on the aluminum surface. Consequently, in this study, the liquid droplet behaviors on the aluminum surface can be used to improve the self-cleaning surface of the heat exchanger such as the air-conditioning system

Keywords: aluminum surface, heat exchanger, contact angle, surface roughness

1. Introduction

The heat exchanger is a device used to transfer heat between solid and fluid. Generally, the heat exchanger system has fins to increase the surface area inside the system. Over time, when operating the heat exchanger, the fins collect the contamination such as dust that reduces the overall efficiency of the system [1]. Excess release dust from the fin hence to improve the solid surface as self-cleaning of the fin, liquid-solid interaction has been studied in this project.

The materials for heat exchanger before use copper with the fin and affected to high price for processing. Moreover, in this case, want to adopt the materials on the fin of the heat exchanger to aluminum. Likewise, the aluminum has a lightweight and approximately density 2.7g/ml and popular use with the

automotive industry. The significance of this study to improve the surface roughness of aluminum such as average surface roughness (Ra) [2] by laser scanning microscope. Moreover, the aluminum surface to test liquid droplets by the contact angle machine.
The contact angle machine of the droplet uses water drop on the aluminum surface. The movement of water droplets on a solid plane; this method was introduced by Thomas Young [3]. The contact angle is related to the surface tension of the angle between solid-liquid, and liquid-vapor. The equation1 that describes the angle of liquid droplets on a flat solid surface in where the vapor, liquid, and solid phase come into contact angle in Figure 1. Can describe this equation

$$\cos\theta = \frac{\gamma_{SV} - \gamma_{SL}}{\gamma_{LV}} \tag{1}$$

In this equation $\gamma_{SV},\gamma_{SL},\gamma_{LV}$are the interfacial tensions [4] of the solid-gas, solid-liquid, and liquid-gas interfaces, and θ is contact angle of liquid droplets.If the liquid wets on the surface called (hydrophilic surface), the value of the contact angle is0 degree $\leq\theta\leq$ 90 degree: on the other hand if the liquid does not wet the surface (hydrophobic surface), the value of the contact angle is 90 degree $\leq\theta\leq$ 180 degree.

On the other hand, Wenzel's [5] demonstrated the contact angle on a rough surface and modified from Young's equation to describe this equation

$$\cos\theta' = \frac{r(\gamma_{SV} - \gamma_{SL})}{\gamma_{LV}} \tag{2}$$

Where r is the roughness factor and θ' is apparent contact angle. In this equation, the value of r=1 is called smooth surface. On the other hand, r<1 the surface is very smooth and r>1 the surface is very roughness. The apparent contact angle will be changed with the roughness factor.

Cassie-Baxter [6] show an equation describe the contact angle at a heterogeneous surface with solid and air. If the contact angle of liquid droplets is more than 90 deg and the surface has a high roughness. The surface in Cassie-Baxter case has air trapped in the depression of the solid surface. Furthermore, the surface area has a wetted surface area fraction f with a water contact angle θ. In a situation, a drop of liquid surrounded by air has got a contact angle of 180 deg can describe this equation.

$$\cos\theta'_{CB} = -1 + f(\cos\theta + 1) \tag{3}$$

Where f is the solid surface fraction, $\cos\theta'_{CB}$ is contact angle of Cassie-Baxter.

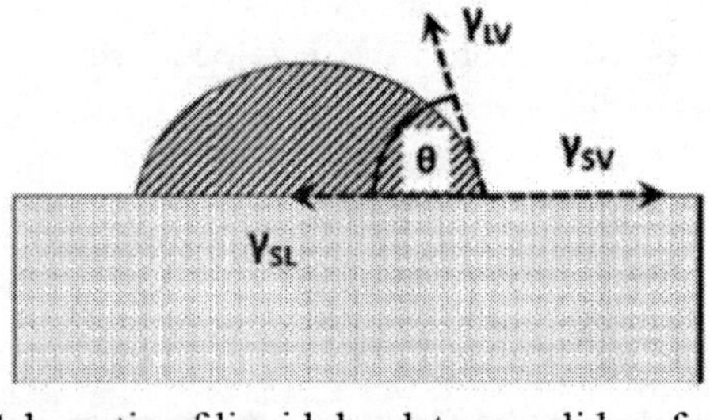

Figure 1. Schematic of liquid droplets on solid surfaces [3]

Figure 1. show the schematic of liquid droplets on solid surfaces related to Young's equation and the angle between the liquid-vapor and solid-liquid interfaces of a solid-liquid-vapor system. The behavior of the water droplets was measured on the foundation of the contact angle. In this work, the various surface roughness of aluminum with different surface roughness to examine the impacts of water droplet contact angle with a different surface roughness on the aluminum

IOP Conf. Series: Materials Science and Engineering **639** (2019) 012025 doi:10.1088/1757-899X/639/1/012025

2. Experiment Procedure

2.1 The Process of preparing the aluminum surfaces

The aluminum metal plate was cut into the sample size of 25 mm by 25 mm, as shown in Figure 2. Then, the aluminum sample was prepared surface used to polished different sandpaper number for varying different surface roughness by the polish machine (STRUER polish machine ROTOPOL-25) in Figure 3. In this experiment, the process to prepare different surface roughness.

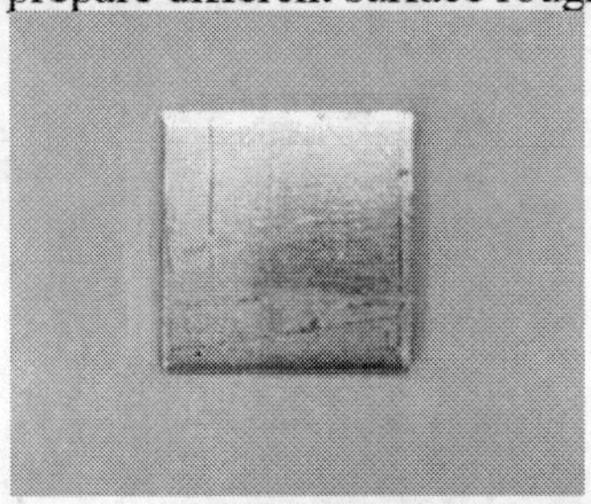

Figure 2. the aluminum surface sample in the preparing process.

Figure 3. polish the aluminum surface by the polish machine (STRUER polish machine ROTOPOL-25)

2.2 The measurement of surface roughness

After polish, in order to observe the effect of surface roughness to the contact angle of water droplets with different surface roughness [7]. The aluminum surface roughness was measured by a laser scanning microscope (Shimadzu lext OLS4500). Further study, the surface layer profile of aluminum surface was observed by AFM(AFM-Seikospa400) with different surface roughness.

2.3 The measurement of the contact angle

The prepared sample surface by polishing, the 5 μL of the liquid water dispense is placed on top surfaces by the contact angle machine. Then, the value of contact angles between surfaces was showed on the screen monitor of contact angle machine to compare the different value of contact angle on different surface roughness of aluminum surfaces to study wettability on the aluminum surfaces as shown in Figure 4.

IOP Conf. Series: Materials Science and Engineering **639** (2019) 012025 doi:10.1088/1757-899X/639/1/012025

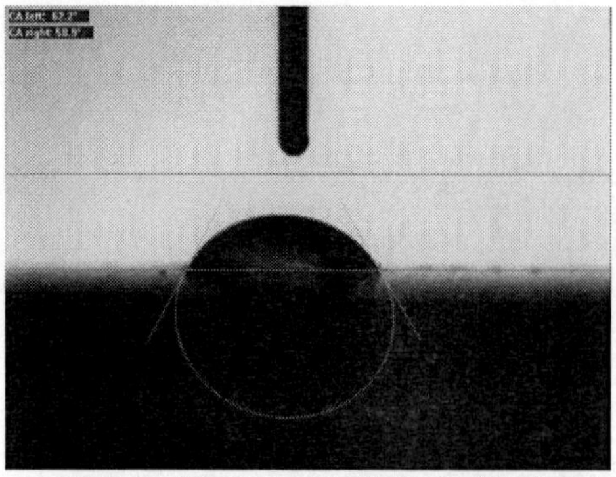

Figure 4. the contact angle water droplets on the aluminum surfaces

2.4 Sliding droplet on aluminum surfaces

The preparing sample aluminum surface to measure sliding droplet on the surface by dropped volume of water droplet 40 μL on the aluminum surfaces. Then, the tilt base was adjusted increase angle from 0degree to 30 degree until water droplet is begun sliding [8] on the surface to study the begin sliding droplet on different surface roughness of aluminum surfaces as shown in figure 5.

Figure 5. the water droplet is begun sliding on the aluminum surfaces

3. Result and Discussion

This experiment was set up to understand the behavior of liquid water droplet on the aluminum surface with a different surface roughness that affects droplet movement on the surface.

3.1 The surface roughness

The aluminum surface profile was observed by laser scanning microscope and AFM as shown in Figure 5. Moreover, Figure 6. Figure 5(A) shows the optical images the roughest surface at Ra=350 nm and Figure 5(B) shows the smoothest surface at Ra =11.3 nm. However, the optical images only review 2D information of the surface layer. For this reason, the AFM was used to investigate the 3D profile of surface for further study as shown in Figure 6. Figure 6 (D) shows a 3D picture of a very smooth surface compare to Figure 6 (A) looking rougher.

Figure 6. Optical images of aluminum surface with different surface roughness from (A) Ra=350nm, (B) Ra=183.33nm, (C) Ra=219.8nm, (D) Ra=122.7nm and (E) Ra=11.3nm).

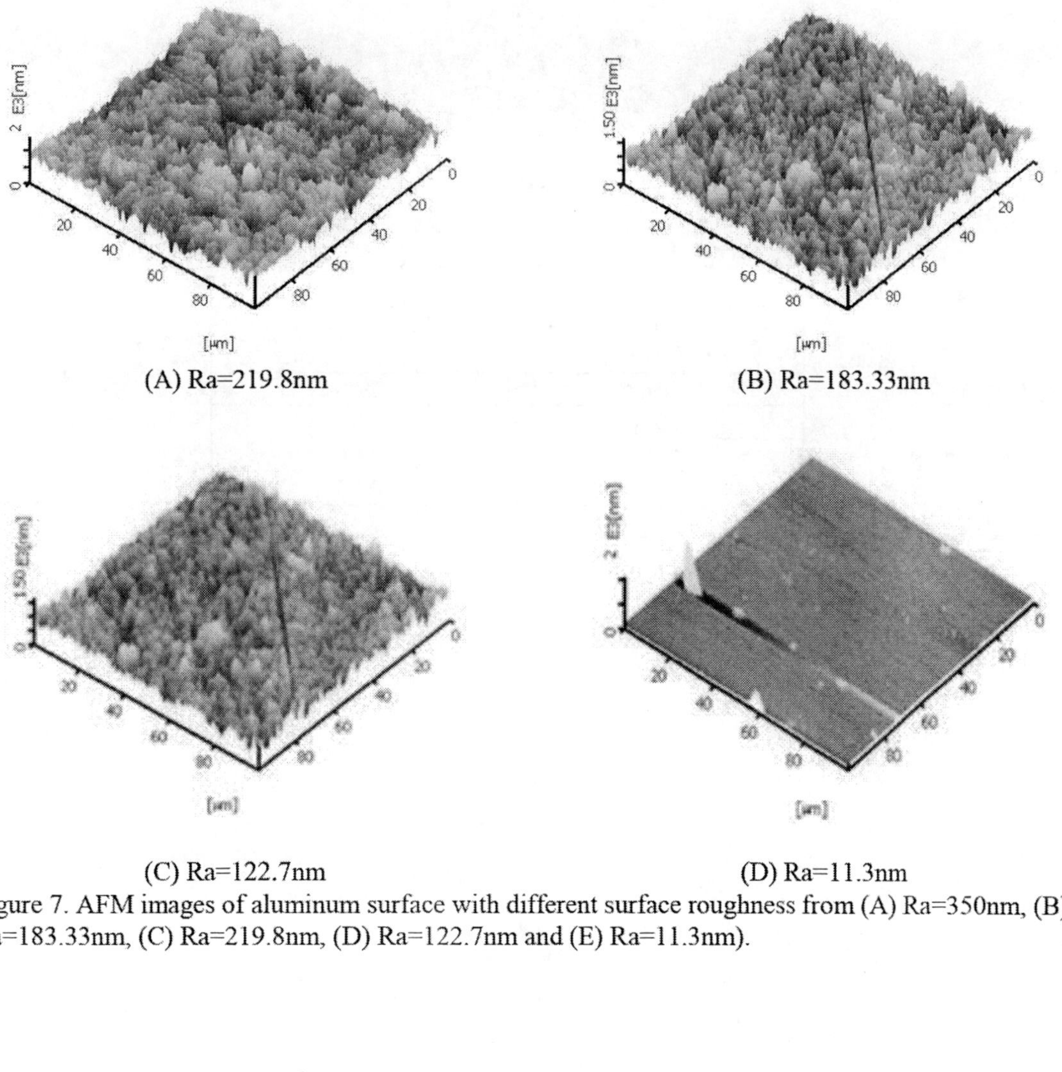

Figure 7. AFM images of aluminum surface with different surface roughness from (A) Ra=350nm, (B) Ra=183.33nm, (C) Ra=219.8nm, (D) Ra=122.7nm and (E) Ra=11.3nm).

3.2 Wettability on the aluminum surface

According to the experiment, the static contact angle of a 5μl liquid water droplets on the aluminum surface with different surface roughness as shown in Figure 7. The static contact angle of the water droplet on the aluminum surface with different surface roughness is shown in Table 1. This table shows the contact angle of water droplets relationship with an average surface roughness (Ra). The result has shown that the value of the contact angle between 62.51 - 90.64 degrees. Furthermore, the result of the contact angle on the aluminum surface is the relationship with a different surface roughness (Ra) shown below Figure 8. This figure shows that surface roughness affected the static contact angle. The result has shown that when the aluminum surfaces are rough the contact angle slight increase. These results are disparity from Wenzel's equation which the static contact angle decrease as surfaces roughness is rougher. Also, the Cassie-Baxter's equation cannot explain these conditions as well.

Figure 8. Water droplet on the aluminum surface with a different value of surface roughness (Ra)

Table1. The relationship between average surface roughness (Ra) and the static contact angle of a water droplet.

Ra(nm)	CA(Degree)
350	80.8
219.8	79.5
183.33	74.4
122.7	72.6
11.3	62.5

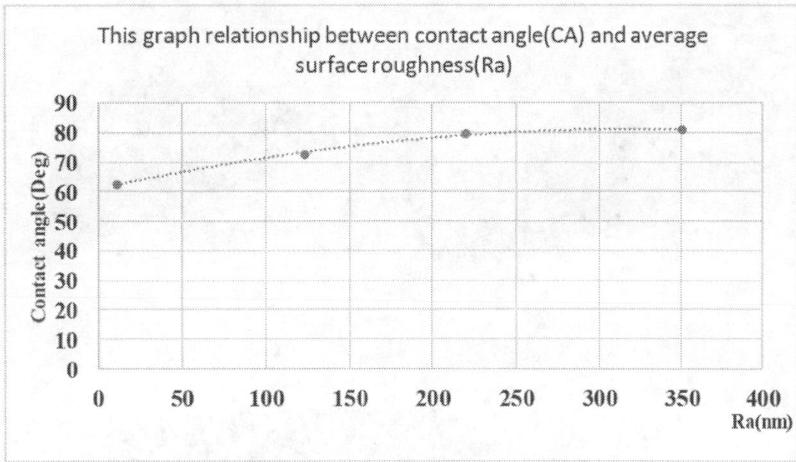

Figure 9. Contact angle vs. average surface roughness (Ra)

3.3 Sliding droplet on aluminum surfaces

In this part, the aluminum surfaces were measured by water repellent with different surface roughness. The tilt base angle of the water droplet on the aluminum surface with different surface roughness has shown the value of contact angle between 15.08-21.48 degrees from Table 2. Furthermore, the tilt base angle with the different surface roughness has shown the figure10.Furthermore, figure11 show tendency of graph relationship between tilt base angle (TB) and average surface roughness (Ra). Consequently, figure 12 shows the tendency of graph contact angle hysteresis and tilt base angle with different surface roughness (Ra). Furthermore, the surface roughness increasing affected to contact angle hysteresis begin sliding in a lower angle. Because the aluminum surface is Ra=11.3nm smoother than Ra= 350nm affected to surface between water and solid surface has low friction force, resulting in sliding angle of water droplet is low. The water droplet is begun sliding low tilt base angle.

These results are similar Wenzel's equation because the water droplet is spreading and wetting on the aluminum surfaces but the result the data from the experiment is a difference from the theory is shown, when the surface roughness is increasing the contact angle is lower angle. The Cassie-Baxter's equation is also cannot explain for this result because the water droplet is not non-wetting or contact angle not more than 90 degrees hence the Cassie-Baxter's equation is cannot explain this result also.

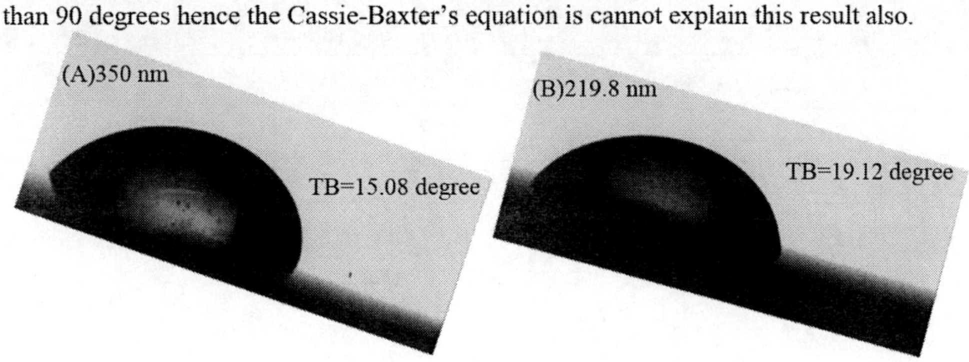

IOP Conf. Series: Materials Science and Engineering **639** (2019) 012025 doi:10.1088/1757-899X/639/1/012025

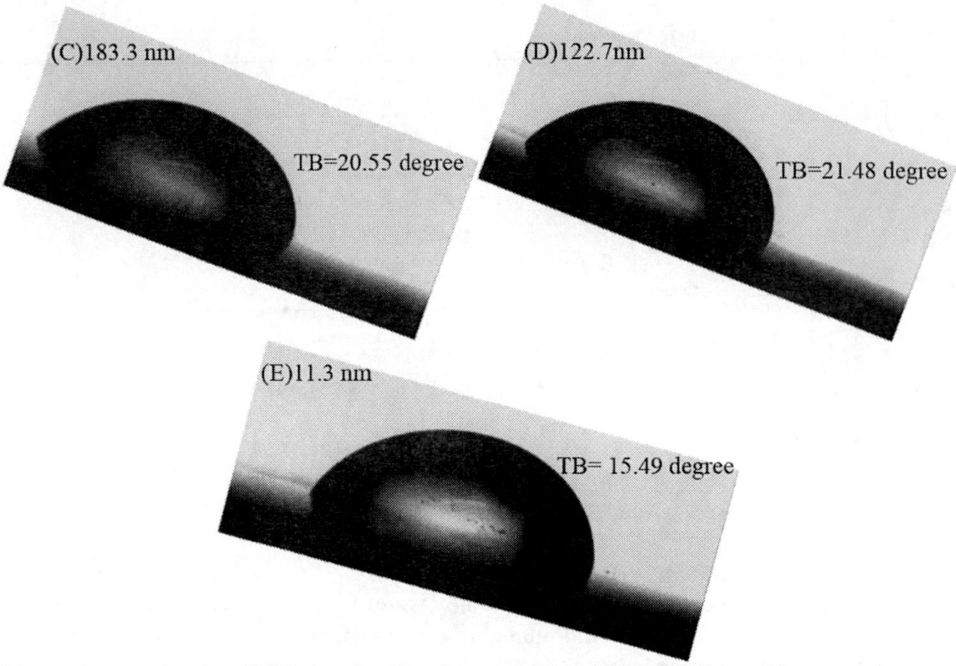

Figure 10. water droplet sliding on the aluminum surface different with surface roughness (Ra)

Table2. The relationship between average surface roughness (Ra) and the tilt base angle (TB) of a water droplet.

Ra(nm)	TB(Degree)
350	15.1
219.8	19.1
183.33	20.6
122.7	21.5
11.3	15.5

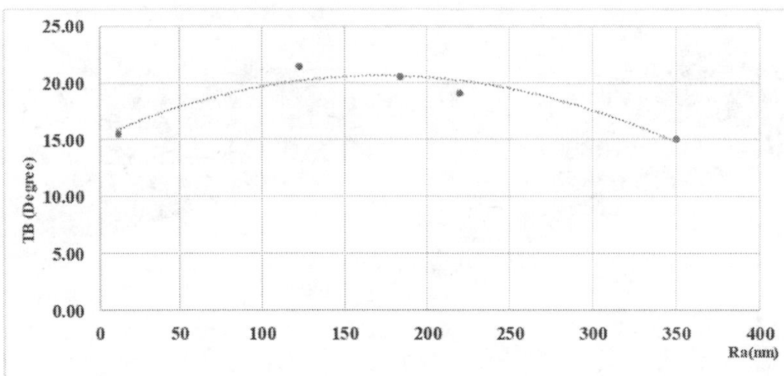

Figure 11. relationship between tilt base angle (TB) and average surface roughness (Ra)

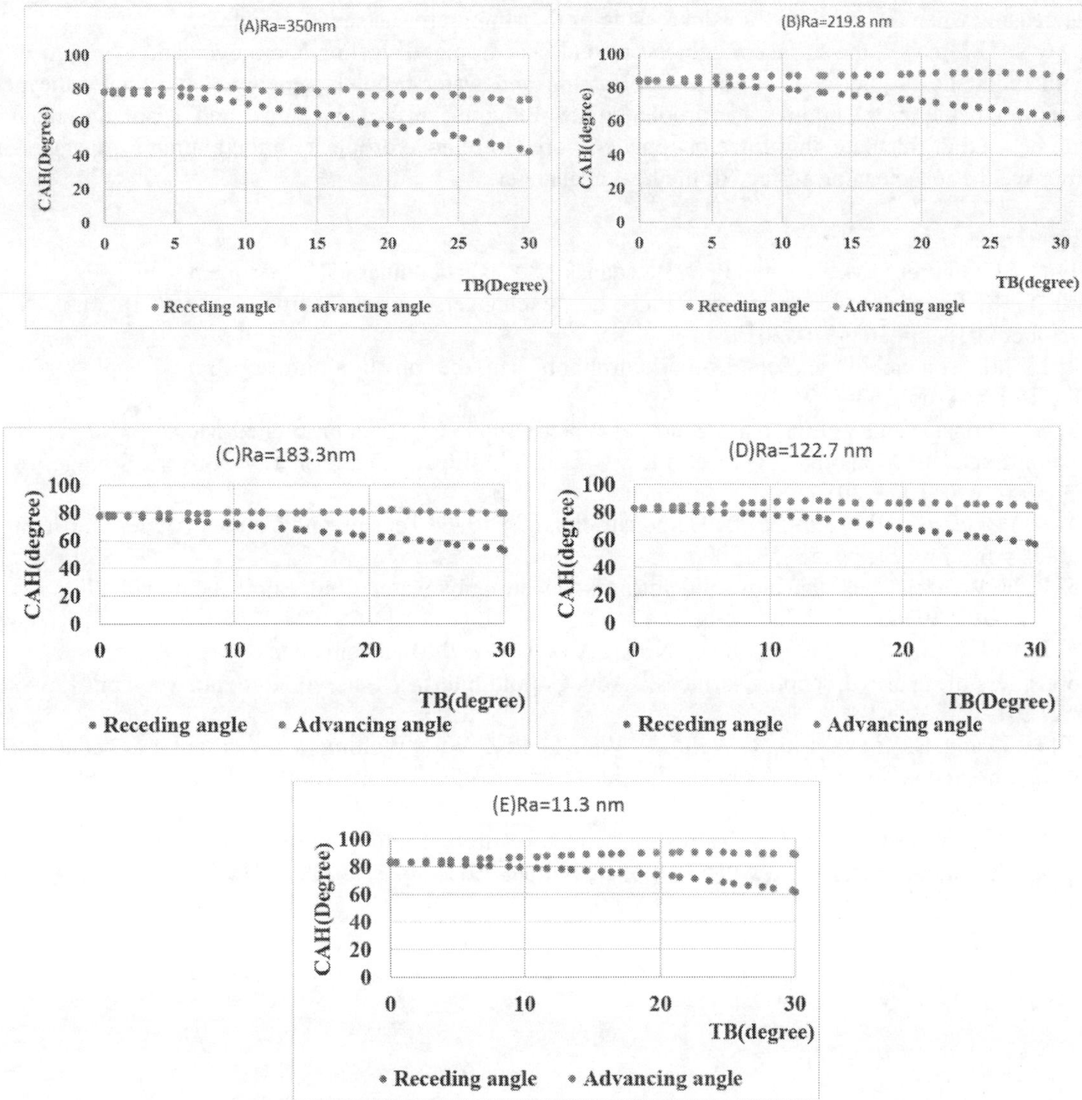

Figure 12. The graph show relationship between contact angle hysteresis (CAH) with tilt base angle on difference surface roughness (A) Ra= 350 nm (B) Ra= 219.8 nm (C) Ra=183.3 nm (D) Ra=122.7 nm (E) Ra=11.3 nm

4. Conclusion

In this study, the liquid water droplet drop behaviors on the aluminum surface with different surface roughness were determined. The value of the average surface roughness of the aluminum surface sample is 11.3-350 nm. The result has shown that surface rough affect the static droplet contact angle. The value of the contact angle is decreasing when the value of Ra is decreasing, or the aluminum surface is smoother. The wettability is related to surface roughness.

In this part, the sliding droplet on the aluminum surface with different surface roughness and adjust tilt base angle from 0 to 30 degree were determined. The value of the tilt base angle begins sliding on surface roughness of aluminum surface between 15.08-21.48 degree. The result has shown that the surface is rough affected to begin the sliding of the water droplet. The value of the tilt base angle is decreasing when the value of Ra is increasing or the aluminum surface is smoother.

Moreover, the surface roughness also affects droplet distortion and movement. Consequently, with the static droplet contact angle information and water droplets repellent investigate, the new equation model for the liquid water droplet on the aluminum surface can be defined. Also, this equation can be used to calculate drag force or repellent force to remove droplets on the aluminum surface and improve the self-cleaning surface of the heat exchanger

References

[1] A. M. González, M. Vaz, and P. S. B. Zdanski, "A hybrid numerical-experimental analysis of heat transfer by forced convection in plate-finned heat exchangers," Appl. Therm. Eng., vol. 148, no. October 2018, pp. 363–370, 2019.

[2] J. Zhu, "A novel fabrication of superhydrophobic surfaces on aluminum substrate," Appl. Surf. Sci., vol. 447, pp. 363–367, 2018.

[3] S. Shimpalee and V. Lilavivat, "Study of Water Droplet Removal on Etched-Metal Surfaces for Proton Exchange Membrane Fuel Cell Flow Channel," J. Electrochem. Energy Convers. Storage, vol. 13, no. 1, p. 011003, 2016.

[4] H. D. Gesser and P. Krause, "A Demonstration of Surface Tension and Contact Angle," J. Chem. Educ., vol. 77, no. 1, pp. 58–59, 2000.

[5] R. N. Wenzel, "Resistance of solid surfaces to wetting by water," Ind. Eng. Chem., vol. 28, no. 8, pp. 988–994, 1936.

[6] L. R. J. Scarratt, U. Steiner, and C. Neto, "A review on the mechanical and thermodynamic robustness of superhydrophobic surfaces," Adv. Colloid Interface Sci., vol. 246, no. May, pp. 133–152, 2017

[7] H. Zhang, L. Yin, X. Liu, R. Weng, Y. Wang, and Z. Wu, "Wetting behavior and drag reduction of superhydrophobic layered double hydroxides films on aluminum," Appl. Surf. Sci., vol. 380, pp. 178–184, 2016.

[8] D. G. Venkateshan and H. V. Tafreshi, "Modelling droplet sliding angle on hydrophobic wire screens," Colloids Surfaces A, vol. 538, no. November 2017, pp. 310–319, 2018.

IOP Conf. Series: Materials Science and Engineering **639** (2019) 012026 doi:10.1088/1757-899X/639/1/012026

Construction project team's competencies influencing contractor business competencies

Kittipos Kawesittisankhun, Jakrapong Pongpeng

Department of Civil Engineering, Faculty of Engineering, King Mongkut's Institute of Technology Ladkrabang, Bangkok 10520, Thailand.
59601034@kmitl.ac.th

Abstract. Consistent arrangement of contractor business competencies and project team's competencies helps improve project operation efficiency. Key participants in the project team are project managers, engineers, and technicians. However, a review of existing literature reveals a lack of research on the causal relationship between project teams' competencies and business competencies. Therefore, this study attempted to identify the relationship through survey contractors' opinions. The data were analyzed using structural equation modeling (SEM). The results suggest the regression weighs presented in parenthesis that business competencies are influenced by the competencies of a project manager (0.50) and engineer (0.49). The study also discovered the influence of competencies among project teams; that is, technician competencies influence engineer competencies (0.87), and engineer competencies influence project manager competencies (0.55). With the causal relationship between the project team's competencies and business competencies, contractors can suitably allocate human resources for increasing their project operation efficiency.

1. Introduction

There are many challenges had been raised in the construction industry due to globalization and technology change. This causes the change in construction techniques, economy and social condition. These changes bring the construction environment to become fiercer competition. To survive, contractors need to adapt their business operation efficiency responding to this changing environment. The key participants in the adaptation process are project teams and their competencies. The project team's key members consist of project managers, engineers, and technicians. The competencies take an important role to develop business effectiveness [1]. Additionally, [2] stated competency is not only concern about performance against an output-based measure but also posses the potential for developing the psychological understanding which is need when selecting and predicting human performance. The construction business also needs to select and predict their business performance. Business competency concerns about demonstrating the characteristic that will develop the business. Business competency can be a major determinant of contractor efficiency. In Turkey, 185 construction companies were surveyed to study business competency aiming to select and predict their business performance [3]. Construction project main works were executed through key project teams: project manager, engineer and technician. As a result, there are possible relationships among these competencies. This result is consistent with [4] (Theory of Hierarchical, Multilevel, Systems), which stated that the performance of higher level (e.g., project manager) is based on the performance of the subordinate level (e.g., engineer). Moreover, [1] mentioned the project team's competencies have impacts on business operation. Thus, this implies a possible relationship between project teams' competencies and business competencies. Although several researchers have studied indicators of project team and contractor business competencies, the influence between competencies has not been identified. As such, the study aimed to identify such influence.

2. Literature review

2.1. Project manager competencies (PMC)

A project manager is responsible for the overall success of the project. The project tasks are distributed from project manager to different divisions which are controlled by an engineer. Then, each division distributes resources to technicians to supervise workers and their works. This character is consistent throughout most contractor organizations.

2.2. Engineer competencies (EC)

An engineer has a role in planning construction sequences, material selection and design. These activities require resources. Thus, to economically utilize project resources, an engineer needs to correspond with a project manager in order to develop an efficient way to use resources.

2.3. A subsection Technician competencies (TC)

The workforce is supervised by a technician. Construction techniques, methods and regular maintenance of tools and equipment are also within the technician roles. A technician has to work under the instruction of engineers to find an effective way to produce physical result from raw materials with acceptable production rate and standard.

2.4. Business competencies (BC)

Risk is unavoidable in the construction business because its performance depends on their project outcomes. If their projects provide uncertain performance, contractors may face many difficulties. Also, [5] mentioned that construction business competes with rivals through project efficiency.

In this study, the literature regarding project team's competencies and business competencies was used as a basis to determine latent and observed variables, which were shown in Table 1.The conceptual framework for the study was illustrated in Figure. 1

Table 1. Summary of latent and observed variables, relationship, and their associated theory.

Latent Variables	**Observed variables**	**Knowledge Base (Theory)**
Project manager competencies (PMC)	knowledge management; strategic implementation; relationship management; decision making; construction management; organization management; risk management	[8, 2, 9, 10]
Engineer competencies (EC)	knowledge; technical skill; communication; problem solving; systematic management; resources management	[11, 12]
Technician competencies (TC)	Learning; social; adaptability; systematic construction work; self-control	[13, 14]
Business competencies (BC)	information management; strategic management; financial management; resource management; relationship management; project management; continuous development	[5, 12, 15]
Multi-level relationships		[4]

In the previous sections, the research shows the project team's competencies were arranged in a three-level structure corresponding to project managers, engineers and technicians. A technician has specific responsibility of workforce production. Then, the technicians are responsible to an engineer. The engineer serves as a mediating role between technicians and a project manager. The engineer outcomes are one of the project manager's performance. This implies that technician competencies have an effect on engineer competencies; then engineer competencies have and effects on project manager competencies. In addition, the contractor business depends on projects and their team. This implies that project manager and engineer competencies likely affect on business competencies.

From the analysis of the literature review, the proposed conceptual framework for the study is illustrated in Figure. 1, and stated as follows:

- H1: Technician competencies have a positive and direct influence on business competencies
- H2: Technician competencies have a positive and direct influence on engineer competencies.
- H3: Engineer competencies have a positive and direct influence on project manager competencies.
- H4: Engineer competencies have a positive and direct influence on business competencies.
- H5: Project manager competencies have a positive and direct influence on business competencies.

3. Research method

The questionnaire was used as a research instrument and divided into two parts. Part 1 consisted of general information. Part 2 contained items related to the competencies of the project team, which are important to competitive competencies. Each item is assessed using a 5-level, Likert type agreement scale showed in Table 2.

Thai contractors' opinions were gathered through personally meet and interview of contractor representatives who participated in well-known infrastructure construction projects. From 300 copies of questionnaires, 216 were obtained (a return rate of 72%). [6] considered this response rate to be excellent, which is sufficient for SEM analysis.

Table 2. Likert type agreement scale.

Level of important	Measurement scale
Less important	1
Little important	2
Medium Important	3
Important	4
Very important	5

Before distributing the questionnaire, the questionnaire results were primarily analyzed to test content validity to ensure validation of the meaning and scope of each competency. Then, questionnaire content was also checked by an in-depth interview with six experts. Four individuals were senior construction managers, and two were vice presidents. All six individuals had more than 20 years of experience in construction projects. Then, the interview results were compared with the literature on competencies to adjust for fit within the construction industry definition. As a result, content validity was confirmed.

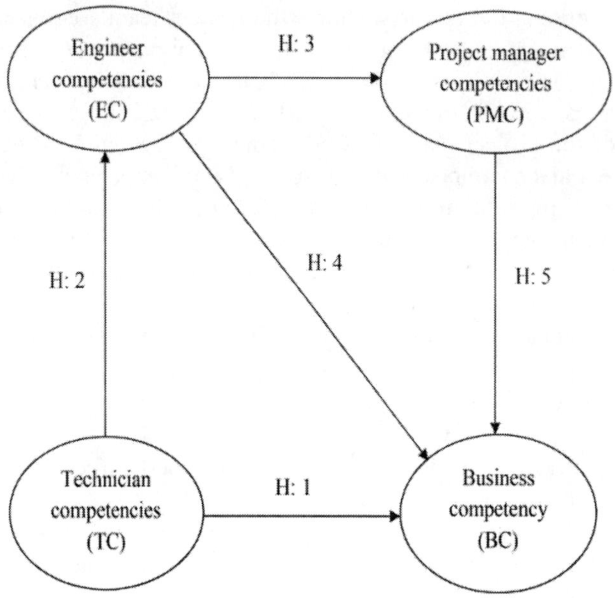

Figure 1. Conceptual framework of the study.

Scale reliability also was tested by the use of Cronbach's alpha (α). The test for the questionnaire's reliability returned results with an average value of 0.924 for each project team's competencies and competitive competencies. Cronbach's alpha value of 0.7 is considered acceptable [7]. Moreover, confirmatory factor analysis (CFA) was used to test the construct validity, which determined by over fit index of each measurement model of PMC, EC, TC and BC as shown in Table 3. In the table, the values of overall goodness-of-fit indices satisfy all recommended criteria values [16]; therefore, the construct validity exists.

Table 3. Criteria and theory of the values of goodness-of-fit(Gof) appraisal for measurement model.

Criteria Index	**Recommended criteria values**	**GoF values of measurement models**			
		PMC	*EC*	*TC*	*BC*
Chi-square (X^2)	$0.05 < p \leq 1.00$	0.531	0.552	0.725	0.457
Relative X^2 $\chi 2$/df	$0 \leq \chi 2 = df \leq 2$	0.374	0.354	0.759	0.935
Goodness of Fit Index (GFI)	$0.09 \leq GFI \leq 1.00$	0.994	0.999	0.988	0.992
Root Mean Square Error of Approximation (RMSEA)	$0 \leq RMSEA \leq 0.08$	0.000	0.000	0.000	0.000
Comparative Fit Index (CFI)	$0.09 \leq CFI \leq 1.00$	1.000	1.000	1.000	1.000
Incremental Fit Index (IFI)	$0.9 \leq IFI \leq 1.00$	1.000	1.000	1.000	1.000
Tucker-Lewis Index (TLI)	$0.9 \leq TLI \leq 1.00$	1.000	1.000	1.000	1.000
Results		passed	passed	passed	passed

4. Result

To develop the structural equation modeling (SEM) for testing the relationships of the latent and observed variables of the respective hypotheses, IBM SPSS AMOS (version 20) software was used. Criteria index and the recommend index values [16] were shown in Table 4. All indices showed

IOP Conf. Series: Materials Science and Engineering **639** (2019) 012026 doi:10.1088/1757-899X/639/1/012026

satisfaction value which means the model fit the data well. The SEM result was shown in Figure 2. Additionally, the hypothesis test results were shown in Table 5.

Table 4. Criteria and theory of the values of goodness-of-fit(Gof) appraisal for SEM model.

Criteria Index	Criteria	Values	Results
Chi-square (X^2)	$0.05 < p \leq 1.00$	0.080	passed
Relative X^2 $\chi 2$/df	$0 \leq \chi 2$=df ≤ 2	1.122	passed
Goodness of Fit Index (GFI)	$0.09 \leq$ GFI ≤ 1.00	0.910	passed
Root Mean Square Error of Approximation (RMSEA)	$0 \leq$ RMSEA ≤ 0.08	0.024	passed
Comparative Fit Index (CFI)	$0.09 \leq$ CFI ≤ 1.00	0.981	passed
Incremental Fit Index (IFI)	$0.9 \leq$ IFI ≤ 1.00	0.982	passed
Tucker-Lewis Index (TLI)	$0.9 \leq$ TLI ≤ 1.00	0.976	passed
Cronbach's alpha	> 0.70	0.924	passed

Table 5. Results of hypotheses testing.

Hypotheses	Coef.	p-value	Results
H1: Technician competencies (TC) have a positive and direct influence on business competencies (BC)	-	> 0.01	Not accepted
H2: Technician competencies (TC) have a positive and direct influence on engineer competencies (EC).	0.87	. < 0.01	Accepted
H3: Engineer competencies (EC) have a positive and direct influence on project manager competencies (PMC).	0.55	. < 0.01	Accepted
H4: Engineer competencies (EC) have a positive and direct influence on business competencies (BC).	0.49	. < 0.01	Accepted
H5: Project manager competencies (PMC) have a positive and direct influence on business competencies (BC).	0.50	. < 0.01	Accepted

Table 6. Standard coefficients of influence equation model of latent variables.

Latent variables	Influence	Latent variables			
		TC	*EC*	*PMC*	*BC*
Technician	DI	-	0.87*	-	
competencies (TC)	II	-	-	0.55	0.43
	TI	-	0.87	0.55	0.43
Engineer competencies	DI	-	-	0.55*	0.49*
(EC)(R^2 = 0.76)	II	-	-	-	0.27
	TI	-	-	0.55	0.76
Project manager	DI	-	-	-	0.50*
competencies (PMC)	II	-	-	-	-
(R^2 = 0.30)	TI	-	-	-	0.50
Business competencies	DI	-	-	-	-
(BC)	II	-	-	-	-
(R^2 = 0.76)	TI	-	-	-	-

Table 6 shows the relationships among the latent variables and their direct influence (DI), indirect influence (II), and total influence (TI) of each construct, with the sum of DI and II referred to as TI. This also shows that engineer competencies (EC) had the highest total influence on business competencies (BC) (TI = 0.76). These influences consist of direct influence (DI = 0.49) and indirect influences (II = 0.27). Project manager competencies (PMC) were also shown to have a direct influence on business competencies (DI = 0.50). Technician competencies shown to have a direct influence on engineer competencies (BC) (DI = 0.87) and indirect influence on project manager competencies (PMC) (II = 0.55) and business competencies (BC) (II = 0.43). These results reflect that project manager skills (MS) had an indirect influence on foreman skills (FS) (II = 0.546).

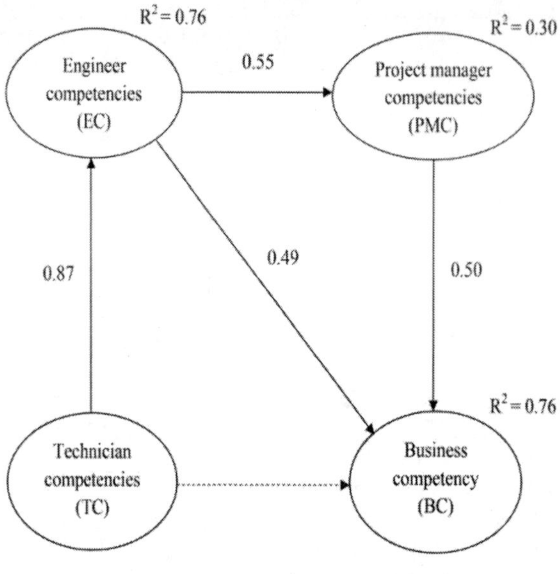

Note. Dash line =Not-significant influence; Solid line = Significant influence

Figure 2. Final result of SEM.

5. Discussion and conclusion.

This study fills a gap in knowledge by exploring the relationships between the construction project team's competencies and contractor business competencies. These relationships imply the characteristics of hierarchical arrangement which was suggested by [4]. This is likely because the need for competency improvement in the lower level staffs can motivate higher level staffs to upgrade their competencies. Moreover, the results indicated that the project manager and engineer competencies have significant influences on business competencies with 0.50 and 0.49 of regression weighs, respectively. This may be because both have a large contribution to the project resources usage, which directly affects the profit of contractor business. This study also finds that technician competencies have no significant influence on business competencies. The possible reason is that a technician has a major role in the supervision of workers to execute works to follow engineer instructions. Thus, most of the technician actions are based on direction from the engineer. However, the model indicates the relationship among their competencies that technician competencies have an influence on engineer competencies (0.87), and engineer competencies have an influence on project manager competencies (0.55), which shows that technician competencies also have an indirect influence on business competencies. These findings comply with [1] who found that the project team's competencies had impacts on business operation.

The model of project team's competencies influencing business competencies can be used by the contractor to develop proper competencies for individual project team leading to the improvement of business competencies.

References

[1] S.M. Takey, M.M. de Carvalho, "Competency mapping in project management: An action research study in an engineering company" Int. J. Proj. Manage, vol. 33, pp. 784-796, 2015

[2] D.K. Ahadzie, D.G. Proverbs, P. Olomolaiye, "Towards developing competency-based measures for construction project managers: Should contextual behavioursbe distinguished from task behaviours?" Int. J. Proj. Manage., vol. 26 pp. 631-645, 2008

[3] Z. Isik, D. Arditi, I. Dikmen, M. T. Birgonul, "Impact of corporate strengths/weaknesses on project management competencies" Int. J. Proj. Manage., 27 269-237, 2009

[4] M.D.Mesarović, D. Macko, Y. Takahara., Theory of Hierarchical, Multilevel, Systems, Volume 68. Elsevier Science, 1970

[5] P.H.K. Ho, "Analysis of competitive environments, business strategies, and performance in Hong Kong's construction industry" J. Manage. Eng., vol. 32(2), 04015044, 2016

[6] P.L. Erdos, A.J. Morgan., Professional mail surveys. New York, NY: McGraw Hill, 1970

[7] M. Tavakol, R. Dennick., "Making sense of Cronbach's alpha" Int. J. of Med. Edu., vol. 2, pp. 53 – 55, 2011

[8] F.T. Edum-Fotwe, R. McCa€er, "Developing project management competency: perspectives from the construction industry" Int. J. Proj. Manage, vol. 18, pp. 111-124, 2000

[9] F. Zhang, J. Zuo, G. Zillante, "Identification and evaluation of the key social competencies for Chinese construction project managers" Int. J. Proj. Manage, vol. 31, pp. 748-759, 2013

[10] H. Sadeghi,• M. Mousakhani,•M. Yazdani, M. Delavari, "Evaluating project managers by an interval decision-making method based on a new project manager competency model" Arab J Sci Eng vol. 39, pp.1417-1430, 2014

[11] G.J. Skulmoski, AACE International Transactions D2A, 2000

[12] T.S. Lee, D. Kim, D.W. Lee., "A competency model for project construction team and project control team" KSCE J. of Civil Eng., vol. 15(5), pp. 781-792, 2011

[13] A. Serpell, X. Ferrada, "A competency-based model for construction supervisors in developing countries" J. Personnel Review vol. 36, pp. 585-602, 2007

[14] V. Shahhosseini, M..H.. Sebt, "Competency-based selection and assignment of human resources to construction projects" Scientia Iranica A, vol. 18(2), pp. 163–180, 2011

[15] J. Lampel, "The core competencies of effective project execution:the challenge of diversity, vol. 19, pp. 471-483, 2001

[16] H. W. Marsh, J. R. Balla, R. P. McDonald., "Goodness-of-fit indices in confirmatory factor analysis: The effect of sample size" Psychol. Bull., vol. 103(3), pp. 391–410, 1988A reference

Relationships of construction project manager, engineer, and foreman skills

Kittipos Kawesittisankhun, Jakrapong Pongpeng

Department of Civil Engineering, Faculty of Engineering, King Mongkut's Institute of Technology Ladkrabang, Bangkok 10520, Thailand.
59601034@kmitl.ac.th

Abstract. Understanding the relationships of construction project manager, engineer and foreman skills can be used by contractors to develop project performance. Although there are widely discussed on construction individual skills, the relationships between individual skills having on others' skills have not been examined in previous studies. Thus, this study examined such relationships. The questionnaire was used to collect data on the important range of skills of the individual to evaluate the relationships among them. The data were then analyzed using structural equation modelling (SEM) to determine the relationships of construction manager, engineer and foreman skills. The final SEM model shows regression weights in parenthesis that construction manager skills have influence on engineer skills (0.65) and engineer skills have influence on foreman skills (0.84). Additionally, the regression weight presents that knowledge management is the most important skill for a construction manager (0.64). Problem-solving is the most important to engineer (0.59). Regarding foreman, physical skill is the most important (0.67). The findings can be used as a guideline for contractors in developing their staffs to increase project performance.

1. Introduction

Today, the changes in construction technique, economy and social condition are factors bring the construction industry to become difficult to survive. These changes cause a competition environment to become fiercer. Thus, contractors have to decide whether they will be outmoded and leave their business or take competitive advantage from technology changes.

Construction is a project-based business which consisted of many functions working under the limited time and resources. In old fashion, a construction project may depend on individual effort; however, to respond with high competition environment, participants need to engage in work through multidisciplinary teams. Contractor executes work through project members who decides which technique and technology should be used. With skilled project members, contractors are easier to adapt to the change in technology. Based on construction project organization structure, the key members of the construction project consist of the construction manager, engineering, and foreman. These key members work under the hierarchical structure. [1] mentioned that a construction manager is responsible for the overall success of delivery the owner's physical development within the constraints of cost, time, quality and safety requirements. An engineer is professional in the knowledge of mathematics and physical sciences, who judges to develop ways to economically utilize resources through design, planning, material selection and workforce management [2]. A construction foreman has a role in overseeing labor production, construction and maintenance of construction facilities (e.g. tools, equipment) under guideline from engineers or managers. A review of the previous literature show the roles of construction project managers, engineers, and foremen. However, the relationships among their skills have not been examined. Hence, the study aim was to examine such relationships.

2. Literature review

2.1. Construction manager skills (MS)

Most of the construction tasks and resources are distributed from a construction manager to every function in order to achieve and deliver the project development under limited resources. Also, a construction manager is responsible for the overall success [1]. Thus, many researchers studied about manager's skills. Also, [3] surveyed the case study from Gas Tank Establishment (GTE) project and suggested that the knowledge and decision-making skills of the project manager have a contribution to project success. Regarding construction manager soft skills that were pointed by [4], these skills are negotiation skill, construction management skill, strategically thinking skill, risk management skill, project organizing skill, and training skill.

2.2. Engineer skills (ES)

Most of the project resources are always utilized by engineers through design, planning, material selection and activities management. To increase project efficiency, an engineer needs to develop economically way to use resources following the guide from the construction manager. Thus, their skills were discussed by many researchers. [5] pointed out the importance of problem-solving skill for a construction engineer, and created construction problem-solving application to assist project participates to execute the project. [6, 7] also determined necessary skills for engineer such as resource management skill, task organization skill, communication skill, technical skill, knowledge applying skill and behavior control skill. [8] mentioned project problem can arise from human factors such as poor interpersonal skills, poor communication, unprofessional interaction and unethical behavior.

2.3. Foreman skills (FS)

A foreman has a role to supervise labors and control their production including maintenance construction tools and equipment in order to keep project acceptable production rate under instruction from an engineer. There are many studies on foremen skills. For example, [9] discussed foremen as persons who are treated as a critical labor function, and have to face the on-site problems frequently. The result from [2] pointed out that not only physical construction skill but also learning skill, problem responding skill, systematic working skill, self-control and work cooperation skill are important characteristics of foremen.

From the analysis of the literature review, the proposed conceptual framework for the study is stated as follows:

- H1: Engineer skills have a positive and direct influence on foreman skills.
- H2: Construction manager skills have a positive and direct influence on engineer skills.

3. Research method

In the construction literature, content analysis regarding skills of construction manager, engineer and foreman have been reviewed. Then, to better understand the influence and importance of construction participants' skills, a structural framework was developed to investigate how construction manager skills affect engineer skills, and how engineer skills affect foreman skills. Then, the structural equation modeling (SEM) was used to test the framework.

A total of 300 copies of the questionnaire were distributed by hand, and 216 were received (a return rate of 72%). According to [10], the return rate is considered as excellent, which is sufficient for SEM analysis. From the survey's Part 1 results (Table 1), 55.55% of the respondents identified themselves as civil engineering. This is in agreement with the 80.10% which indicated they had 10 or more years' experience, and the entire sample who told they participated in the construction field. It was noted that a large majority of the respondents (62.20%) were associated with infrastructure project. There are 88.89% engage with large project with average project value over 30 million USD. In questionnaires part 2, each item about construction manager skills, engineer skills and foreman skills is assessed using a 5-level Likert type agreement scale.

Table 1. Respondent characteristics (N=216).

Position	Frequency	%
Construction senior manager	14	6.49
Construction project manager	82	37.96
Civil engineer	120	55.55
Experience		
Less than 5 years	15	6.94
5-10 years	28	12.96
10-20 years	91	42.14
Over 20 years	82	37.96
Work characteristics		
Strategic management	21	9.73
Construction project management	175	81.02
Construction control	20	9.25
Range of average project value per year		
0.3–3 million USD	22	10.18
3–30 million USD	41	18.98
Over 30 million USD	153	70.83
Project characteristics		
Infrastructure	135	62.20
Low-rise building (< than 8 stories)	9	4.16
High-rise building (> than 8 stories)	72	33.36
Total employees		
500-1000 individuals	24	11.11
Over 1000 individuals	192	88.89

Table 2. Likert type agreement scale.

Level of important	Measurement scale
Less important	1
Little important	2
Medium Important	3
Important	4
Very important	5

Before distributed, the questionnaire was tested to confirm its reliability and validity. Cronbach's alpha was used to test the internal consistency reliability of the scale. The questionnaire's reliability returned results with an average value of 0.904, which a Cronbach's alpha value of 0.7 is considered as acceptable [11]. In addition, construct validity was tested to confirm that latent variables and observed variables are consistent by confirmatory factor analysis (CFA) method. This was determined by the overall fit index of each measurement model of construction manager skills (MS), engineer skills (ES) and foreman skills (FS) as shown in Figures. 1, 2, and 3, respectively. The recommended criteria [13] were used to justify construct validity. In the test, the values of overall goodness-of-fit indices satisfy all recommended criteria values as shown in Table 3; therefore construct validity exists.

In addition, discriminant validity can be measured by analyzing relationships between latent variables [12]. Individual item was examined by looking at the loadings or correlations of each indicator on its respective construct (Table 4). In the table, all the latent variables are highly correlated; thus, the discriminant validity is confirmed.

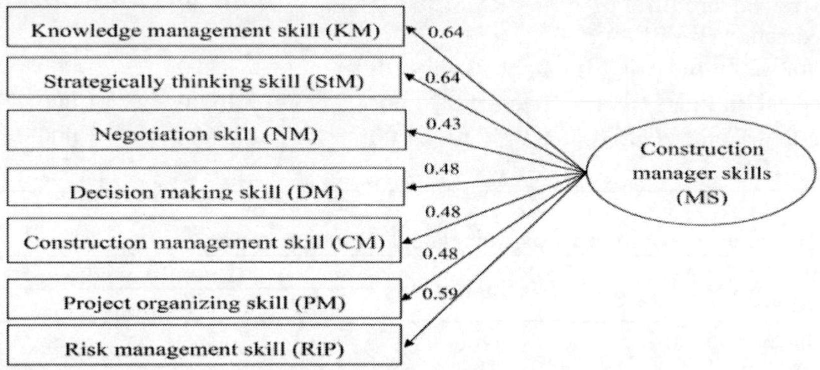

Figure 1. Result of measurement model for construction manager skills.

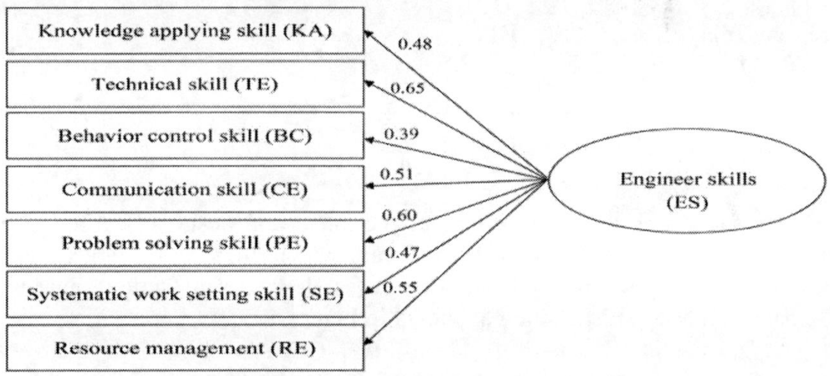

Figure 2. Result of measurement model for engineer skills

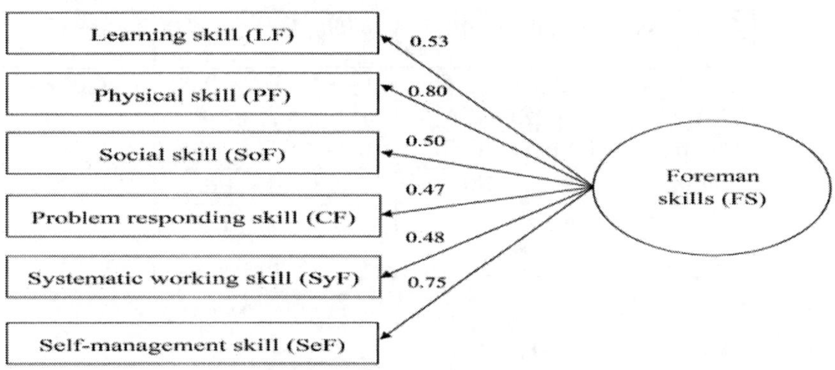

Figure 3. Result of measurement model for foreman skills

Table 3. Criteria and theory of the value of goodness-of-fit for measurement models.

Criteria Index	Recommended criteria values	GoF values of measurement models		
		MS	*ES*	*FS*
Chi-square (X^2)	$0.05 < p \leq 1.00$	0.197	0.392	0.169
Relative X^2 $\chi2$/df	$0 \leq \chi2 = df \leq 2$	1.322	1.056	1.453
Goodness of Fit Index (GFI)	$0.09 \leq GFI \leq 1.00$	0.980	0.988	0.982
Root Mean Square Error of Approximation (RMSEA)	$0 \leq RMSEA \leq 0.08$	0.039	0.016	0.046
Comparative Fit Index (CFI)	$0.09 \leq CFI \leq 1.00$	1.000	1.000	1.000
Incremental Fit Index (IFI)	$0.9 \leq IFI \leq 1.00$	1.000	1.000	1.000
Tucker-Lewis Index (TLI)	$0.9 \leq TLI \leq 1.00$	1.000	1.000	1.000
Results		passed	passed	passed

Table 4. Correlation coefficients between latent variables.

Latent variables	Influence	Latent variables		
		MS	*ES*	*FS*
Manager skills (MS)	DI	-	0.65*	-
	II	-	-	0.55
	TI	-	0.65	0.55
Engineer skills (ES)	DI	-	-	0.84*
($R^2 = 0.42$)	II	-	-	-
	TI	-	-	0.84
Foreman skills (FS)	DI	-	-	-
($R^2 = 0.71$)	II	-	-	-
	TI	-	-	-

4. Result

To develop SEM for testing the relationships of the latent and observed variables of the respective hypotheses, IBM SPSS AMOS (version 20) software was used. Criteria index and the recommend index values [13] were shown in Table 5. All indices showed satisfaction value which means the model fit the data well. The SEM result was shown in Figure 4.

Table 5. Criteria and theory of the value of goodness-of-fit for SEM.

Criteria index	recommended values	GoF values
Chi-square (X^2)	$0.05 < p \leq 1.00$	0.211
Relative X^2	$\chi2$/df $0 \leq \chi2 = df \leq 2$	1.095
Goodness of Fit Index (GFI)	$0.09 \leq GFI \leq 1.00$	0.937
The Root Mean Square Error of Approximation (RMSEA)	$0 \leq RMSEA \leq 0.08$	0.021
Comparative Fit Index (CFI)	$0.09 \leq CFI \leq 1.00$	0.989
Incremental Fit Index (IFI)	$0.9 \leq IFI \leq 1.00$	0.990
Tucker-Lewis Index (TLI)	$0.9 \leq TLI \leq 1.00$	0.985
Results		passed

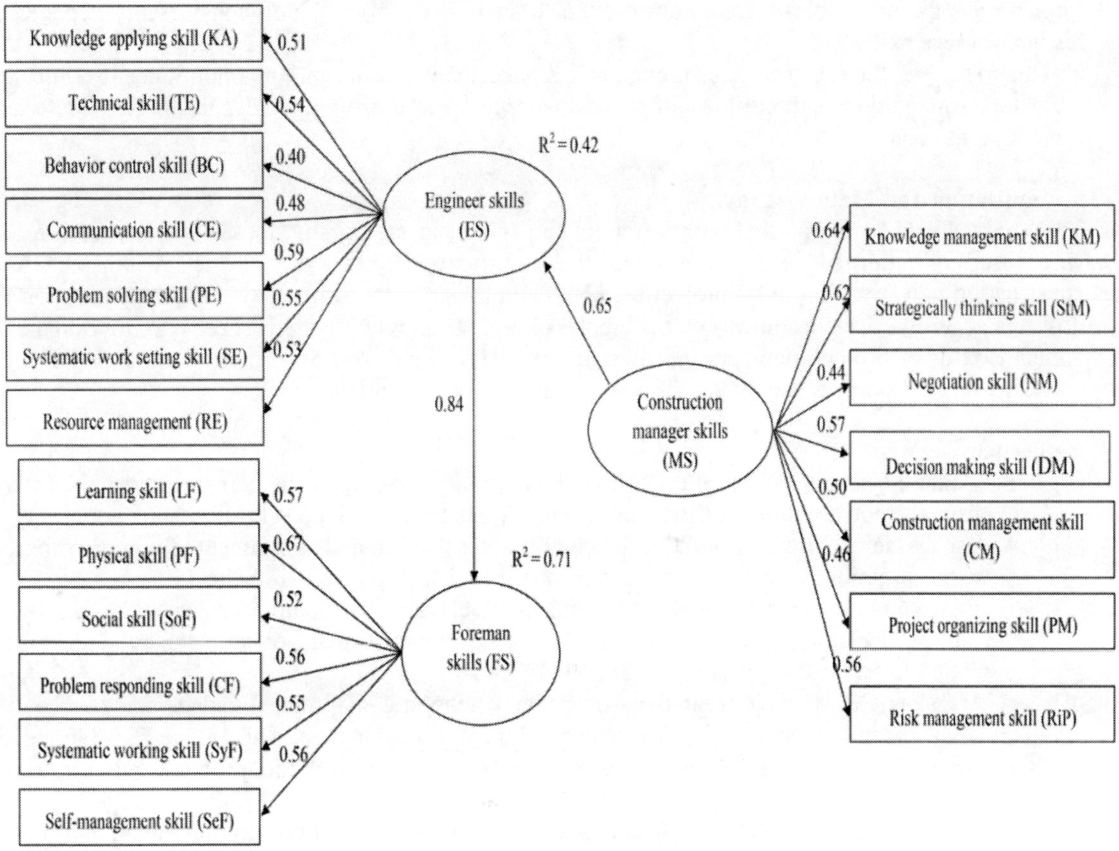

Figure 4. Result of final structural equation modeling for influence among construction project participant's skills

SEM analysis shows the results of hypothesis testing. The p-values of all hypotheses are less than 0.01, indicating statistical significance, so H1 and H2 were accepted. Table 6 additionally shows the relationships among the latent variables and their direct influence (DI), indirect influence (II), and total influence (TI) of each construct, with the sum of DI and II referred to as TI. This table also shows that engineer skills (ES) had a direct influence on foreman skills (FS) (DI = 0.84). Project manager skills (MS) were also shown to have a direct influence on engineer skills (DI = 0.65). This results reflect that project manager skills (MS) had an indirect influence on foreman skills (FS) (II = 0.546).

5. Discussion and conclusion.

The model of participant skills can be arranged into a hierarchical structure of construction manager skills, engineer skills, and foreman skills corresponding to most of project organization structure. However, few research works have explored the relationships, showing a knowledge gap. To fill this knowledge gap, this study aimed to find the relationships among construction manager skills, engineer skills, and foreman skills. The results show that construction manager skills significantly influence engineer skills (a regression weight of 0.65) and engineer skills significantly influence foreman skills (a regression weight of 0.84).This is likely because in working environment supervisor need to train their subordinates. This result is consistent with [4]. In addition, the result shows that knowledge management skill is the most important for construction manager skill (0.64). However, unexpected problem can occur during construction. This may be the reason for problem-solving skill is the most important for an engineer (a regression weight of 0.59), which complies with [5]. Physical skill is the most important for foreman (a regression weight of 0.67). Also, [9] mentioned that a foreman has the main role in field construction to supervise labors, which considerably depends on his/her physical skill.

Understanding the relationships among project participants' skills and the importance level of their indicators can guide contractors to improve the project participants' skills, leading to increasing project performance.

6. Limitation and further study.

This study should be interpreted with limitations in some aspects. Firstly, the SEM framework model was based on international literature but the statistical survey was collected within the Thai construction industry. Secondly, project members and their related skills may vary, depending on the differences in each environment in different culture. Thus, future study needs to investigate or compare skills of project members in different types of various construction environments. Different construction environment may have different opinions on skills and circumstances.

References

[1] F.T. Edum-Fotwe, R. McCa€er, "Developing project management competency: perspectives from the construction industry" Int. J. Proj. Manage, vol. 18, pp. 111-124, 2000
[2] V. Shahhosseini, M..H.. Sebt, "Competency-based selection and assignment of human resources to construction projects" Scientia Iranica A, vol. 18(2), pp. 163–180, 2011
[3] H. Sadeghi,• M. Mousakhani,•M. Yazdani, M. Delavari, "Evaluating project managers by an interval decision-making method based on a new project manager competency model" Arab J. Sci. Eng., vol., 39, pp.1417-1430, 2014
[4] S.M. Takey, M.M. de Carvalho, "Competency mapping in project management: An action research study in an engineering company" Int. J. Proj. Manage., vol. 33, pp. 784-796, 2015
[5] P. Chang, W. Yu, "Developing a general model for construction problem solving for an engineering consulting firm" KSCE J. Civil Eng., vol. 20(6), pp. 2143-2153, 2016
[6] T.S. Lee, D. Kim, D.W. Lee., "A competency model for project construction team and project control team" KSCE J. of Civil Eng., vol. 15(5), pp. 781-792, 2011
[7] G.J. Skulmoski, AACE International Transactions D2A, 2000
[8] O.J. Alkhatib,• A. Abdou., "An Ethical (Descriptive) Framework for Judgment of Actions and Decisions in the Construction Industry and Engineering–Part I" Sci., Eng., Ethics, 2017
[9] A. Serpell, X. Ferrada, "A competency-based model for construction supervisors in developing countries" J. Personnel Review, vol. 36, pp. 585-602, 2007
[10] P.L. Erdos, A.J. Morgan., Professional mail surveys. New York, NY: McGraw Hill, 1970
[11] M. Tavakol, R. Dennick., "Making sense of Cronbach's alpha" Int. J. Med. Edu., vol. 2, pp. 53 – 55, 2011
[12] J. Henseler, C.M. Ringle, M. Sarstedt., "A new criterion for assessing discriminant validity in variance-based structural equation modeling" J. Acad. Marketing Sci., vol. 43(1), 116-135, 2015
[13] H. W. Marsh, J. R. Balla, R. P. McDonald., "Goodness-of-fit indices in confirmatory factor analysis: The effect of sample size" Psychol. Bull., vol. 103(3), pp. 391–410, 1988S.M. Takey, M.M. de Carvalho, "Competency mapping in project management: An action research study in an engineering company" Int. J. Proj. Manage, vol. 33, pp. 784-796, 2015

IOP Conf. Series: Materials Science and Engineering **639** (2019) 012028 doi:10.1088/1757-899X/639/1/012028

The Enhancement of Pool Boiling Heat Transfer by Passive Technique using Rotating Blades Applied on the Nucleation Sites

A. Suriyawong[1], S. Saisorn[2], A. Kawahara[3], S. Wongwises[4]

[1] Department of Mechanical Engineering, Faculty of Engineering, King Mongkut's Institute of Technology Ladkrabang, Bangkok 10520, Thailand
[2] Department of Engineering, King Mongkut's Institute of Technology Ladkrabang Prince of Chumphon Campus, Pathiu District, Chumphon 86160, Thailand
[3] Faculty of Advanced Science and Technology, Kumamoto University, Kumamoto, Japan
[4] Fluid Mechanics, Thermal Engineering and Multiphase Flow Research Lab. (FUTURE), Department of Mechanical Engineering, Faculty of Engineering, King Mongkut's University of Technology Thonburi, Bangmod, Bangkok 10140, Thailand

E -mail: adirek.su@kmitl.ac.th

Abstract. This paper was aimed to investigate the nucleate boiling phenomena with the presence of the rotating blades above the heat transfer surface. The boiling process at 1 atm was carried out by allowing heat transferred from a copper surface to distilled water. The 4-blade rotor made from copper was vertically installed at different distances above the heat transfer surface (L_{SB}). The flow visualization during phase-change phenomena under heat flux values of 160-610 kW/m^2 was feasible by high-speed camera. The results showed that pool boiling at the heat flux of 160 kW/m^2 corresponded to isolated bubble regime, and at the value of 610 kW/m^2, the column regime was observed. In addition to the two-phase flow pattern, the heat transfer coefficient data was reported in this work. The heat transfer results were found to depend on the distance at which the rotor was installed.

1. Introduction

The pool boiling is a sort of heat transfer with the processes of phase changing between liquid and gas under vaporization. The investigation of nucleate pool boiling has been needed to develop and control the steam generation process. The major techniques, i.e., passive method and active method, applied on boiling process to enhance heat transfer have been investigated by different researchers in the past years. The tasks dealing with improved heat transfer surface, different working fluids, and secondary motion over the heat transfer surface have been proposed. The following are the discussions based on different previous works available for nucleate boiling phenomena.

Jaikumar and Kandlikar [1] reported the pool boiling behaviors affected by sintered open micro-channel. The heating surfaces which were improved based on nano-technology were presented by Shi et al [2] and Kruse et al [3]. They pointed that the nano-coated surfaces were able to significantly improve the heat transfer performance.

Peng et al [4], Heris [5] and Amiri et al [6] investigated the effect of nanofluids on the nucleate boiling phenomena. It was reported that nanofluids resulted in significant heat transfer enhancement.

Kim et al. [7] and Bartoli and Baffigi [8] studied the enhancement of pool boiling heat transfer with installation of the ultrasonic vibration on the heating surface. The results indicated that an ultrasonic vibration was major factor enhancing heat transfer rate. Sheikhbahai et al [9] studies the pool boiling heat transfer on the heating surface under the electrohydrodynamic conditions. Although the electric field has advantages for the electrohydrodynamic technique producing agitation near the heat transfer surface, it needs the external power source to support the system. In this work, the passive technique using the rotating blades without external energy was applied to enhance the heat transfer performance. The pool boiling curves as well as heat transfer results are reported in the present paper.

2. Experimental arrangement

The experimental method is briefly presented in this section. It should be noted that the detailed apparatus can be found in Suriyawong et al [10]. The apparatus was designed in order to study two-phase flow pattern and heat transfer phenomena. In this work, the 4-blade rotor was employed to induce a commotion of the water near the heat transfer surface. The copper rotor was machined to have a diameter of 30 mm, a length of 50 mm, an angle of 90°, and a core of 5 mm. It was adjusted to be vertically installed at different distances (L_{SB}) above the heating surface, i.e., 5 mm, 15 mm, and 25 mm, respectively. The experimental conditions were controlled under a pressure of 1 atm and heat flux values ranging from 160 to 610 kW/m^2.

3. Data reduction

Experimental studies were carried out, based on the flat plate heating surface, to observe the pool boiling heat transfer characteristics. The heat fluxes, q'' (W/m^2), can be calculated from the following equation:

$$q'' = \frac{4VI}{\pi D^2} \tag{1}$$

where V is the voltage (V), I is the direct current (A), and D is the heating surface diameter. The average boiling heat transfer coefficient, h_b (W/m^2·°C), is calculated from the equation below:

$$h_b = \frac{q''}{T_s - T_{sat}} \tag{2}$$

where T_{sat} is the saturation temperature of water (°C), and T_s is the average heater surface temperature (°C) calculated from

$$T_s = \frac{\sum_{i=1}^{4} T_i \sum_{i=1}^{4} x_i^2 - \sum_{i=1}^{4} x_i \sum_{i=1}^{4} T_i x_i}{4\sum_{i=1}^{4} x_i^2 - \left(\sum_{i=1}^{4} x_i\right)^2} \tag{3}$$

where x_i is the position (m) and T_i is the local temperature (°C) which is measured at the test section. Equation (3) was obtained by the linear least squares method using the four thermocouple installed in the test section with their respective points and considering the origin ($x = 0$) at the heating surface.
The uncertainty of heat flux $\delta q''$ (W/m^2) can be calculated from the following equation:

$$\frac{\delta q''}{q''} = \sqrt{\left(\frac{\delta V}{V}\right)^2 + \left(\frac{\delta I}{I}\right)^2 + \left(\frac{2\delta D}{D}\right)^2} \tag{4}$$

where δV is uncertainty of voltage (V), δI is uncertainty of direct current (A) and δD is uncertainty of heating surface diameter (m) The uncertainty of heat transfer coefficient δh_b (W/m^2·°C) can be calculated from the following equation:

$$\frac{\delta h_b}{h_b} = \sqrt{\left(\frac{\delta q}{q}\right)^2 + \left(\frac{\delta T_s}{T_s - T_{sat}}\right)^2 + \left(\frac{\delta T_{sat}}{T_s - T_{sat}}\right)^2} \tag{5}$$

where δT_{sat} is uncertainty of liquid saturation temperature (°C) and δT_s is uncertainty of surface temperature (°C) can be calculated from the following equation:

$$\frac{\delta T_s}{T_s} = \frac{1}{T_s}\sqrt{\sum_{i=1}^{4}\left(\frac{\partial T_s}{\partial T_i}\delta T_i\right)^2 + \sum_{i=1}^{4}\left(\frac{\partial T_s}{\partial x_i}\delta x_i\right)^2} \tag{6}$$

where δT_i is uncertainty of local temperature (°C) and δx_i is uncertainty of position of temperature measurement (m).

The uncertainties of the heating surface temperature, heat flux and heat transfer coefficient are ±0.6%, ±2.6%, and ±7.8% respectively.

4. Results and discussion

In this research, the effects of the L_{SB} on the nucleate pool boiling curve, heat transfer coefficient and boiling phenomenon of distilled water at atmospheric pressure were studied. The horizontal circular plate heating surface was made from copper with surface roughness of 0.2 μm.

The pool boiling curve was developed and presented with the nucleate pool boiling phenomena induced by the presence of the 4-blade rotor rotating above the heat transfer surface. The L_{SB} varied between 5, 15 and 25 mm, while the number of blade was kept at 4 blades. It was found from the results that, the increase in the L_{SB} caused an increase in the T_s-T_{sat} when compared to the non-blade configuration. The corresponding pool boiling curves are illustrated in figure 1.

Figure 2 represents the dependence of heat transfer coefficient on the heat flux. It was found that the increase in heat flux resulted in the increased heat transfer coefficient. Moreover, the decrease in L_{SB} also caused the heat transfer coefficient increased. The heat transfer enhancements corresponded to 6.35 %, 9.45%, and 13.01%, for the L_{SB} at 25 mm, 15 mm, 5 mm, respectively.

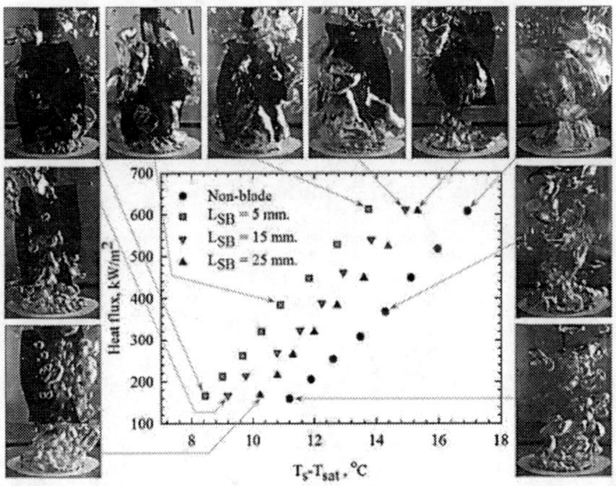

Figure 1. Distilled water pool boiling curves with different L_{SB} configurations.

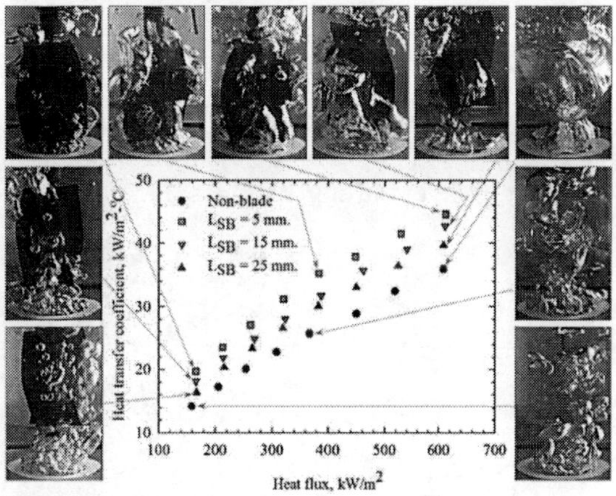

Figure 2. Heat transfer data for different L_{SB} configurations.

Figure 3 shows images of bubble formation on heating surface for the different L_{SB}, during saturated nucleate pool boiling with three different heat flux conditions (167, 385 and 610 kW/m^2). The heat flux of 610 kW/m^2 induced larger bubbles than that of 385 and 167 kW/m^2. In this work, the L_{SB} with 5 mm tended to be strongly driven by the bubbles. The kinetic energy seems to decrease with increasing the L_{SB}.

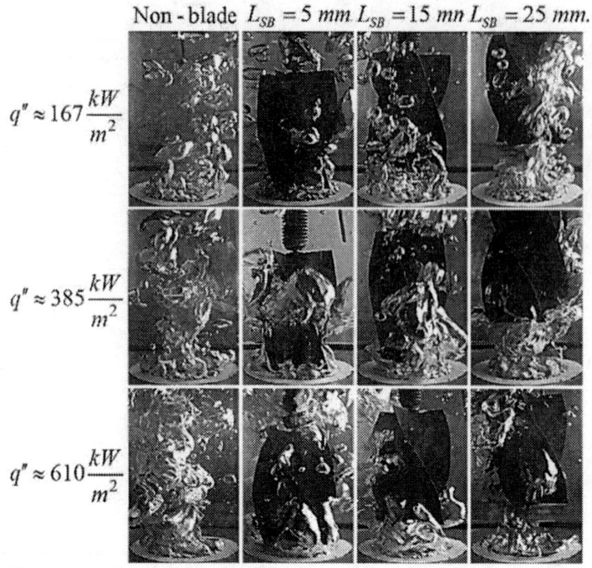

Figure 3. Photographs of the distilled water pool boiling phenomena with different L_{SB} configurations.

The images of nucleate pool boiling behaviors taken at various time points by an Aos Promon Studio camera are shown in figures 4 and 5. The video frames of nucleate pool boiling heat transfer of distilled water at atmospheric pressure in the apparatus without the rotating blades are shown in figures 4. The observations showed that the bubble nucleates on the heating surface, grew in size, and eventually departed from the heating surface. The bubble size for the heat flux of 167 kW/m^2 was smaller than that obtained from the heat flux values of 385 and 610 kW/m^2, respectively.
Figure 5 showed the video frames of nucleate pool boiling heat transfer of distilled water at atmospheric pressure in the apparatus installed with rotating blades with the L_{SB} of 5 mm. The L_{SB} with 5 mm was experimented during saturated nucleate pool boiling under three different heat flux conditions (167, 385 and 610 kW/m^2). The heat flux of 167 kW/m^2 resulted in the vapor bubbles smaller than those generated by the heat flux values of 385 and 610 kW/m^2. In addition, the heat flux of 610 kW/m^2 seemed to cause the bubbles transmitting more energy to the rotating blades, and hence, the secondary motion of the distilled water near the heating surface was highly disordered. The higher the heat flux, the stronger is the secondary motion caused by the rotating blades.

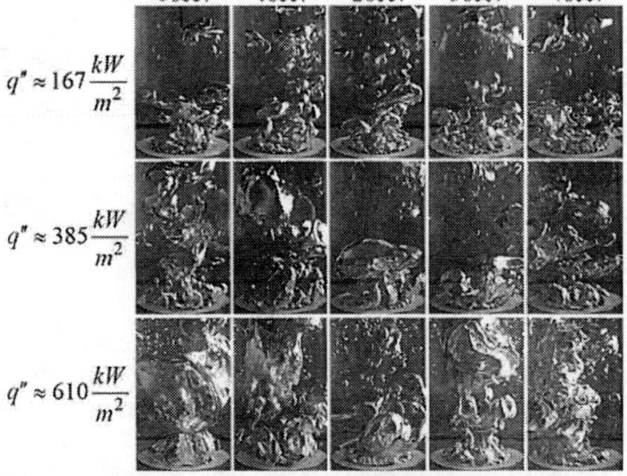

Figure 4. Video frames of nucleate pool boiling phenomena without the rotating blades.

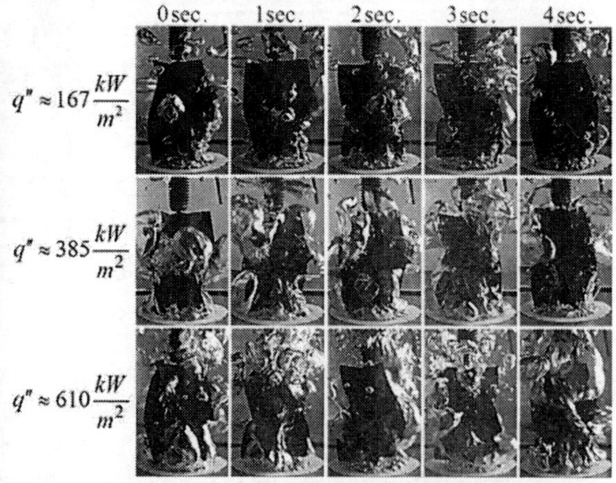

Figure 5. Video frames of the pool boiling phenomena with L_{SB} = 5 mm configuration.

5. Conclusions

The experimental investigation was carried out to explore the nucleate boiling behaviors with the presence of the rotating blades positioned at different locations above the heat transfer surface. Isolated bubble regime was observed at low heat flux, and was developed to be the column regime when heat flux became higher. With the presence of the 4-blade rotor vertically installed above the surface at a distance of 5 mm, the heat transfer enhancement was up to 13.01% when compared to the case without the rotor.

References

[1] Jaikumar A and Kandlikar S G 2015 *Int. J. Heat Mass Transfer* **88** 652-61
[2] Shi B, Wang Y and Chen K 2015 *Appl. Therm. Eng.* **75** 115-21
[3] Kruse C M, Anderson T, Wilson C, Zuhlke C, Alexander D, Gogos G and Ndao S 2015 *Int. J. Heat Mass Transfer* **82** 109-16
[4] Peng H, Ding G, Hua H, Jiang W, Zhuang D and Wang K 2010 *Int. J. Refrigeration* **33** 347-58.
[5] Heris S Z 2011 *Int. Commun. Heat Mass Transfer* **38** 1470-3
[6] Amiri A, Shanbedi M, Amiri H, Heris S Z, Kazi S N, Chew B T and Eshghi H 2014 *Appl. Therm. Eng.* **71** 450-9
[7] Kim H Y, Kim Y G and Kang B H 2004 *Int. J. Heat Mass Transfer* **47** 2831-40
[8] Bartoli C and Baffigi F 2011 *Exp. Thermal Fluid Sci.* **35** 423-32
[9] Sheikhbahai M, Esfahany M N and Etesami N 2012 *Int. J. Thermal Sciences* **62** 149-53
[10] Suriyawong A, Saisorn S and Wongwises S 2017 *Exp. Thermal Fluid Sci.* **87** 109-16

Acknowledgement

The research is financially supported by King Mongkut's Institute of Technology Ladkrabang and King Mongkut's University of Technology Thonburi.

IOP Conf. Series: Materials Science and Engineering **639** (2019) 012029 doi:10.1088/1757-899X/639/1/012029

An Improvement on Heat Reduction in a Lightweight Composite Roof Material

Amphon Jarasjarungkiat[1], Phuridech Lertrattanaporn[2]

[1,2]Department of Civil Engineering, Faculty of Engineering, King Mongkut's institute of technology Ladkrabang, Bangkok, Thailand.

[1]amphon.ja@kmitl.ac.th

Abstract. The objective of this research is to improve heat reduction in a lightweight composite roof material. This research has studied and compared roofing material with opening for ventilation and without opening. A comprehensive model is simulated from computational fluid dynamics simulation of a finite element software. The simulation considered three aspects of heat transfer which are heat conduction, heat convection and radiation. The simulation result showed that the roofing material with opening for ventilation can reduce temperature in the composite roof about 20%

1. Introduction

Roofing materials are made of various kinds of materials by depending on people demand for example zinc, metal sheet, concrete tile and ceramic as well as composites.

Composite materials are made from two or more materials with different properties combined to produce a new material. Physical and chemical properties of each of the constituent materials remain distinct in the new material. These constituent materials work synergistically to produce a composite material that has improved properties when compared with the individual constituent materials.[1] Therefore, properties of the composite material cannot be predicted from any type of material, The constitutive equations in the finite element software can be used to simulate deflection behaviors of a linear composites material.[2]

Heat transfer is a movement of thermal energy from one object to another object of different temperature. There are three different ways the heat can be transferred by conduction (through direct contact), convection (through fluid movement) and radiation (through electromagnetic waves).

1. The concept of conduction is carried out in the case of the two objects in contact. Heat is passed from the higher temperature to another lower temperature. Conductive heat losses that occurred on a material sheet are due to thermal gradients between the material sheet and ambient surrounding condition that the solar panel connected [3]. In a rectangular coordinate, Heat transfer in one dimension can be calculated by Equation (1)

$$Q = kA\frac{T_h - T_c}{\Delta x} = -kA\frac{T_c - T_h}{\Delta x} = -kA\frac{\Delta T}{\Delta x} \quad (1)$$

Where: Q = Heat transferred per times (W)

k = Thermal conductivity (W/m°C)

A = Heat transferred area (m^2)

T_h = Maximum Temperature (°C)

T_c= Minimum Temperature (°C)

ΔT= Temperature difference between (°C)

Δx= Thickness (m)

2. Convection heat transfer is the transfer of heat by movement of a fluid from higher temperature region to the lower temperature one. Generally, liquid and gas are the dominant form of this heat transfer. Convection heat transfer can be occurred either natural convection or forced convection. The natural convection is triggered by density differences within fluid that happened due to temperature gradient and without any external source of support. Forced convection is movement of the fluid that is produced by an external source. By Newton's law of cooling, the rate of convective heat transfer can be calculated [4] as given by Equation (2)

$$Q = hA(T_s - T_f) \tag{2}$$

Where: Q = Heat transferred per times (W)

h = Heat transferred coefficient (W/m°C)

A = Cross section area (m^2)

T_S= Temperature of the solid (°C)

T_f= Temperature of the fluid (°C)

3. Radiation heat transfer is not similar as the concept of conduction and convection. Both conduction and convection is the transfer of heat via matter. While, radiation transfers the heat in form of electromagnetic waves. Radiation is the heat transfer from the body according to the temperature; it increases as the body temperature increases. When the heat reaches to another surface of the body, they may be absorbed, reflected or transmitted. The emitted energy by a blackbody [5] is given by Stefan-Boltzmann law as Equation (3)

$$Q = \varepsilon A \sigma T_A^4 \tag{3}$$

Where: Q = Heat transferred per times (W)

σ = Stefan - boltzmann constant (W/m^2K4)

ε = Infrared emissivity

T_A= Effective radiating temperature (K)

This heat energy can be transferred away by the fact that heat is transferred to the conduction via matter from particle to particle. Convective heat transfer of the entire roofing material caused by wind. Solar radiation reaches on roofing material as shown in Figure 1.

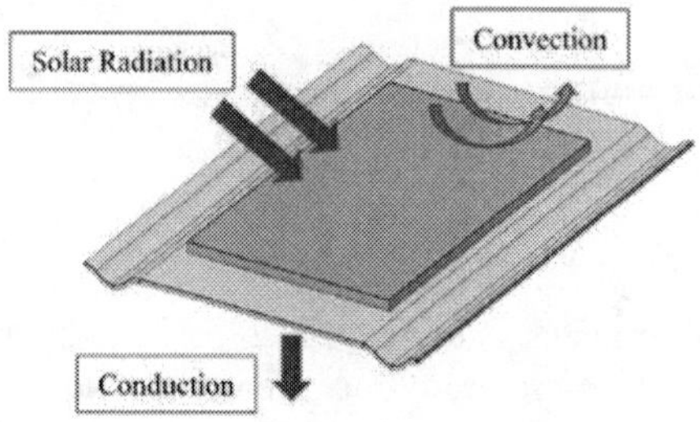

Figure1. Heat transfer behaviors of a roofing material.

The aim of this work is studied on heat transfer behaviors of a roofing material. This research is divided into two models; Firstly, Roofing material with opening for ventilation. Secondly, Roofing material without opening for ventilation the heat transfer behaviors is analyzed from a finite element software in computational fluid dynamics simulation.

2. Method

2.1. Geometry Model of Finite Element.

The roofing material model consists of two composite materials. The composite material A and B as shown in Figure2. The composite material A has size of 88×60×4cm and material B has size of 120x99x1.25cm. The demonstration divided into two models; Firstly, Roofing material with opening for ventilation below the composite material A as shown in figure3. Secondly, Roofing material without opening for ventilation as shown in figure4.

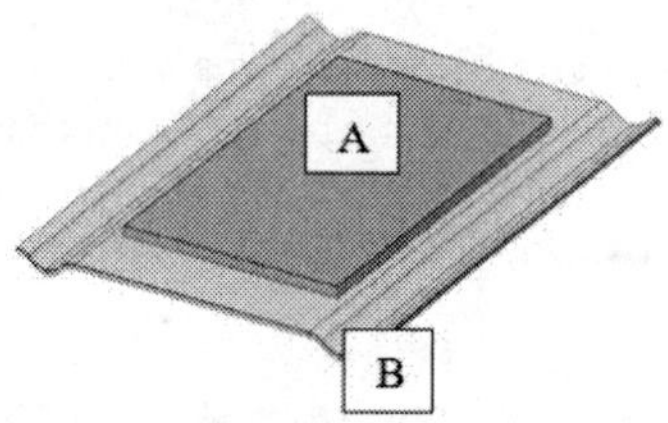

Figure 2. Pattern of composite roofing material

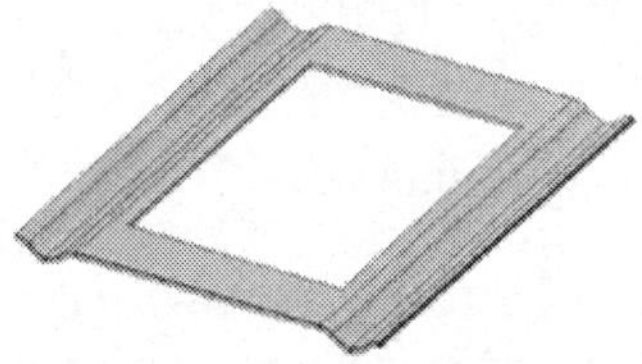

Figure 3. Roofing material with opening for ventilation

IOP Conf. Series: Materials Science and Engineering **639** (2019) 012029 doi:10.1088/1757-899X/639/1/012029

Figure 4. Roofing material without opening for ventilation

2.2. Materials Properties

The material properties of the considered specimen consist of Density (ρ), Thermal Conductivity (K), Specific Heat (Cp) following table I

TABLE 1. MATERAIL PROPERTIES

Materials	*Thermal Conductivity, K (W/m°C)*	*Specific Heat ,Cp (J/kg°C)*	*Density ρ (kg/m2)*
A	1.8	500	3000
B	0.42	921	1430

2.3. Boundary Condition

Finite Element Model for analysis heat transfer behaviors of composite roofing material is prepare in a software for computational fluid dynamics simulation.

Both the initial temperature of composite material the model and ambient temperature are defined at 30 °C.

Beside, no-slip in boundary condition between composite material is prescribed.

By using mesh generation of the composite roof with type of element of SOLID186 and SOLID187, The model of roofing material with opening for ventilation and without opening have the number of elements 920,652 and 929,813 respectively as shown in figure5 and figure6.

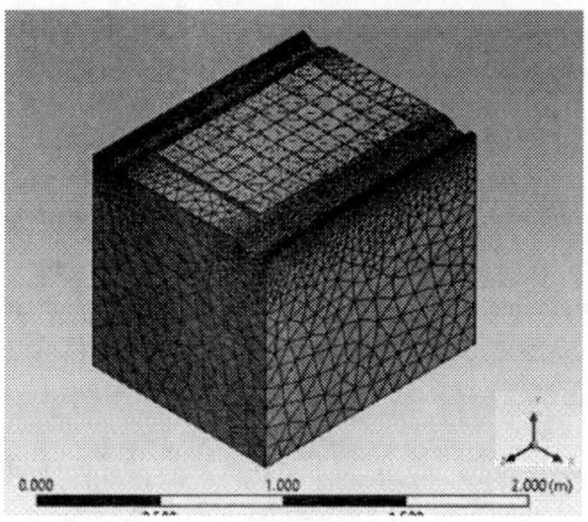

Figure 5. Roofing material with opening for ventilation model was meshing.

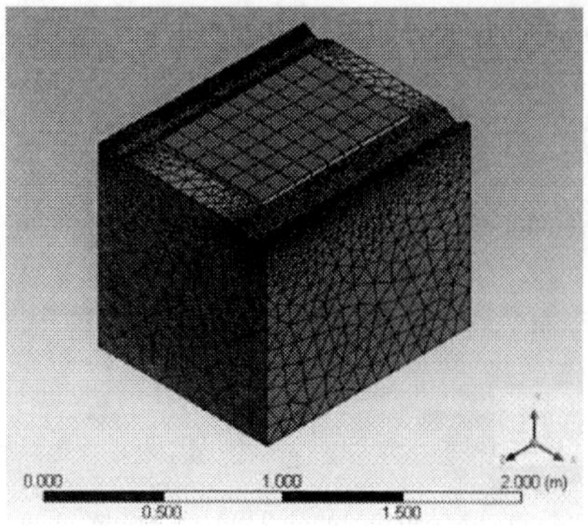

Figure 6. Roofing material without opening for ventilation model was meshing.

According to the computational fluid dynamics simulation, the model was specified two surfaces which normal toZ axis to be an inlet and outlet surface. Beside, other surfaces being normal to X axis as specified as wall condition as written in figure7.

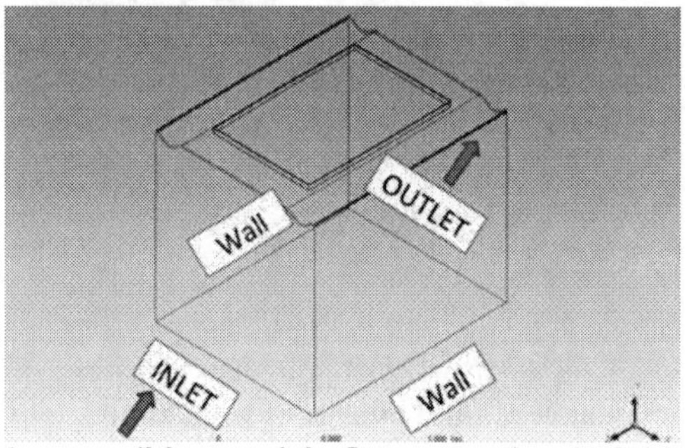

Figure 7. Boundary condition model of Computational Fluid Dynamics Domain.

2.4. Load Input

The initial temperature from solar radiation has 60°C at top surface of composite material A and B.

Applied wind with velocity 1m/s and ambient temperature 30 °C flows beneath composite roofing material, the wind passes through the inlet and outlet as shown in figure7.

3. Result

The result from simulation of the A composite material with opening for ventilation has maximum temperature at top surface about 42.3 °C as shown in figure8. and at bottom surface about 42.1 °C as shown in figure9. The result from stimulation test of the A composite material without opening for ventilation has maximum temperature at top surface about 52.7°C as shown in figure10. and at bottom surface about 52.6°C as shown in figure11. The figure12 shown comparing the temperature of composite material A with opening and without opening.

IOP Conf. Series: Materials Science and Engineering **639** (2019) 012029 doi:10.1088/1757-899X/639/1/012029

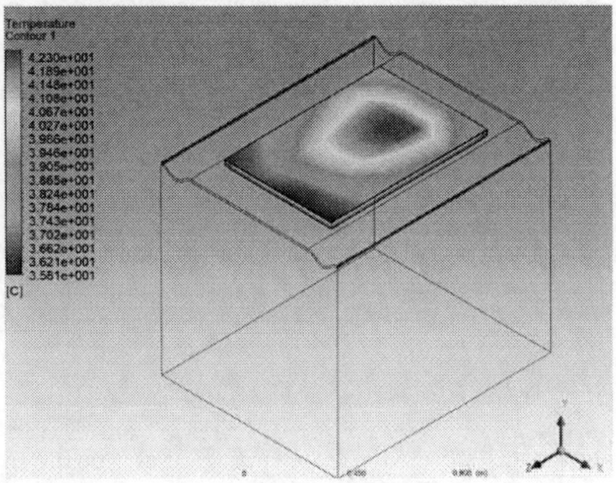

Figure 8. Temperature distribution at top surface of composite material A of roofing material with opening for ventilation

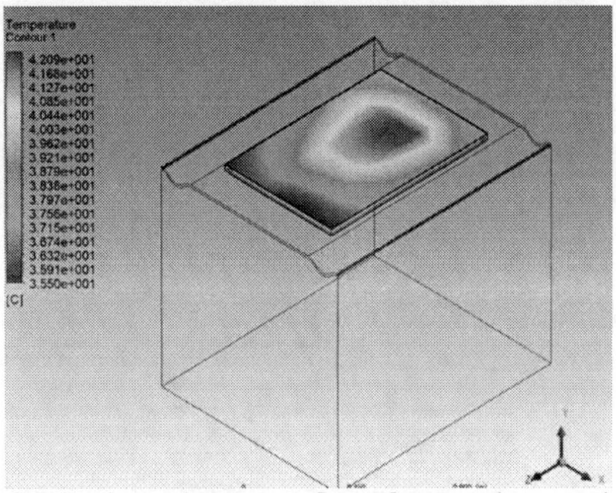

Figure 9. Temperature distribution at bottom surface of composite material A of roofing material with opening for ventilation

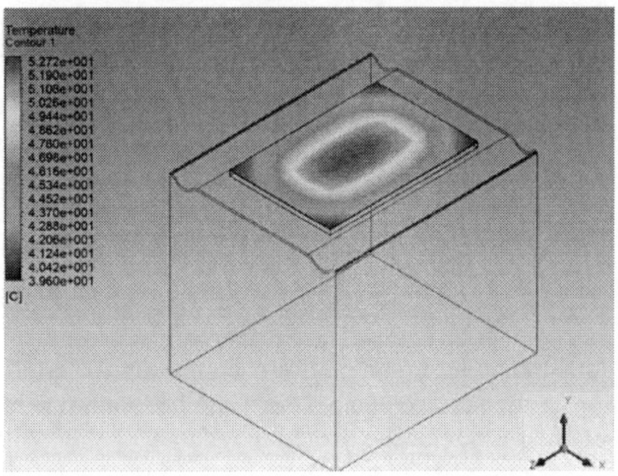

Figure 10. Temperature distribution at top surface of composite material A of roofing material without opening for ventilation

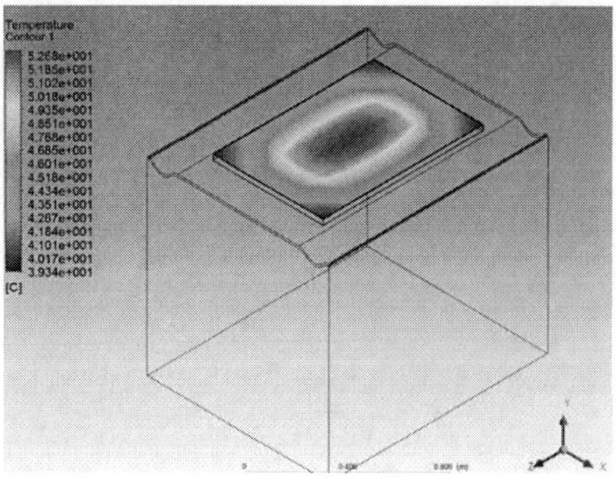

Figure 11. Temperature distribution at bottom surface of composite material A of roofing material without opening for ventilation

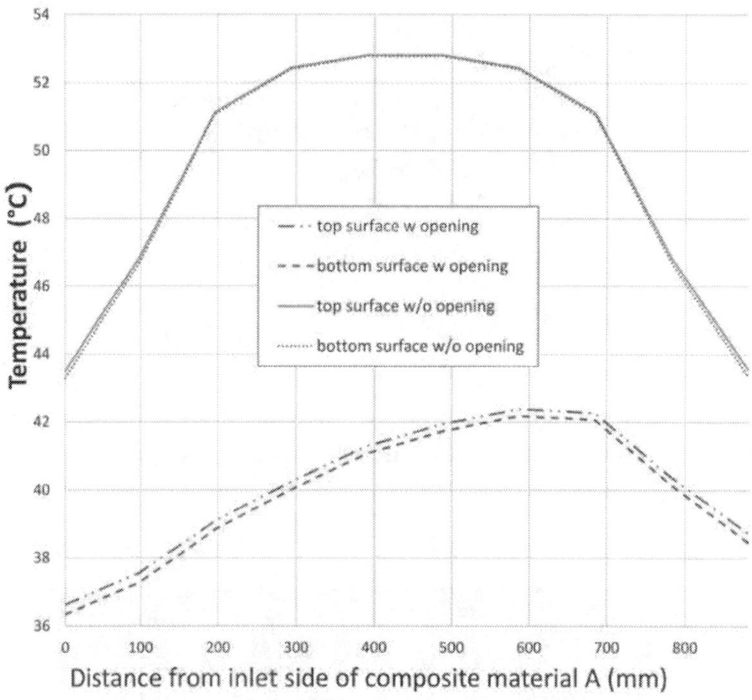

Figure 12. Graph comparing the temperature of composite material A.

4. Conclusion

The result from simulation model of the A composite material with opening for ventilation has maximum temperature at top surface about 42.3 °C and at bottom surface about 42.1 °C The result from stimulation test of the A composite material without opening for ventilation has maximum temperature at top surface about 52.7 °C and at bottom surface about 52.6 °C .To summarize as the

IOP Conf. Series: Materials Science and Engineering **639** (2019) 012029 doi:10.1088/1757-899X/639/1/012029

result, that roofing material with opening ventilation could reduce temperature in the A composite material about 20% under specified conditions.

This paper gives a guideline to develop composite material roofing in the future.

References

[1] A. Kelly, A. Mortensen. Composite Materials: Overview. Eds. Amsterdam: Elsevier. 2001pp.1361-1371Another reference

[2] Amphon Jarasjarungkiat, Phuridech Lertrattanapor. A comparison study for deflection of a foam composite material with results from the finite element method", Proceedings of the 22nd National Conference on Civil Engineering (NCCE22), Nakornrachasima, Thailand, July 18-20, 2017 (STR) CDROM.

[3] Don W Green, Robert H Perry. Perry's Chemical Engineers' Handbook, Eighth Edition. HEAT TRANSFER BY CONDUCTION. McGraw-Hill Professional. 2008

[4] J Vlachopoulos, D Strutt. Heat Transfer. SPE Plastics Technicians Toolbox. 2002: 21-33.

[5] John R Howell, Robert Siegel, M Pinar Mengu9. Thermal Radiation Heat Transfer. Fifth Edition. New York: Taylor & Francis Group.

Computing the maximum value engineering for a building construction costs using linear programming applications.

K Siripob[1] **and C Vuttichai** [2]
[1] Ph.D. Student, Civil Engineering Department, Faculty of Engineering King Mongkut's Institute of Technology Ladkrabang , Bangkok, Thailand.
[2] Asst. Prof. Dr , Civil Engineering Department, Faculty of Engineering King Mongkut's Institute of Technology Ladkrabang , Bangkok, Thailand.

E-mail: siripob.k@gmail.com

Abstract. Value Engineering is an effective technique for reducing costs, increasing productivity and improving quality. Value engineering is a technique directed toward analyzing the functions of an item or process to determine best value or the best relationship between worth and cost functions. Building construction costs comprises 2 major components as materials cost and labor cost. Linear Programming (LP) is a quantitative technique by means of a mathematical modeling technique designs to solve allocation problem (resource allocation) to achieve the maximum profit or lowest cost. Therefore the authors bringing the advantages of both as value engineering and linear programming combined to work together in order to be able to effectively reduced building construction costs or create maximum value engineering that can be best used in the construction projects. Aim of this paper shows the reduction of building construction cost as to achieved the targets for reducing building construction cost or maximum value engineering as much as possible that leads to efficiency of cost reduction by used of principles of value engineering to apply with using linear programming for the reduction of building construction costs based on the material costs and labor costs to serve construction resource reallocation by consideration is made only 3 type construction materials as concrete, rebar and formwork because of 3 type are main component of building construction materials.
Keywords — Value engineering, Linear programming, Construction Costs , Cost function , Worth function .

1. Introduction

Value engineering is a systematic, interdisciplinary examination of factor affecting the cost of a product so as to devise means of achieving the specified purpose at the required standard of quality and realiability at the target cost. At present, for current construction project in several countries it has applied the value engineering popularly, it can reduce the construction costs. The major aim of value engineering is to systematic identification and elimination of unnecessary cost of building construction cost or businesses whereas those products or work can have the same quality, function, and reliability just like before the value engineering is applied. Relative equation is as below.

$$\text{Value} = \text{Function} / \text{Cost} \tag{1}$$

Equation(1) illustrated for value engineering is a technique directed toward analyzing the functions of an item or process to determine best value or the best relationship between worth and cost functions of items or products.

1.1. Definition of cost function and worth function and their relation.

Function cost is price that can perform function. Emphasis is placed on basic functions or sometimes function worth can be defined as one man's price is another man's cost. Function worth is the lowest cost to provide a given function. Emphasis is placed on basic functions or the lowest cost that can contribute basic functions.

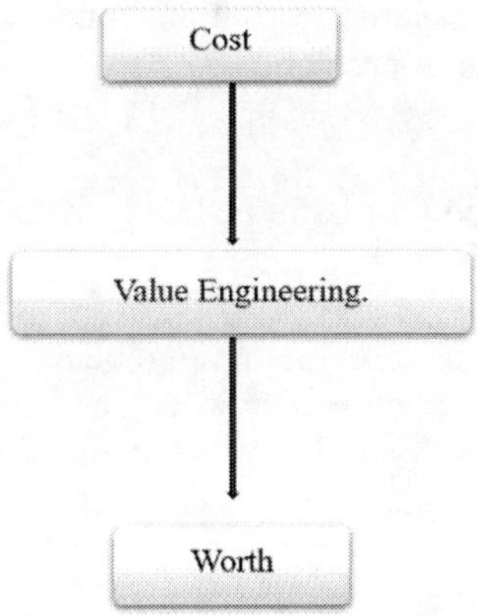

Figure 1. The relationship between cost function with worth function in value engineering process.

Figure1. that illustrated to the relationship between cost and worth function by applying Value Engineering. Cost function which represents the initial costs of the building project which included material costs and labor costs after that will be applied value engineering with various conditions. The resulted will be worth function which meaning to the construction cost is reduced but the functions are remains the same.

1.2. Linear programming.

Linear programming (LP) has proven its efficiency to mathematically model many problems aiming at the maximization or minimization of certain function (objective function) that is linearity dependent on set of variables related to each other through a set of linear constraints. Linear Programming (LP) is a quantitative technique by means of a mathematical modeling technique designs to solve allocation problem (resource allocation) to achieve the maximum profit or lowest cost. The linear programming can be applied with many types of work such as budget allocation, production cost reduction planning, investment and including to construction. Structure of a linear programming model that consists of one objective which is a linear equation that must be maximized or minimized. From constraint equation that specifies restriction of objective function, optimal solution comprises values of decision variables; for example, x1 x2 x3, to xN ,etc. The values of decision variables are non-negative and the objective function must be in the form of a linear equation and each constraint must be a linear equation or a linear inequality. Components in Linear programming are shown to below. :

1.2.1 Objective function.

The objective of a linear programing model will be to maximize or minimize some numerical value; for example, the best outcome from maximum investment or optimal profit or maximum saving for this paper that meaning to maximum value engineering.

1.2.2 Decision variables.

Determination of decision variables is essential because a problem may be specified with various variables. It can be said that decision variables are the variables which will decide output such as x1, x2, x3, x4,.....xN when xN refers to the quantity of each type of materials which are required for this paper.

1.2.3 Constraints.
The objectives of problem can be achieved the optimum or minimum more or less depend on constraints or restriction such as resource restrictions, budget or cost for construction , requirements rule, regulation, policy or condition of problems.

1.2.4 Non Negativity Restriction.
All variables in a linear programming must not be less than zero.

2. Method
This paper presents realistic linear Programming models that define the variables and problem constraints also objective function for maximize value engineering of building costs . Firstly, it should understand what is operation research and linear programming where it's mathematical modeling is a cornerstone to solve more problems and making a decision before a final decision can be reached.

2.1. Linear programming applications.
Therefore, it is advisable to describe maximum value engineering of building construction cost or optimization materials as a mathematical programming problem, and to solve it using LINGO (release17.0).

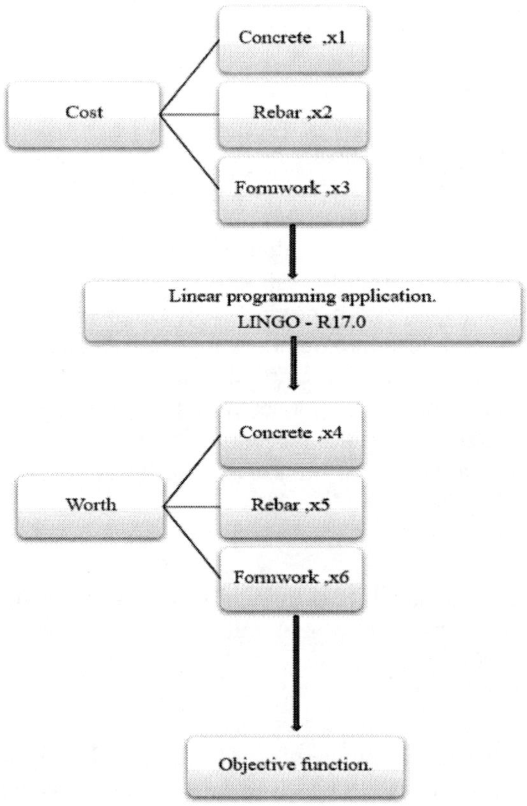

Figure2.explained the procedure of linear programming applications to finding the objective function consisting of the 6 variable that is called the cost function with the concrete quantity (x1) ,rebar quantity (x2) and formwork quantity (x3). Besides, it has used Lingo-R17 software to solve this problem , and the gaining result is the worth function with the concrete quantity(x4) , rebar quantity (x5) and formwork quantity(x6). Therefore, the most important answer is the objective function that is the maximize value engineering of building costs.

IOP Conf. Series: Materials Science and Engineering **639** (2019) 012030 doi:10.1088/1757-899X/639/1/012030

2.1.1 Decision variables. Which are as follows.

Table1. Cost function variables

No.	Details	Unit
x1	Concrete quantity	(m^3.)
x2	Rebar quantity	(kg.)
x3	Formwork quantity	(m^2.)

Table2. Worth function variables.

No.	Details	Unit
x4	Concrete quantity	(m^3.)
x5	Rebar quantity	(kg.)
x6	Formwork quantity	(m^2.)

2.1.2 Objective function. This integrated model is developed to maximize value engineering a construction building costs , as described below.

2500*(x1-x4)+ 22X(x2-x5)+500*(x3-x6) +350*(x1-x4)+6*(x2-x5)+150*(x3-x6) (2)

Equation (2) represent to the objective function which illustrated will be able to maximum value engineering for building construction costs under all constraints and only 3 types of construction material as concrete, rebar and formwork by that including material costs and labor costs which have a unit costs for material cost and labor cost, that are the assumption datas used in table 3 as follow:

Table3 . Unit costs of construction.

Materials	Unit	Material (THB)	Labor (THB)
Concrete quantity	(m^3.)	2500	350
Rebar quantity	(kg.)	22	6
Formwork quantity	(m^2.)	500	150

2.1.3 Constraints. This model, we have constraints all 16 equations as follows.

2500*x1+22*x2+500*x3 <= 1,000,000 (3)
350*x1+6*x2+150*x3 <= 500,000 (4)
2500*x4+22*x5+500*x6 <= 500,000 (5)

$$350*x4+6*x5+150*x6 \quad <= 100{,}000; \tag{6}$$
$$x1/x3 <= 6; \tag{7}$$
$$x1/x2 <= 0.008; \tag{8}$$
$$x2/x3 <= 20.8; \tag{9}$$
$$x4 >= 0.7*x1; \tag{10}$$
$$x5 >= 0.7*x2; \tag{11}$$
$$x6 >= 0.7*x3; \tag{12}$$
$$x1,x2,x3,x4,x5,x6 >=0; \quad \text{(Non negativity)} \tag{13}$$

Where ;
(3) is the material cost (THB.) of building construction costs which is assumption data before value engineering process.
(4) is the labor cost (THB.) of building construction costs which is assumption data before value engineering process.
(5) is the material worth (THB.) of building construction costs which is assumption data.
(6) is the labor worth (THB.) of building construction costs which is assumption data.
(7) is the relationshiped between concrete quantity with formwork quantity before value engineering process which is cost function.
(8) is the relationship between concrete quantity with rebar quantity before value engineering process which is cost function.
(9) is the relationship between rebar quantity with formwork quantity before value engineering process which is cost function.
(10) is the value engineering of the concrete quantity which will be reduced no more than 30% of original quantity which is worth function.
(11) is the value engineering of the rebar quantity which will be reduced no more than 30% of original quantity which is worth function.
(12) is the value engineering of the formwork quantity which will be reduced no more than 30% of original quantity which is worth function.
(13) are the all variables to be nonnegative.

3. Results.

This problem solving by the mathematical model using LINGO (Release17.0) software.

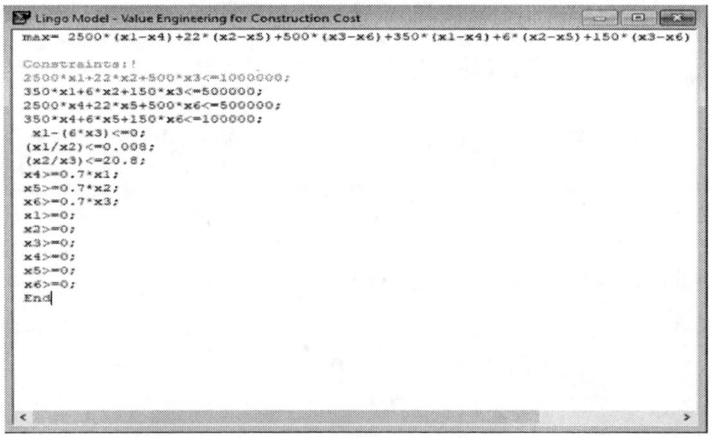

Figure3. is all equations which is input in Lingo from the software release(17.0) for solve the objective function that is the maximize value engineering of building costs under all constraints are concern cost function and worth function.

Table 4. The resulted of x1, x2, x3 are cost function variables.

No.	Materials	Quantity	Unit
x1	Concrete	71.37	(m^3.)
x2	Rebar	8,922.13	(kg.)
x3	Formwork	428.949	(m^2.)

Table 5. The resulted of x4, x5, x6 are worth function variables.

No.	Materials	Quantity	Unit
x4	Concrete	49.96	(m^3.)
x5	Rebar	6,245.49	(kg.)
x6	Formwork	300.26	(m^2.)

From table4 and table5 are a resulted of all decision variables that shows the resulted of worth variables (x4,x5,x6) are reduced from the cost variables (x1,x2,x3) according to the all constraints and represents the difference between the cost and worth variables in Table 6.

Table 6. As show the materials quantity reduced between cost function and worth function.

No.	Materials	Quantity	Unit
x1-x4	Concrete	21.41	(m^3.)
x2-x5	Rebar	2,676.64	(kg.)
x3-x6	Formwork	128.69	(m^2.)

Table6. illustrated the materials quantity reduced between cost function and worth function that can be reduced the quantity of construction materials as concrete, rebar and formwork b using linear programming application that is able to reduce the amount of concrete 21.41 m3 (x1-x4) reduced rebar was 2,676.64 kg.(x2-x5) and reduced formwork be 128.69 m3 respectively. The objective function is 219,618.4 THB. as figure.5 which meaning to the maximum value engineering for a building construction costs under all constraints ,that value displayed to a benefit for reduced of building construction costs and including construction materials that are reduced the quantity of materials as concrete, rebar and formwork which benefit to resources and costs management for building construction.

4. Conclusions.

The objectives function can be meet the targets which depend on all constraints and decision variables. The resulted of materials quantity will be worth function which meaning to the construction cost is reduced but the functions are remains the same. Therefore ,the value Engineering is an effective technique for reducing of building construction cost by apply with linear programming which resulted shown to achieved the targets for reducing building construction cost as much as possible that leads to efficiency of maximum value engineering and the resulted of all decision variables which is benefit for resource management of construction projects. The future research in the field of value engineering of construction for ,that can used to goal programming to be applied with the value engineering for construction to solved the best results of all goals under the corresponding conditions.

5. References.

[1] Barry Render ,Ralph M.Stair,Jr , Micheal E.Hanna. and Trevor S.Hale, *Quantitative Analysis for Management,* **12th** edition ,(published by Pearson Education @ 2015.)

[2] Alphonse Dell'Isola,PE,,*Value Engineering: Practical Application for Design, Construction , Maintenance & Operations*, (Construction Publisher&Consultants)

[3] Wayne L. Winston , *Operations Research Applications and Algorithms* **3rd** Edition, (Publisher: Duxbury Press)

[4] Robin Cooper and Regine Slagmulder, *Target Costing and Value Engineering*., Copy right@1977 by The IMA Foundation for Applied Research,Inc. (published by Productivity Press, a division of Productivity , Inc.)

[5] Alphonse J.Dell'Isola , *Value Engineering in the Construction Industry* 2**nd** edition, (Copy right@1974 by Construction Publishing Company)

[6] Hamdy A.Taha, *Operations Research An Introduction*, **9th** Edition , (publishing as Prentice Hall, Copyright@2011 Pearson Education, Inc., One Lake Street,Upper Saddle River, New Jersey , The United States of America.)

[7] John Kelly and Steven Male ,*Value Management in design and construction, The economic management of projects,* **1st** edition 1993

[8] Anderson, Sweeney and Williams , *An Introduction to Management Science, Quantitative Approaches to decision Making* , **7th** edition, (copyright@1994 by West publishing company.)

[9] S.Christian Albright ,Wayne L. Winston and Christopher Zappe , *Data Analysis & Decision Making with Microsoft Exce*l , **3rd** edition (copyright@2006 by The Thomson Corporation.)

Effect of Throat Size on Performance of Microbubble Generator and Waste Water Treatment

Xayalak Vilaida[1], Sengratry Kythavone[1] and Toshio Iijima[2]

[1]Department of Mechanical Engineering, National University of Laos, Vientiane, Laos

[2]Professor emeritus, Tokai University, Kanagawa, Japan

E-mail: x.vilaida@nuol.edu.la

Abstract. Microbubble can be widely applied to many fields such as waste water treatment from community, industry and agricultural. This study focus on set up experimental apparatus of microbubble generation to examine the optimum ratio between air to water for generating the best microbubble by using three different sizes of venturi with throat diameters of 3mm, 4mm and 5mm, respectively; and also for apply microbubble to treat community waste water. The venturi with throat diameter of 3mm can generate more microbubble than the other. When applied microbubbles to treat the sampling of the waste water, chemical oxygen demand (COD) of waste water was considered.

1. Introduction

Microbubbles are bubble smaller than one millimetre in diameter, but larger than one micrometre. A microbubble is smaller than a human hair (70 micrometres). They have widespread application in industry, life science, agriculture, and medicine. The composition of the bubble shell and filling material determines important design features such as buoyancy, crush strength, thermal conductivity and acoustic properties [1, 2, 3, 4].
Microbubbles actually may be used for drug delivery, biofilm removal, membrane cleaning, biofilm control and waste water treatment purposes. When pressured water is introduced into the generator, the velocity around the spherical body becomes extremely high compared to that at a generator exits; thus from the energy conversion principle the pressure a little downstream of the body becomes negative. With the aid of the negative pressure, air is automatically sucked through to small hole drilled around the pipe wall, and the air sucked is broken into a huge number of microbubbles by high shear introduced by water flow around the body. Thus, the microbubble generator can discharge a water jet with microbubble form the nozzle exit [5, 6, 7, 8].
The purpose of this research are to present and validate a physical model to predict the microbubble generator performance such as air to water suction rate and testing microbubble to waste water treatment form community.

2. Experimental apparatus and experimental procedure

2.1 Experimental Apparatus

Figure 1 shows the experimental apparatus that composes of following parts, 1; main water tank, 2; circulation water tank, 3; water pump, 4; venturi (main microbubble generator), 5; water regulation valve, 6; water flow meter of floater type, 7; air flow meter of floater type, 8; pressure gage, 9; electric circuit breaker, 10; vacuum gage, 11; bypass valve, 12; air regulating valve, 13; over flow pipe. The capacity of main water tank and circular water tank are 45 liters. Centrifugal pump power is 0.37kW with maximum pressure of 2 bar and maximum flow rate is 40 l/min. Venturi was placed at H= 0.15 m for all size of venturis. Figure 2 shows venturi with three different dimeter d.

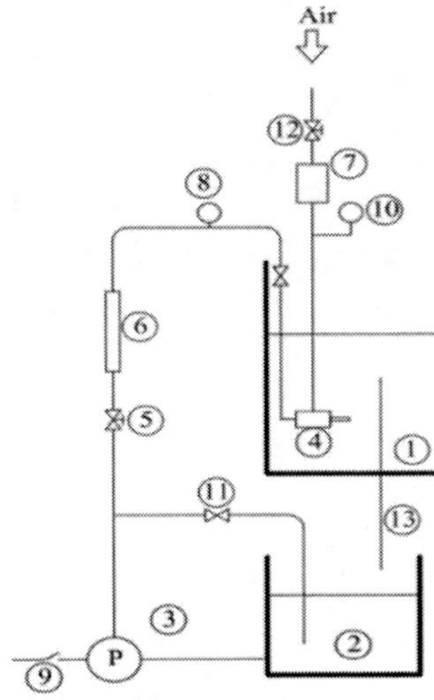

Figure 1 Experimental apparatus

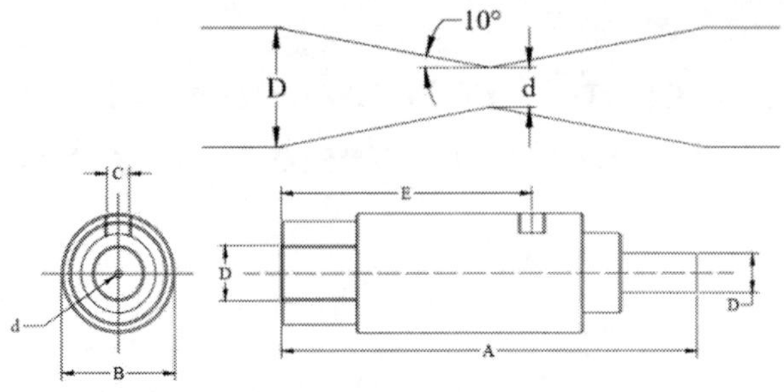

Figure 2 Venturi with three different dimeter d

2.2 Experimental Procedure

In the present research work, experimental was conducted in the fish aquarium with maximum depth of 0.65 m. The venturi was placed at 0.15m for all size. At each venturi, water was discharge from a centrifugal pump at 9 -20 l/min. Three different sizes of venturi were designed with throat diameter of 3mm, 4mm, and 5mm respectively. Two flow meters are installed for both water and air flow measurements with pipe fitting system.

Table1 Specifications of venturi

Total length A (mm)	Outside dia. B (mm)	Air Section dia. C (mm)	Throat dia. d (mm)	Water inlet dia. D (mm)	Air suction pos. E (mm)
100	30	12	3	12	52
100	30	12	4	13	52
100	30	12	5	14	52

The microbubble generator test and their specification are shown in Figure 2 and Table 1. The three microbubble generators had the same total length, outside diameter, suction air diameter and air suction position, but different air throat diameter namely d=3mm, 4mm and 5mm are used.

3. Results and discussions

Table 2 shows the measurement result of mass of air in water. The two venturis with 3 mm of throat's diameter is applying to water. The result was tested after 1 hour and 2 hours with varying water flow rate (9l/min – 16 l/min) and air suction rate (1.35 l/min – 4 l/min). Figure 3 shows the measurement result by using pure water. The two venturis with 3 mm, 4mm and 5mm of throat diameter were applying to water. The result was measured after 1 hour and 2 hours with varying water flow rate (9l/min – 16 l/min) and air suction rate (1.35 l/min – 4 l/min). Hence, the 15% of air to water ratio can generate up to 13.735 mg/l for two hours. The mass of air in water decreases when air to water ratio increase to 25% and it can generate around 2.773 mg/l.

Table2 Mass of air in water

T (h)	Q_w (l/m)	Q_a (l/m)	Q_a/Q_w	M_w (mg)	M_a (mg)	M_a/M_w (mg/l)
				9.8432[a]	0	0
1	9	1.35	0.15	9.7884	0.0548	5.567
2	9	1.35	0.15	9.7080	0.1352	13.735
1	12	2.4	0.20	9.8099	0.0333	3.381
2	12	2.4	0.20	9.7845	0.0587	5.964
1	16	4	0.25	9.8159	0.0273	2.773
2	16	4	0.25	9.7762	0.0670	6.807

[a]Raw sample

The result of application via applying four microbubble generators for treatment of waste water from community. The sample was measured in every hour. Figure 3 compares mass of air in water for the air to water ratio 15%, 20% and 25%, respectively. The mass of air in water increase with time series. The air to water ratio 15% shows best result. The mass of air was increased from 10.311 mg/l after one hour and it reach up to 22.852 mg/l after six hours. It means the quality of waste water is improved. The quality of waste water that 25% of air to water ratio is lower than 20% of air to water ratio. The mass of air is increased form 4.575 mg/l after one hour and it reach up to 20.027 mg/l after six hours.

The application of venturi for 3 mm of throat diameter were shown in the figure 5. Community waste water was used as substances and the experiment was used 4 venturi together for 15% of air to water ratio. The result shown that the colour of waste water was changed from black to clay colour after 2 hours. It means oxygen in waste water increase whereas chemical oxygen demand (COD) reduce. In addition, the quality of waste water was improved steadily with time series. The experimental of treated waste water operates up to six hours.

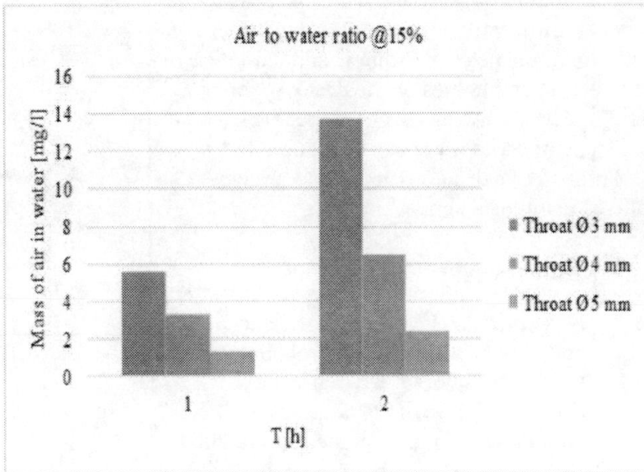

(A) Air to water ratio is 15%

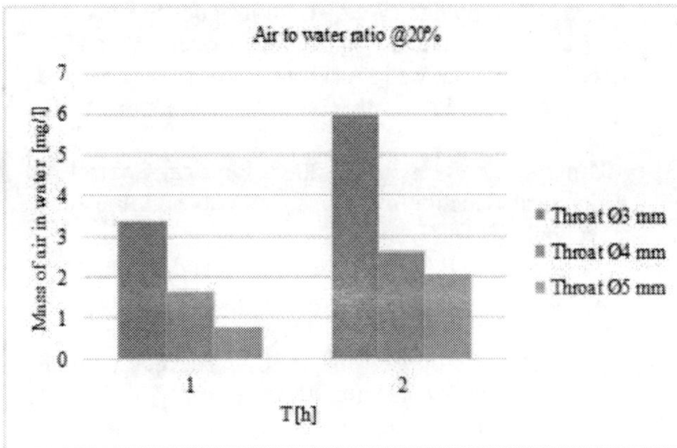

(B) Air to water ratio is 20%

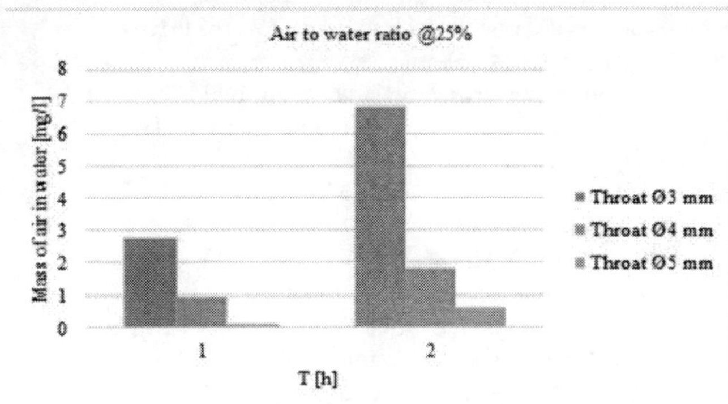

(C) Air to water ratio is 25%

Figure 3 Mass of air in water for each nozzle size and ratio of air to water

4. Conclusions

From the Figure 3 and Figure 4 all size of throat diameter of venturi for air to water ratio of 15% can generate more microbubbles, and throat diameter of venturi with 3mm can generate more microbubble compare with other ones. For 2 hours of experiment the mass of microbubble or air in the water generated by venturi with throat diameter of 3mm, 4mm, and 5mm for air to water ratio of 15% are 0.1352mg (equivalent to 13.735mg/l), 0.0638mg (equivalent to 6.481mg/l), and 0.0236mg (equivalent to 2.397mg/l), respectively for volume of water 9.8432mg, therefore for applying microbubbles to treat the community waste water venturi with throat diameter of 3mm, and air to water ratio of 15% were used. Figure 5 shows experimental result before and after treatment.

When applied microbubbles to treatment of the community waste water, COD of waste water significantly reduces. The venturi with throat diameter of 3 mm is applied to generate microbubbles with air to water ratio of 15%, 20%, and 25%, respectively and COD of waste water can be reduced by 55%, 43.5%, and 29.5% for air to water ratio as mentioned above.

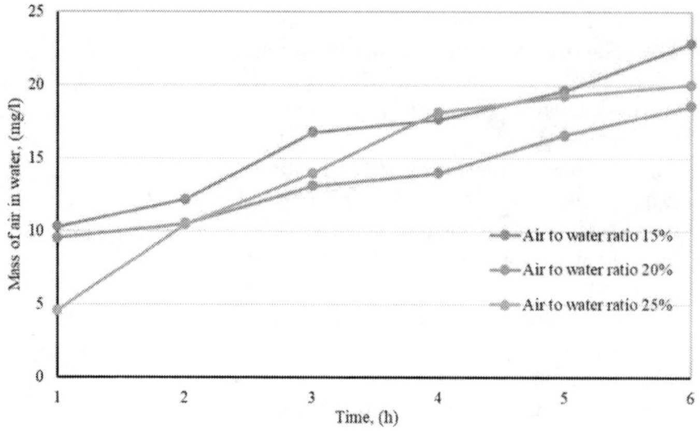

Figure 4 Mass of air in water with 3mm throat diameter

(A) Waste water

(B) Treated wast water

Figure 5 Waste water treatment by using 4 venturis of 3 mm throat diameter

5. References

[1] Serizawa A. Flow characteristics and application of microbubble containing bubble two-phase flow, Proc. Of the 16 int. Congress of Chem. And Process Eng. – Prague, Czech. 2004.

[2] Eriksson, J.C. and Ljunggren S., 1999, On the Mechanically Unstable Free Energy Minimum of a Gas Bubble which is Submerged in Water and Adheres to a Hydrophobic Wall, Colloid and Surface A: Physicochemical and Engineering Aspects, 159: 159–163J. Clerk Maxwell, A Treatise on Electricity and Magnetism, 3rd ed., vol. 2. Oxford: Clarendon, 1892, pp.68–73.

[3] Kimura, T. and Ando, T., 2002, Physical Control of Chemical Reaction by Ultrasonic Waves, Ultrasonic Technology, 14: 7–8.

[4] Kobayashi, F., Ikeura, H., Tamaki, M. and Hayata, Y., 2009a, Application of CO Micro- and Nano-Bubbles at Lower Pressure and Room Temperature to Inactivate Microorganisms in Cut Wakegi (*Allium wakegi* Araki), Acta Horiculturae, 875: 417-424.

[5] Kobayashi, F., Hayata, Y., Ikeura, Y., Tamaki, M., Muto, N. and Osajima, Y., 2009b, Inactivation of Escherichia coli by CO Microbubbles at a Lower Pressure and Near Room Temperature, American Society of Agricultural Engineers, 52: 1621-1626.

[6] Y. Yorozu, M. Hirano, K. Oka, and Y. Tagawa, "Electron spectroscopy studies on magneto-optical media and plastic substrate interface," IEEE Transl. J. Magn. Japan, vol. 2, pp. 740–741, August 1987 [Digests 9th Annual Conf. Magnetics Japan, p. 301, 1982].

[7] Pengphol S., J. Uthaibutra, O.A. Arqueron, N. Nomura and K. Whangchai. 2012. Reduction of residual chlopyrifos on harvested bird chilies (Capsicum frutescenes Linn.) using ultrasonication and ozonation. Thai Journal Agricultural Science 44(5): 182-187.M. Young, The Technical Writer's Handbook. Mill Valley, CA: University Science, 1989.

[8] Zhang, H., L. Duan and D. Zhang. 2006. Absorbtion kinetics of ozone in water with ultrasonic radiation. Ultrasonics Sonochemistry 14: 552-556.

Acknowlegement

We sinrely thank to Daido Kogyou Cooporation, Japan and Professor Dr. Katsumi Aoki of Tokai University for giving us very important information microbubble generation experiment setup.

IOP Conf. Series: Materials Science and Engineering **639** (2019) 012032 doi:10.1088/1757-899X/639/1/012032

Effect of Climate Change on Groundwater Age of Thailand's Lower Chao Phraya Basin

Pinit Tanachaichoksirikun[1] **and Uma Seeboonruang**[2]

[1] Ph.D. Student, Department of Civil Engineering, Faculty of Engineering, King Mongkut's Institute of Technology Ladkrabang, Bangkok, Thailand
[2] Associate Professor, Department of Civil Engineering, Faculty of Engineering, King Mongkut's Institute of Technology Ladkrabang, Bangkok, Thailand

57601070@kmitl.ac.th; uma.se@kmitl.ac.th

Abstract. Water is indispensable for life, including, groundwater that the largest fresh water source. Groundwater sustainability is threatened by many factors, one is climate change which lead to flood and drought. Groundwater age is an indicator that can be applied for vulnerability assessment and sustainable groundwater management. This research focus on the changing of groundwater age under the effect of climate change using the steady-state three-dimensional mathematical simulation including, MODFLOW-2000 and MODPATH. In addition, the climate scenario including, IPSL-CM5A-MR that consist the increasing of carbon dioxide (Representative Concentration Pathways; RCP) between 2.6, 4.5 and 8.5 in the period 2017-2036. The results revealed groundwater age was highly distributed between 140 to 177,505 years with the average 18,665 years due to the distribution of groundwater recharge and pumping in the basin. In addition, the average groundwater age with RCP 2.6, 4.5 and 8.5 were decreased to 17,217, 15,960, 16,286 years from base case (18,665 years), respectively. Because the quantity of rainfall which contribute the hydraulic head, hydraulic gradient and velocity was changed. In conclusion, the groundwater sustainability in the Lower Chao Phraya basin was consistent during this period because the groundwater age was decreased by rainfall. However, the old groundwater, non-renewable groundwater could be conservative due to the distribution of groundwater age was increased.

1. Introduction

Water resource is the important resource for alive. Especially, Groundwater is the largest fresh water source in the world but the increasing of carbon-dioxides gas in the atmosphere affect the global and local climate change. Additionally, the global climate change affects the world the agricultures and water resource (Ali et al., 2012 [1]; Kumar, 2014 [2]; Pholkern et al., 2018 [3]). So, plants, food and water are become low quality and may be facing the shortage which is the anxious situation.

The global climate change highly effects the tropical climate in term of groundwater quality and quantity, sustainability and vulnerability assessment (Green et al., 2011 [4]; Taylor and Callist, 2012 [5]; Chris et al., 2012 [6]; White and Tony, 2012 [7]). Thailand located in the tropical climate. Then, groundwater resource in Thailand may be facing these problems. Several researches investigated the climate change effect on groundwater resources for example, Saraphirom et al. (2013) [8], Srisuk and Nettasana (2017) [9]; Tanachaichoksirikun et al. (2018) [10]. Although, the research about the groundwater management for groundwater sustainability under the global climate change is very important, the indicators of groundwater management are not approached.

Groundwater age is the time elapsed since the water molecule entered the groundwater basin until the water molecule located in the interesting position (Cook and Herczeg, 2000 [11]; Kazemi et al., 2006 [12]) which is useful such as, infer recharge rates (e.g. Sanford et al., 2004 [13]) and to sustainably exploit groundwater resources, analyze aquifer vulnerability (Manning and Solomon, 2005 [14]; Bethke and Johnson, 2008 [15]; Molson and Frind, 2012 [16]; Zhu et al., 2016 [17]; Sonnenborg et al., 2016 [18]) and understand the complex flow system (Troldborg et al., 2008 [19]; Ebert et al., 2012 [20]).Tanachaichoksirikun and Seeboonruang (2015) [21] reported the average age of groundwater was inclined, although the young and old groundwater age was mixed. Because the quantity of older groundwater was higher than the younger groundwater. however, the effect of groundwater pumping decreased groundwater age, the old groundwater may be was extracted. Although, the groundwater age is excellent indicator of groundwater, it never used to describe the effect of climate change.

Thailand's Lower Chao Phraya (LCP) basin was selected for this research by three reasons 1) it is the largest groundwater basin in Thailand and also located in the central part of Thailand that cover Bangkok (Capital City of Thailand). 2) there are a lot of groundwater monitoring stations, the calibration and verification are efficient. 3) this area faced the groundwater over-extraction, its effect contributed to the land subsidence. These reasons show that the LCP basin is important for groundwater management. Then, if the correlation between climate change and groundwater age is determined, the groundwater planning for vulnerability and sustainability will be efficient.

Then, this current research investigated the groundwater age under the effect of variable climate change of the Thailand's Lower Chao Phraya Basin to evaluate the changing and sensitivity of groundwater age and suggest the preliminary groundwater management under the effect of climate change.

2. Material and method

2.1. Modelling framework

The solution approach to steady-state groundwater flow equation and particle tracking are present here. The general form of the equation for groundwater flow over the physical dimension is

$$\frac{\partial}{\partial x}\left[K_x\frac{\partial h}{\partial x}\right]+\frac{\partial}{\partial y}\left[K_y\frac{\partial h}{\partial y}\right]+\frac{\partial}{\partial z}\left[K_z\frac{\partial h}{\partial z}\right]+W=0 \quad (1)$$

where Kx, Ky, Kz are the hydraulic conductivity in x, y, z dimension, respectively (LT^{-1}), $h(x, y, z, t)$ is the hydraulic head (L), W is sink or source (L^3T^{-1}/L^3).

Additionally, the solution of the particle tracking equation over the physical dimension is

$$(x_p)_{t_2} = x_1 + \frac{1}{A_x}\left[(v_{x_p})_{t_1} e^{(A_x\Delta t - v_{x_1})}\right] \quad (2a)$$

$$(y_p)_{t_2} = y_1 + \frac{1}{A_y}\left[(v_{y_p})_{t_1} e^{(A_y\Delta t - v_{y_1})}\right] \quad (2b)$$

$$(z_p)_{t_2} = z_1 + \frac{1}{A_z}\left[(v_{z_p})_{t_1} e^{(A_z\Delta t - v_{z_1})}\right] \quad (2c)$$

where x_p, z_p, z_p are the last position of particle (L), x_1, y_1, z_1 are the initial position of particle (L), A_x, A_y, A_z are velocity gradient,$v_x, v_y, v_z$are velocity in x, y, z dimension, respectively.

In this research, Modular Finite-Difference Ground-water Flow Model-2000; MODFLOW-2000 (Harbaugh et al., 2000 [22]) was selected to determine the groundwater flow field. Because the application and determination described groundwater flow in the complex problem. In addition, A Particle-Tracking Model for MODPATH version-6 (Pollock, 2012 [23]) was carried out to report the moving and age of groundwater due to the conservative of groundwater age and efficient for determine the changing of groundwater age.

2.2. Study area

Thailand's Lower Chao Phraya (LCP) basin locates in the central plain, which cover 21 provinces including, Bangkok. The shape like triangle which the tip is in Chai Nat province and widespread to Chao Phraya delta region near the Gulf of Thailand. The plain is approximately 230 km from the east-west and about 300 km for the north-south, the total area of 43,317 km^2 (Fig. 1.). The western bounds the Tenasserim hills. The northern is connected to the Upper Chao Phraya basin which Chao Phraya river connected. The eastern part is series of small hill. The south side is near to the Gulf of Thailand. The area consists 4 major rivers i.e., Chao Phraya, Pa Sak, Tha Chin and Mae Klong rivers. The basin is relatively flat with water flow from the north to south.

The hydrogeological of the Lower Chao Phraya basin is sedimented from the Tertiary to Quaternary strata of Lower Central Plains. The groundwater system consists of sand, gravel and lensed of clay of Pliocene-Pleistocene-Holocene age. The aquifers are predominantly old flood-plain 8 principal confined aquifers with 30-70 m thick. The top subsurface is overlain by the Bangkok Clay with 10-20 m. The total depth of the groundwater system is approximately 500 meters. The 2nd (Phra Pradeang; PD), 3rd (Nakorn Luang; NL), 4th (Nonthaburi; NB) aquifers that the major groundwater pumping wells for consumption agriculture and industry were focused in this research. Since the groundwater was over-extracted, the subsequence land subsidence was occurred in this area.

The climate records of Lower Chao Phraya monitoring stations during 1983-2014 indicated that the average annual rainfall had increased from 1,191 mm (1983-1998) to 1,208 mm (1999-2014) and the average temperature had increased from 27.1 °C (1983-1998) to 27.4 °C (1999-2014).

The projected climate scenario weather data during 2017-2036 was generated by Wattanasetpong et al. (2015) [24] based on 30 years weathering records of Thailand's Lower Chao Phraya monitoring stations. The six criteria which consist the annual rainfall, average temperature, pressure, humidity, evaporation and groundwater discharge were selected because they are the important climate change factor. In this research, Ruangrassamee et al. (2015) [25] described the climate scenario IPSL-CM5A-MR that was developed by the Institute Pierre Simon Laplace offered the minimum root mean square error and minimum bias in the Chao Phraya watershed. So, this model was chosen. Additionally, the Representative Concentration Pathways (RCP) was selected between 2.6, 4.5 and 8.5 because the quantity of carbon dioxide gives the contrasted of the quantity of rainfall.

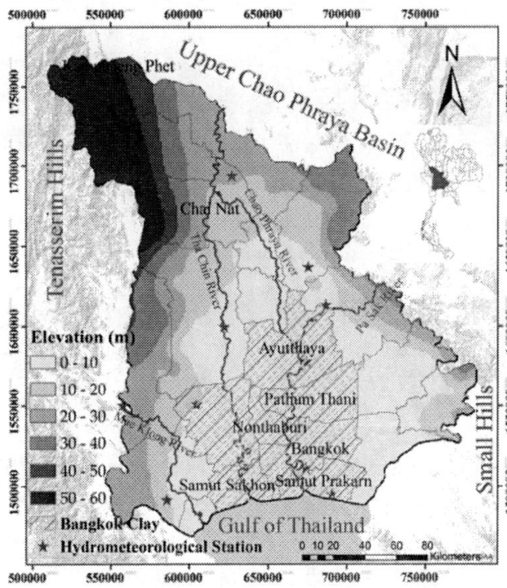

Figure 1. Topographical map of LCP basin (Tanachaichoksirikun et al., 2018 [10])

2.3. Methodology

2.3.1 Groundwater simulation

2.3.1.1 Model construction and boundary conditions
The LCP basin was distributed to nine strata including, eight aquifers and the top stratum was Bangkok clay (BKC) which individually discretized to 350 × 300 (row × column) with grid size of 1 × 1 km^2. The constant head was applied in the south for the Gulf of Thailand. The north was constant for groundwater flux from Upper Chao Phraya basin. The river boundary was applied for 4 rivers. the others were no-flow boundary. In addition, the climate was treated as recharge boundary and the groundwater extraction was treated as pumping boundary.

2.3.1.2 Groundwater parameters
The groundwater parameter is an important, but the numerical groundwater model cannot determine directly. So, the inverse simulation needs to be performed the groundwater parameters; effective porosity (Φ), horizontal hydraulic conductivity (K_h) and vertical conductivity (K_v). The groundwater parameters were shown in the Table 1.

Table 1. Groundwater parameter in the LCP basin

Aquifers	**Φ**	**K_h (m/s)**	**K_v (m/s)**
Bangkok Clay	0.03	$1x10^{-8}$-$1x10^{-9}$	$1x10^{-8}$-$1x10^{-9}$
Bangkok	0.2-0.3	$5x10^{-5}$-$6x10^{-6}$	$5x10^{-9}$-$6x10^{-10}$
Phra Pradeang	0.25-0.35	$2x10^{-5}$-$7x10^{-6}$	$2x10^{-9}$-$7x10^{-10}$
Nakorn Luang	0.2-0.35	$1x10^{-4}$-$6x10^{-6}$	$1x10^{-8}$-$6x10^{-10}$
Nonthaburi	0.3-0.35	$2x10^{-4}$-$7x10^{-6}$	$2x10^{-8}$-$7x10^{-10}$

2.3.1.3 Calibration and Verification

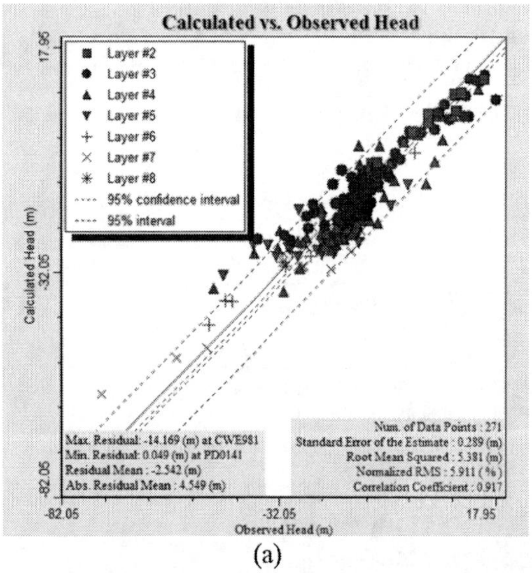

(a)

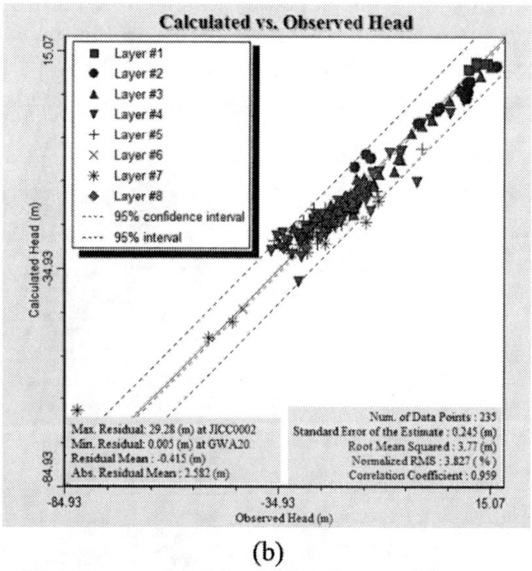

(b)

Figure 2. Calculated vs observed head (a) calibration model (b) verification model

The calibration and verification of the groundwater model were illustrated in the Figure 2. In the calibration model, the simulated hydraulic heads were compared to the measured hydraulic heads in the period 2009-2014. The absolute residual mean error was 4.59 m. and the root mean squared error was 5.38 m. and the normalized root mean squared error was 5.91%. In addition, the verification model, the simulated hydraulic heads were compared to the measured hydraulic heads in the period 2007-2008.

The absolute residual mean error was 2.58 m and the root mean squared error was 3.77 m and the normalized root mean squared error was 3.82%.

2.3.2 Modelling scenarios
The climate scenario, IPSL-CM5A-MR consists the Representative Concentration Pathways (RCP) 2.6, 4.5 and 8.5. The average annual rainfall of the scenarios based on the period 2017-2036 were 1,182, 1,258 and 1,227 mm, under RCP 2.6, 4.5 and 8.5, respectively, while the annual rainfall has the increasing trend by 1.85, 2.64 and 12.78 mm/year, respectively. Then IPSL-CM5A-MR with RCP 4.5 has the greatest annual rainfall while IPSL-CM5A-MR with RCP 8.5 has the highest trend of annual rainfall in this period.

The base case period (2009-2014) was used as the control model for check the effect of climate change on the groundwater age of Thailand's LCP basin. Additionally, the research was suggested the groundwater sustainability assessment in the basin. In the climate scenario which is IPSL-CM5A-MR of RCP 2.6, 4.5 and 8.5, monthly precipitation was average during 2017-2036. The assessment was analyzed with regard to the reaction of groundwater age. In addition, the groundwater extraction in the future was assumed the same pumped at the base case. Because the area has the low increasing pumping rate during this period.

2.3.3 Particle tracking
The particles which assume as the water molecule were randomly assigned in the PD NL and NB layers. The backward tracking was analyzed because it easy to compare with the existing research that groundwater was tracked at the observation location. In addition, the predicted groundwater age was tracked by varied the climate scenarios and backward tracking method. Because the changing of groundwater age was affected the groundwater location of and velocity.

3. Results and discussion

3.1. Observed and modelled age
Table 2. shows the simulated groundwater age in the Lower Chao Phraya basin and comparing with Sanford and Buapeng (1996) [26] age. the average groundwater age in the PD, NL and NB aquifers were 6,031, 12,466 and 33,325 years and the average age was 18,665 years. The average groundwater age can comparable with Sanford and Buapeng (1996) [26] that the absolute error 7.58%. In addition, groundwater age which computed by MODPATH-6 (Pollock, 2012 [23]) was higher distributed than Sanford and Buapeng (1996) [26] age because the starting point of particles were randomly distributed all of the model domain while the starting point of Sanford and Buapeng (1996) [26] age located at the observation well in the model.

Table 2. Simulated groundwater age in the LCP basin

Aquifers	**Groundwater age (years)**		
	Sanford and Buapeng (1996) [26] model	MODPATH-6	Average age
PD	4300-35000	180-31281	6031
NL	10500-40000	140-74959	12466
NB	12100-34700	215-177505	33325
Average	20195		18665

3.2. Projected groundwater age
Table 3. shows the predicted groundwater age in the Lower Chao Phraya basin under the base case (2009-2014) and the variable climate change (IPSL-CM5A-MR) including RCPs 2.6, 4.5 and 8.5 with the PD, NL and NB aquifers. the finding revealed the groundwater age were decreased from 18,665 to 17,217, 15,960 and 16,286 years under the IPSL-CM5A-MR with RCPs 2.6, 4.5 and 8.5, respectively.

Because the largest average rainfall which contribute the hydraulic head, hydraulic gradient and groundwater velocity. the velocity direct effects on the groundwater age. In addition, the distribution of groundwater in PD, NL and NB aquifers was decreased due to the distribution of groundwater recharge and pumping in each aquifer. Moreover, the average groundwater age in each aquifer was decreased from base case indicating, the groundwater in this period still sustainable.

Table 3. Predicted groundwater age in the LCP basin

Climate	**Groundwater age (years)**			
	PD	NL	NB	Mean
Base Case	180-31281	140-74959	215-177505	18665
Average	6031	12446	33325	
IPSL RCP 2.6	189-30838	124-58148	212-152495	17217
Average	6347	12308	34792	
IPSL RCP 4.5	192-19713	123-72311	214-151586	15960
Average	5756	11593	32210	
IPSL RCP 8.5	191-20385	122-72040	213-144167	16286
Average	5834	11684	33064	

4. Conclusion

This research has investigated the effect of variable climate on the groundwater age and the preliminary suggestion of Thailand's Lower Chao Phraya (LCP) basin. The climate scenario which IPSL-CM5A-MR was varied Representative Concentration Pathway (RCP) between 2.6, 4.5 and 8.5 during periods 2017-2036. The simulations were performed using MODFLOW-2000 and MODPATH in steady-state condition. The findings revealed that the average groundwater age was decreased. Specifically, in the simulation IPSL-CM5A-MR RCP 4.5 had the lowest average age as a result of the largest average rainfall which contribute the hydraulic head, hydraulic gradient and groundwater velocity. the velocity direct effects on the groundwater age. In addition, the Lower Chao Phraya basin had the higher distribution of groundwater age due to the distribution of groundwater recharge and pumping in the area. Moreover, the average groundwater age was lower than the groundwater age in the past indicating, the groundwater in the basin still sustainable during this period (2017-2036). However, the old groundwater, non-renewable groundwater could be conservative due to the distribution of groundwater age was increased. Additionally, this research was caries the particle tracking (MODPATH) model that dispersion and diffusion were ignored.

5. References

[1] Ali R, McFarlane D, Varma S, Dawes W, Emelyanova I, Hodgson G and Charles S 2012 Potential climate change impacts on groundwater resources of south-western Australian Australia *J. Hydrol.* **475** 456–472

[2] Kumar C P 2014 Impact of climate change on groundwater resources *Handbook of research on climate change impact on health and environmental sustainability* 196–221

[3] Pholkern K, Saraphirom P and Srisuk K 2018 Potential impact of climate change on groundwater resources in the Central Huai Luang Basin, Northeast Thailand *Sci. Total Environ.* **633** 1518–1535

[4] Green,T R, Taniguchi M, Kooi H, Gurdak J J, Allen D M, Hiscock K M, Treidel,H and Aureli A 2011 Beneath the surface of global change: Impacts of climate change on groundwater *J. Hydrol.* **405** 532–560

[5] Taylor R and Callist T 2012 The impacts of climate change and rapid development on weathered crystalline rock aquifer systems in the humid tropics of sub-Saharan Africa: Evidence from south-western Uganda *Clim. Chang. Eff. Groundw. Resour. A Glob. Synth. Find. Recomm.* 17–32

[6] Chris M H, Harm D, Diana M A and Dirk K 2012 Groundwater recharge and storage variability in southern Mali *Clim. Chang. Eff. Groundw. Resour. A Glob. Synth. Find. Recomm.* (London: United Kingdom)
[7] White I and Tony F 2012 Reducing groundwater vulnerability in Carbonate Island countries in the Pacific *Clim. Chang. Eff. Groundw. Resour. A Glob. Synth. Find. Recomm.* 75–110
[8] Saraphirom P, Wirojanagud W and Srisuk K 2013 Potential Impact of Climate Change on Area Affected by Waterlogging and Saline Nagarajan Nagarani, Arumugam Kuppusamy Kumaraguru, Velmurugan Janaki D *EnvironmentAsia* **77** 19–28
[9] Srisuk K and Nettasana T 2017 Climate change and groundwater resources in Thailand *J. Groundw. Sci. Eng.* **5** 67–75
[10] Tanachaichoksirikun P, Seeboonruang U and Saraphirom P 2018 Impact of climate change on the groundwater sustainability in the Lower Chao Phraya Basin, Thailand *Proc. 4th Int. Conf. Eng. Applied Sci. Technol.* 119–122
[11] Cook P G and Herczeg 2000 *A environmental tracers in subsurface hydrology* Kluwer Academic Publishers (Boston: USA)
[12] Kazemi G A, Lehr J H and Pierre P 2006 *Groundwater Age*. John Wiley & Sons
[13] Sanford W E, Plummer L N, Mcada D P, Bexfield L M and Anderholm S K 2004 Hydrochemical tracers in the middle Rio Grande Basin, USA:2. Calibration of a groundwater-flow model *Hydrogeol. J.* **12** 389–407
[14] Manning A H and Solomon D K 2005 An integrated environmental tracer approach to characterizing groundwater circulation in a mountain block *Water Resour. Res.* **41** 1-18
[15] Bethke C M and Johnson T M 2008 Groundwater age and groundwater Age dating *Annu. Rev. Earth Planet. Sci.* **36** 121–152
[16] Molson J W and Frind E O 2012 On the use of mean groundwater age, life expectancy and capture probability for defining aquifer vulnerability and time-of-travel zones for source water protection *J. Contam. Hydrol.* **127** 76–87
[17] Zhu Y, Shi L, Wu J, Ye M, Cui L and Yang J 2016 Regional quasi-three-dimensional unsaturated-saturated water flow model based on a vertical-horizontal splitting concept *Water* **8**
[18] Sonnenborg T O, Scharling P B, Hinsby K, Rasmussen E S and Engesgaard P 2016 Aquifer vulnerability assessment based on sequence stratigraphic and 39Ar transport modeling *Ground Water* **54** 214–230
[19] Troldborg L, Jensen K H, Engesgaard P, Refsgaard J C and Hinsby K 2008 Using environmental tracers in modeling flow in a complex shallow aquifer system *J. Hydrol. Eng.* **13** 1037–1048
[20] Eberts S M, Böhlke J K, Kauffman L J and Jurgens B C 2012 Comparison of particle-tracking and lumped-parameter age-distribution models for evaluating vulnerability of production wells to contamination *Hydrogeol. J.* **20** 263–282
[21] Tanachaichoksirikun P and Seeboonruang U 2015 Preliminary study on exposure time model for groundwater age in Lower Chao Phraya Basin *Proc. The 6th National Convention on Water Resour. Eng. (NCWRE6)* 1–12
[22] Harbaugh A W, Banta E R, Hill M C and McDonald M G 2000 *MODFLOW-2000, The U.S. Geological Survey modular ground-water model-user guide to modularization concepts and the ground-water flow process*
[23] Pollock D 2012 *User guide for MODPATH version 6: a particle tracking model for MODFLOW: Tech. Methods 6–A41* 58
[24] Wattanasetpong J, Charoenvaravut P and Laosinwattana W 2012 Downscaling climate models in Thailand by artificial neural network method *Thesis of Civil Eng. KMITL* (Bangkok: Thailand)
[25] Ruangrassamee P, Khamkong A and Chuenchum P 2015 Assessment of precipitation simulations from CMIP5 climate models in Thailand Proc. *The 3rd EIT Int. Conf. on Water Resour. Eng. (ICWRE3)* 1-9
[26] Sanford W E and Buapeng S 1996 Assessment of a ground water flow model of the Bangkok Basin, Thailand, using carbon 14 based ages and paleohydrology *Hydrogeol. J.* **4** 26–40

IOP Conf. Series: Materials Science and Engineering **639** (2019) 012032 doi:10.1088/1757-899X/639/1/012032

Acknowledgement
The authors would like to express the special thanks of gratitude to the Royal Golden Jubilee PhD Program (Grant No. PHD/0189/2556) from Thailand Research Fund for the financial support who gave the golden opportunity to do the wonderful project and sincere appreciation to the Thailand Meteorological Department for the Meteorological data, Thailand Land Development Department for the topographic map and Thailand Department of Groundwater Resources for the data and many things.

IOP Conf. Series: Materials Science and Engineering **639** (2019) 012033 doi:10.1088/1757-899X/639/1/012033

Uncertain External Pressure Act on Thick Cylinder in Fuzzy Context

W Yukunthorn[1] **and J Tariboon**[2]

[1]Faculty of science and technology, Kanchanburi Rajabhat University, Kanchanaburi, Thailand
[2]Department of mathematics, faculty of applied science, King Mongkut's University of Technology North Bangkok, Bangkok, Thailand

E-mail: w.yukunthorn@kru.ac.th

Abstract. In this paper, we study a thick cylinder with uncertain external pressure. The traditional model was extended in fuzzy context. We use the governing equation which was rewritten as delta operator equation. The interpretations of external pressure are based on triangular, trapezoidal and Gaussian fuzzy numbers. By Adomian Decomposition Method gives an approximate solution. The maximal and minimal radial stresses are obtained, corresponding to the level cut. Finally, we get the methodology for simulating uncertain problems in fuzzy context and utility of each membership function.

1. Introduction

Many theoretical mechanics are interpreted by concept of calculus. In solid mechanics, calculus is used to describe deformation of solid. The equilibrium of forces can be written as governing equation with fact conditions, called initial value problems or boundary valued problems. But, most of models are based certain system.

In fact, we avoid the uncertainty in nature by design and control our experiments. The other method, we may use probability theorem to understand the randomness of existence of events. In addition to both previous concepts, considering the possibility of the fact is an alternative way. It is well known as fuzzy systems.

Mass flow in tube is one of the problem, which concerns uncertainty of system and environment [1], [2]. The governing equation of the thick cylinder with radius $r \in [a, b]$ subjects to internal pressure p_i and external pressure p_0 is form of differential equation: $u''(r) + u'/r - u/r^2 = 0$ subjects to boundary conditions, $\sigma(u(a)) = -p_i$ and $\sigma(u(b)) = -p_o$, where $u(r)$ is displacement and $\sigma(r)$ is stress at the radius r. If atleast one of parameters in previous problem is vague and we solve it in fuzzy context. This problem becomes fuzzy differential equation.

Fuzzy differential equation is a concept of differential equation on fuzzy space. The quantities are indicated by a grade of membership function in sense of Zadah's fuzzy set and logic [3]. The researchers still continue to develop theory and methods of fuzzy differential equations.

Recently, the governing equation of thick cylinder was developed from derivative to delta operator, defined by $\delta u(r) = ru'(r)$. In this paper, we organize concepts and methods for simulating thick cylinder with uncertain external pressure by delta operator in Fuzzy Context. In the next section, we recall delta operator model and its computational method, then we extend the problem in real space to fuzzy space under triangular, trapezoidal and Gaussian membership functions. Their simulation results are presented in section 4. Finally, we conclude and discuss our results in this work.

2. Mathematical model and its computation

Through out this section, the mathematical and theoretical framework are omitted. We refer some results which is used to investigate the simulation, see [4] for more details.

2.1. Delta operator model

Let $u : [a, b] \to R$ and its second order delta derivative exists. The stress distribution in thick cylinder can be written as following delta operator equation:

$$\delta^2 u(r) = u(r) \tag{1}$$

subject to

$$\delta u(a) - \left(\frac{\nu}{1-\nu}\right) u(a) = -k_1 \text{ and } \delta u(b) - \left(\frac{\nu}{1-\nu}\right) u(b) = -k_2,$$

where,

$$k_1 = ap_i(1+\nu)(1-2\nu)/(E(1-\nu)), k_2 = bp_o(1+\nu)(1-2\nu)/(E(1-\nu)). \tag{2}$$

2.2. n-terms ADM approximation

The closed-form of the n-terms ADM approximated solution of 1, given by

$$u^n = D_1^n \left(\sum_{k=0}^{n} \frac{\ln^{2k+1}(r/a)}{(2k+1)!} \right) + D_2^n \left(\sum_{k=0}^{n} \frac{\ln^{2k}(r/a)}{(2k)!} \right) = D_1^n S_O^n(r) + D_2^n S_E^n(r).$$

Moreover, 2-terms ADM approximations are close to the exact solution. Here, we use $u^2 \approx u(r)$.

3. Fuzzy Model

In this section, we assume that the external pressure is uncertain quantity ($p_2 \in E$). Note that E is fuzzy set. We describe this vague quantity into 3 types of fuzzy number: 1) Triangular Fuzzy Number (TFN), 2) Trapezoidal Fuzzy number (TrFS) and 3)Gaussian Fuzzy Number (GFN)[5]. Let $u : [a, b] \to E$ be a d monotone fuzzy function and its second order δ derivative exists. The stress distribution in thick cylinder with uncertain external pressure is represented as follow:

$$\delta^2 u(r) = u(r) \tag{3}$$

subject to

$$\delta u(a) - \left(\frac{\nu}{1-\nu}\right) u(a) = -k_1 \in E \text{ and } \delta u(b) - \left(\frac{\nu}{1-\nu}\right) u(b) = -k_2 \in E.$$

Applying t-cut representation, the fuzzy solution is in form of $u(r) = [\underline{u}(r), \bar{u}(r)] \approx [\underline{u}^2, \bar{u}^2]$ where,

$$\underline{u} \approx \underline{u}^2 = \underline{D}_1^2 \left(\ln(r/a) + \frac{\ln^3(r/a)}{3!} + \frac{\ln^5(r/a)}{5!} \right) + \underline{D}_2^2 \left(1 + \frac{\ln^2(r/a)}{2!} + \frac{\ln^4(r/a)}{4!} \right).$$

Its arbitrary constants D_1^2 and D_2^2 can be updated by solving linear system which is replaced by the boundary conditions. Note that $D_i^2 = [\underline{D}_i^2, \bar{D}_i^2]$ corresponding to $[\underline{k}_i, \bar{k}_i]$ for $i \in \{1, 2\}$. The approximation of $\bar{u}(r)$ is obtained in the same manner.

3.1. Triangular Fuzzy Number

In a case of triangular fuzzy number, let $p_2 = [p_m, \bar{p}, p_M]$. Then t-cut representation of p_2 is given by

$$[\underline{p}_2, \bar{p}_2] = [p_m + (\bar{p} - p_m)t, p_M - (p_M - \bar{p})t] \tag{4}$$

Suppose that $p_2 = (100, 200, 300)$ MPa, the grade of membership function of p_2 is shown in Fig. 1.

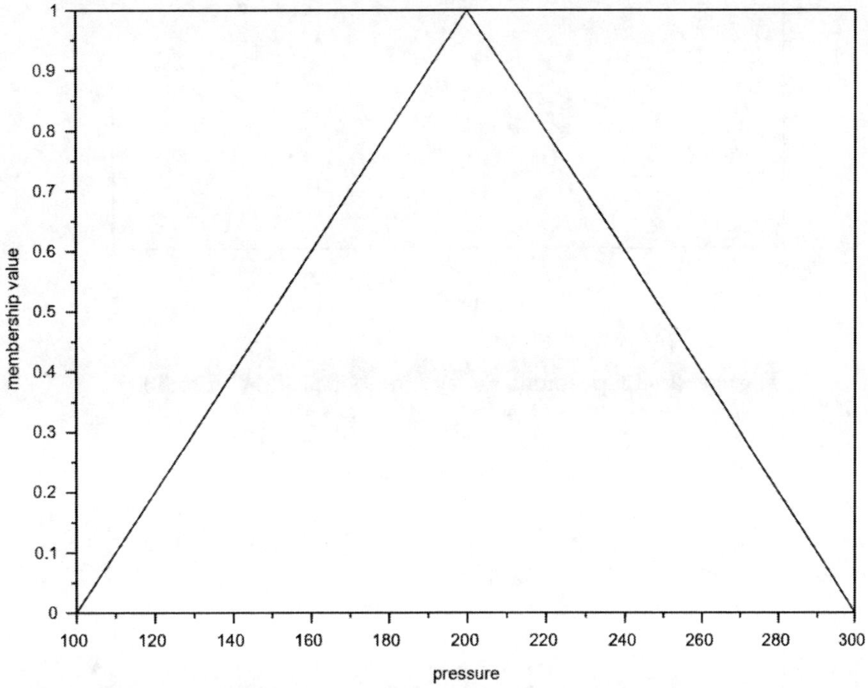

Figure 1. Triangular Fuzzy Number (100,200,300)

3.2. Trapezoidal Fuzzy Number

The trapezoidal fuzzy number $p_2 = [p_m, p_u, p_U, p_M]$. The t-cut approach

$$[\underline{p}_2, \bar{p}_2] = [a + (p_u - p_m)t, p_M - (p_M - p_U)t] \tag{5}$$

Fig. 2 is the trapezoidal fuzzy number of external pressure $(100, 175, 225, 300)$ MPa

3.3. Gaussian Fuzzy Number

The external pressure p_2 is defined by Gaussian Fuzzy Number, $[\bar{p}, \sigma_l, \sigma_r]$. Its t-cut represents

$$[\underline{p}_2, \bar{p}_2] = \left[\bar{p} - \sqrt{-\frac{\ln t}{\lambda}}, \bar{p} + \sqrt{-\frac{\ln t}{\lambda}}\right], \tag{6}$$

where $\lambda = 1/(2\sigma^2)$ Assume that the external pressure distribute as Gaussian fuzzy number $(200, 50, 100)$ MPa, see its shape in Fig 3.

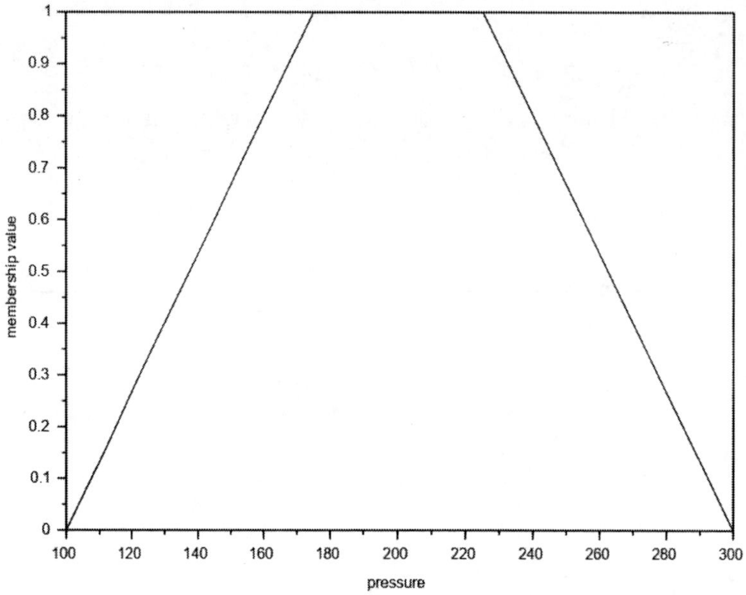

Figure 2. Trapezoidal Fuzzy Number (100,175,225,300)

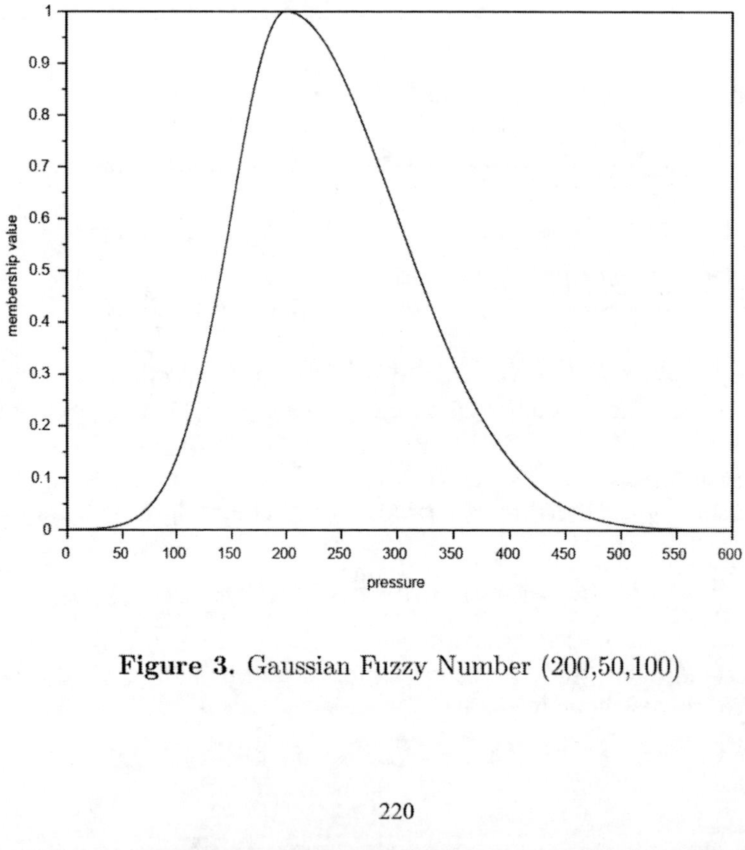

Figure 3. Gaussian Fuzzy Number (200,50,100)

IOP Conf. Series: Materials Science and Engineering **639** (2019) 012033 doi:10.1088/1757-899X/639/1/012033

4. Numerical results and simulations

We illustrated numerical solutions and simulations of the cylinder which is 50 cm mean diameter with 15 cm thick. Suppose that internal pressure (p_i = 35 MPa) and allowed outside pressure is decided by t-cut of fuzzy number in Fig.1, Fig. 2 and Fig. 3. Young modulus and poisson's ratio of material are $210MN/m^2$ and 0.3 respectively.

Table 1 shows numerical results of minimal and maximal stress beloging to t-cut at the radius r. The Fig. 4 is radial stress simulation of 3 and 4 when $t = 0, 0.5, 1.0$ and similarly for trapezoidal shape 5. Its simulation is presented in Fig. 5. But $t = 0$ cannot be evaluated in case of GFN. Then we set $t = 0.025, 0.5, 1.0$ and plots Gaussian fuzzy solution in Fig. 6.

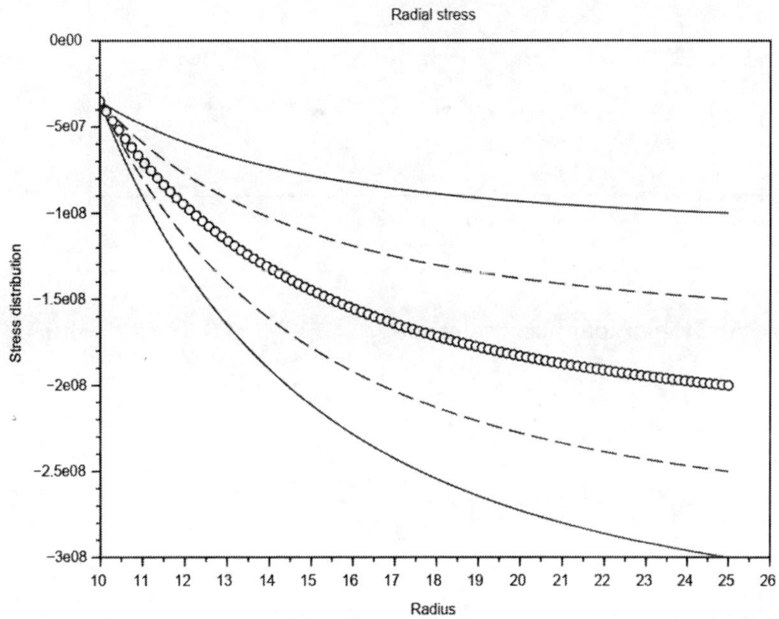

Figure 4. Triangular fuzzy solutions [$t = 0$,'-'],[$t = 0.5$,'–'],[$t = 1.0$,'o']

5. Conclusion and discussion

In this work, we studied the thick cylinder subjects to internal and external pressures such that inside is certainty but not outside. We used concept of fuzzy set to indicate the grade of membership function. Consequently, the traditional deformation become fuzzy deformation. Three types of fuzzy number are demonstrated and investigated simulations of radial stress distribution. Firstly, $t = 0$ of triangular fuzzy number is the support and $t = 1$ is core of TFN. For $t \in [0, 1)$, TFN yields minimal and maximal radial stresses while TrFN yields radial stress interval for all $t \in [0, 1]$. GFN is the last one that we used to interpret external pressure. Since the support of GFN is $(-\infty, \infty)$, its shape should be established from density and distribution of external pressure.

We observed that if the quantity is certain, the fuzzy interval is singleton set, see internal pressure and core of TFN. This mean that TFN is suitable when the best selection of external pressure is fixed. On the other hand, TrFN is the better if best selection still vague. But if we want to decide the possibility from data, determine Gaussian fuzzy shape and apply GFN.

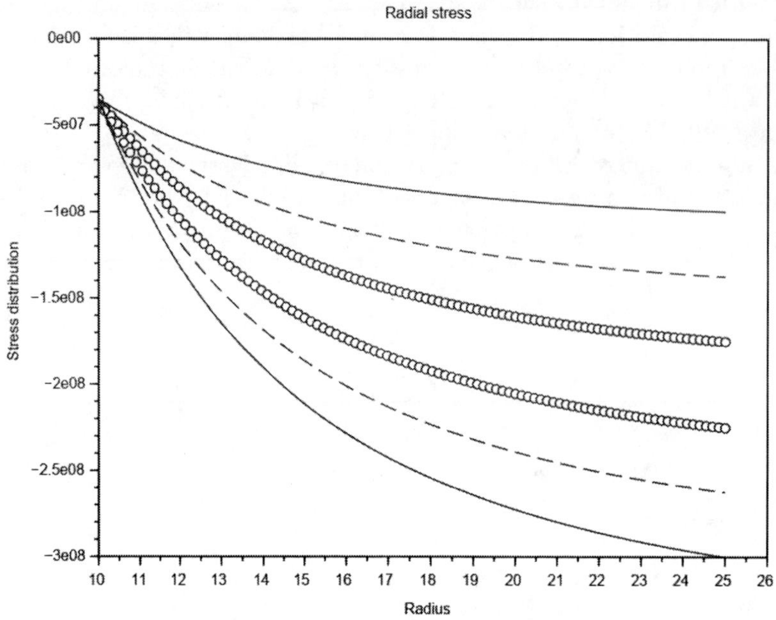

Figure 5. Trapezoidal fuzzy solutions [$t = 0$,'-'],[$t = 0.5$,'–'],[$t = 1.0$,'o']

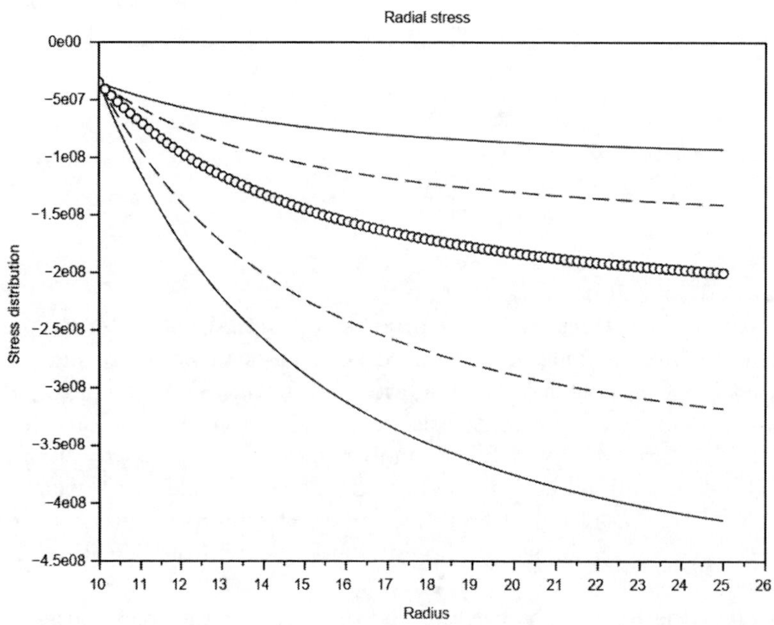

Figure 6. Gaussian Fuzzy solutions [$t = 0.025$,'-'],[$t = 0.5$,'–'],[$t = 1.0$,'o']

Table 1. Approximation of radial stress intervals by 2-terms ADM

t-cut	Radial stress(MPa)			
	$r = 10$	$r = 15$	$r = 20$	$r = 25$
TFN				
$t = 0$	[-35,-35]	[-78.30,-211.4]	[-93.32,-272.6]	[-100,-300]
$t = 0.5$	[-35,-35]	[-111.6,-178.1]	[-138.2,-227.8]	[-150,-250]
$t = 1.0$	[-35,-35]	[-144.8,-144.8]	[-183,-183]	[-200,-200]
TrFN				
$t = 0$	[-35,-35]	[-78.30,-211.4]	[-93.32,-272.6]	[-100,-300]
$t = 0.5$	[-35,-35]	[-103.3,-186.4]	[-126.9,-239]	[-137.5,-262.5]
$t = 1.0$	[-35,-35]	[-128.2,-161.5]	[-160.6,-205.4]	[-175,-225]
GFN				
$t = 0.025$	[-35,-35]	[-54.47,-325.6]	[-61.21,-426.5]	[-64.19,-471.6]
$t = 0.5$	[-35,-35]	[-105.7,-186.4]	[-130.2,-239]	[-141.1,-262.5]
$t = 1.0$	[-35,-35]	[-144.8,-144.8]	[-183,-183]	[-200,-200

Hence the utility of fuzzy differential equation can support decision making for planing and control under uncertainty of static and dynamics problems with soft computing. Our concepts maybe develop and apply to some engineering problems such as a pipeline subjects to internal pressure (water) and external pressure (traffic and earth)[6].

5.1. *Acknowledgments*

The financial support provided by Kanchanaburi Rajabhat University under budget bureau.

6. References

[1] Kim H K 2017 *Procedia structural Integrity* B **5** pp 63–68.
[2] Reis M L C C, Mello O A F and Chisaki M 2004 *Engenharia Térmica(Thermal Engineering).* **3** pp 122–126.
[3] Nakamura S, Senoh M, Nagahama S, Iwase N, Yamada T, Matsushita T, Kiyoku H and Sugimoto Y 1996 *Fuzzy Sets and Systems* **23** pp 301–317.
[4] Yukunthorn O 2018 *Proceeding of URU International Conference on Science and Technology 2018* **2018** pp 13–19.
[5] Chakraverty S, Tapaswini S and Behera D 2016 *Fuzzy arbitrary order system:Fuzzy Fractional Differential Equations and Application.*
[6] Robert D J, Rajeev P, Kodikara J and Rajanib B 2016 *Canadian Geotechnical Journal* **53(8)** pp 1315–1331

Development of Effective Lightweight Material for Construction Building.

Amphon Jarasjarungkiat[1], Setsawat Chuaisrinual[2]

[1,2]Department of Civil Engineering, Faculty of Engineering, King Mongkut's institute of Technology Ladkrabang, Bangkok, Thailand.

[1]Corresponding author; E-mail address: amphon.ja@kmitl.ac.th

Abstract. Due to an increasing trend of building construction in Thailand. Lightweight concrete has been recently introduced into the structural engineering applications. This study was aimed to investigate two aspects; the first one is studying the properties of Cellular Lightweight Concrete (CLC) and verifying that the lightweight concrete satisfies Thai Industrial Standard (TIS Standard) [12]. To verify with the TIS Standard, investigated properties of CLC are dry unit weight, water absorption and compressive strength. Researcher prepared admixtures with sand cement ratio (S/C), water cement ratio (W/C) and percentage of foam to collect statistical data and analyse. Verified by the standard, specimens that pass the standard criteria will be chosen for proper applications in manufacturing of prefabricated construction materials.

1. Introduction

Lightweight concrete is a concrete that has less unit weight than general concrete. Lightweight concrete is a new product produced from natural raw materials include portland cement, lime sand, gypsum, water, air entraining admixtures and design admixtures in a unique formula. Normally, the concrete has a unit weight around 2300 kg/m^3 in dry conditions.

Nowadays, most of production technology and equipment are imported from abroad, such as Germany and Australia. Lightweight concrete is a new construction material with special properties which are lightweight materials, heat resistant, sound insulation, easy to trim. More benefits are working quickly, reduce construction time, construction structural costs and environmental impact. (Krisada Sutthiphan, 2014) [1]

Lightweight concrete is widely used because it is an alternative material with advantages are lightweight, heat insulation and other features that are suitable for use rather than brick. Development of foamed concrete properties has become more various by using a new aggregate material for enhancing other construction requirement. This development of new innovations can reduce both direct and indirect costs within the household

2. Method

This study compares properties of cellular lightweight concrete. Specimen size is 5 cm wide, 5 cm long, 5 cm high. Specimens have admixtures which are water and cement ratio (w/c) at 0.4, 0.5 and 0.6, sand and cement ratio (s/c) at 0.5, 1.0 and 2.0 and percent foam per volume of specimen at 40, 60, 80 and 100% as shown in table 1. Then comparing with CLC standard (TIS. 2601 - 2556) is carried over for select admixture which pass the criteria. Lightweight prefabricated products in various parts of the household such as walls and roofs.

Ratio		Percentage of Foam			
S/C	W/C	40	60	80	100
0.5:1	0.4	12	12	12	12
	0.5	12	12	12	12
	0.6	12	12	12	12
1.0:1	0.4	12	12	12	12
	0.5	12	12	12	12
	0.6	12	12	12	12
2.0:1	0.4	12	12	12	12
	0.5	12	12	12	12
	0.6	12	12	12	12

Table 1 Number of samples of each of the admixture

2.1. The manufacturing process of lightweight concrete.

1. Mixing cement, sand and water with the specified ratio.

2. Produce CLC foam from foaming agent with foam generator machine and air pump. The foam can be mixed with mortar. The foam must be smooth and must be used immediately while the foam leaves from foam generator machine because foam's quality will be decreased with increasing of time.

Fig 1 Foam for CLC

3. Mixes mortar and foam together in a temperature control room. Due to the high temperatures, water in the mortar will be evaporated more rapidly that accelerates the hydration reaction between water and cement and reduce workability.

4. Cast the mortar into the formwork to make the mortar fit well, then finishing the surface with a trowel, and then sealing the surface with plastic seal to prevent water evaporation

5. Curing in the formwork for 24 hours and then remove the formwork. Then curing the samples by soaking them in water.

6. By the age is 28 days old, the specimens were brought out of the water. Wiping the skin to dry, then measuring the weight of the condition at saturated dry skin. Then take the sample to bake for 24 hours and then measure the weight that is completely dry. And measure the size on both sides of the length with the vernier caliper and then tested to compressive strength.

2.2. Unit weight and water absorption.

The specimens were brought out of the water when it completely cured and dry the skin then measure the weight of the samples to collect the saturated condition weight. Then dry the samples in the oven for 24 hours and collect the absolute dry condition weight. After that calculated the water absorption from the formula below.

$$\%W = \left[\frac{SW-D}{DW}\right] \times 100\ \% \tag{1}$$

- SW : Saturated weight, kg
- DW : Dry weight, kg
- %W : Water absorption, %

Then measure the sample size with Vernier Caliper. And calculate the density of the sample from the formula below

$$\rho = \left[\frac{DW}{V}\right] \tag{2}$$

- V : Volume = cm × 10^{-4} = m^3
- ρ : Unit weight, kg/m^3

2.3. Compressive Strength

1. Taking the finished curing samples measure in width, length and height by vernier caliper.

2. Weighing samples and taking notes.

3.Taking samples into the concrete compression testing machine. Prepare to perform compression tests.

4. Applying the compression to the specimens at a constant rate in the range of 0.14 - 0.34 Newton per square millimeter per second to the point of failure. And collect compressive strength data to calculate compressive stress from formula below. [TIS 409-2525][11]

Calculation

$$\sigma = F / (L \times W) \tag{3}$$

- W : Width of the sample(cm.)
- L : Length of the sample(cm.)
- H : Height of the sample (cm.)
- F : Compressive force (kg.)
- σ : Compressive strength (ksc.)

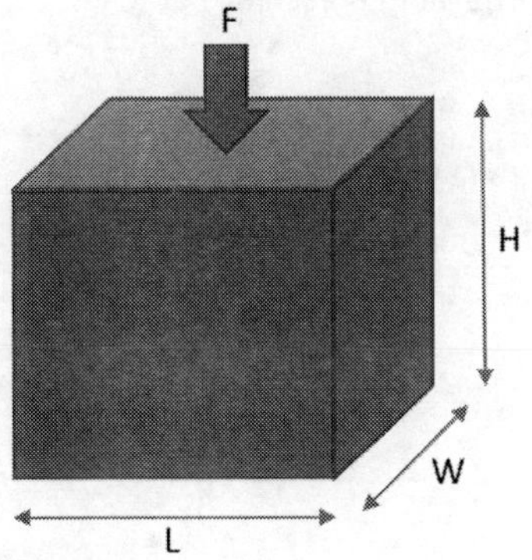

Fig 2 Dimension of specimen

2.4. *Analysis results*

From the experiments, the properties were compared with CLC criteria (TIS. 2601 - 2556). Specimens which pass the standard shall be considered for engineering construction.

3. Result

From the testing results, the density of specimens show that density is inversely proportional to percentage of foam.

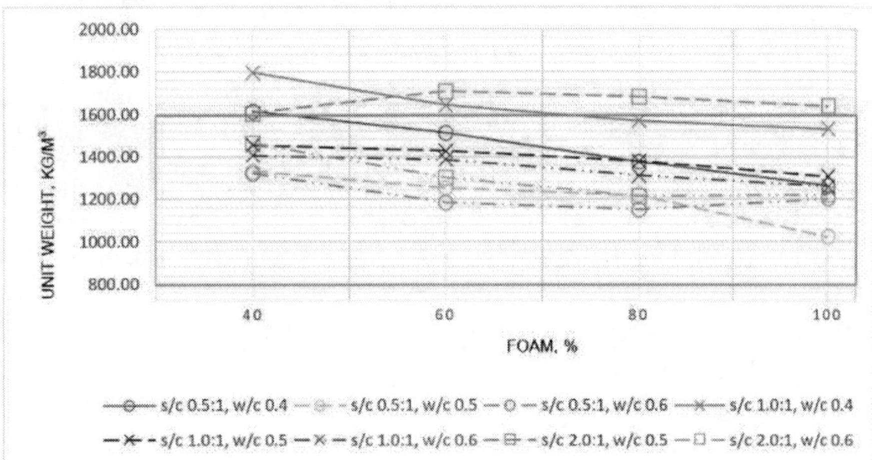

Fig 3 Relationship between average unit weight and admixtures

From Figure 3, the average unit weight and foam (%), w/c and s/c relationship as shown. The standard density is 800 to 1600 kg / m^3, i.e. s/c 0.5 at w/c 0.4, 0.5 and 0.6, s/c 1.0 at w/c 0.5 and 0.6.

From the results of the experiment to find compressive strength was found that compressive strength is inversely proportional to percentage of foam.

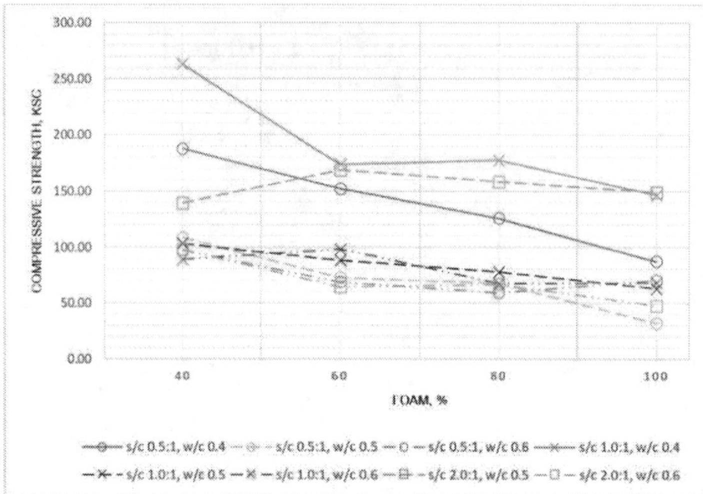

Fig 4 Relationship between average compressive strength and admixtures

Figure 4 describes average compressive strength and foam (percent), w/c and s/c relationship. For density range 801 - 1200 kg/m^3 the compressive strength values according to standards is not less than 25.5 ksc. and density range 1201 - 1600 kg/m^3 values is not less than 51.0 ksc. The results show that every specimen has average compressive strength is greater than the standard. Consequently, all specimens pass this criterion.

From the experiment, water absorption is proportional with percentage of foam.

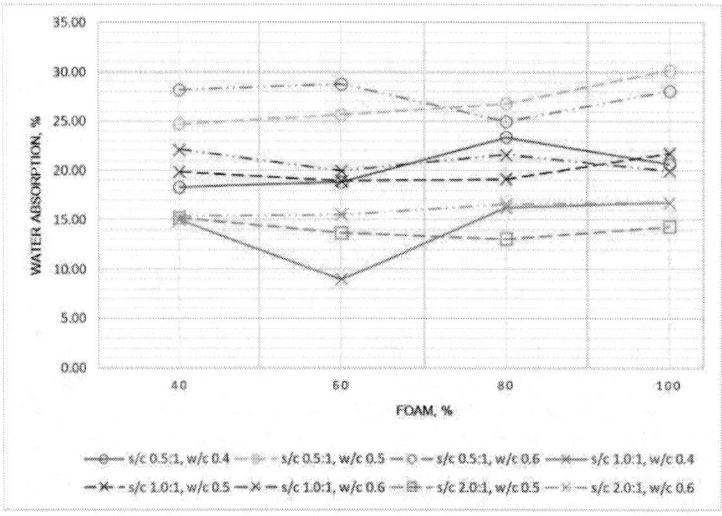

Fig 5 Relationship between average water absorption and admixtures

This tendency is elaborated in figure 5, the average water absorption, foam (percent), w/c and s/c relationship. The standard of water absorption is not greater than 23% by volume between the density 801 - 1200 kg/m^3 and not greater than 20% by volume between the density 1201 - 1600 kg/m^3. All specimens except a specimen at s/c 0.5 at w/c 0.5 and 0.6 has average water absorption over 25%. The remaining sample passes the criteria because it is lower than 23%.

4. Result

By comparison, the three CLC properties ;unit weight, compressive strength and water absorption. The specimens which pass all criteria were labeled in the table below.

Type	Average Dry Unit weight, (kg/m^3)
C6	501 - 600
C7	601 - 700
C8	701 - 800
C9	801 - 900
C10	901 - 1000
C12	1001 - 1200
C14	1201 - 1400
C16	1401 - 1600

Table 2 CLC classification by TIS Standard

Ratio	Percentage of foam				
S/C	W/C	40	60	80	100
0.5:1	0.4	C16	C16	-	C14
	0.5	-	-	-	-
	0.6	-	-	-	-
1.0:1	0.4	-	-	C14	C14
	0.5	C16	C16	C14	-
	0.6	-	C14	-	C14
2.0:1	0.5	-	-	-	-
	0.6	C16	C14	C14	-

Table 3 Admixtures which classified by TIS standard (1. unit weight, 2. compressive strength and 3. water absorption

The relationship between the admixture of CLC and the properties. Can classify specimens with average dry density by table 2.

As TIS standard, specimens were classified to C14 and C16, which C14 will be applied as a wall structure with the advantages are sound and heat insulation. And C16 will be applied as concrete roof structure due to the sample type C16 has compressive strength around 250 ksc. Therefore, this is a good option to be applied in engineering because it has a light weight and has a close air-void pore pattern to preventing water permeability.

5. Conclusion

The specimens which passed all criteria can be classified in CLC standard. From the analysis, the specimens have unit weight ranges between 1201 - 1400 kg/m^3 are C14 type and 1401 - 1600 kg/m^3 are C16 type. The specimens at s/c 0.5 are light weight samples, but average water absorption is over the standard. Therefore, these samples were rejected. From the table 3 displaying exclusively specimen passes all criterion. To summarize, specimens passing all standard criteria are suitable for prefabricated constructions as shown within this research.

References

[1] Krisada Sutthiphan. (2014). Light weight concrete for the construction of physics and engineering. Retrieved September 2, 2014, from http://www.dss.go.th/images/st-article/pep-11-2557-Concrete.pdf.

[2] Nattawoot Tengsiritum. (2015). An Optimization Model for a Prediction on mix proportion of Cellular Lightweight Concrete (Master's thesis, Faculty of Engineering, King Mongkut's Institute of Technology Ladkrabang).

[3] Zaler Mundher Yaseen. (2018). Predicting Compressive Strength of Lightweight Foamed Concrete Using Extreme Learning Machine Model. (Science direct research paper, Faculty of Engineering and Built Environment, University Kebangsaan Malaysia).

[4] Tuan Nguyen. (2018). 3D Meso-Scale Modelling of Foamed Concrete Based on X-Ray Computed Tomography. (Science direct research paper, Department of Infrastructure Engineering, The University of Melbourne, Melbourne, Australia).

[5] Kai Tai Wan. (2018). Development of Low Drying Shrinkage Foamed Concrete and Hygro-Mechanical Finite Element Model for Prefabricated Building Façade Applications. (Department of Civil and Environmental Engineering, Brunel University London, UK).

[6] Prasert Damrongchai. (1999). Construction Materials. (1th ed.). Khonkaen: Department of Civil Engineering Faculty of Engineering Khonkaen University.

[7] Winit Chowichian, Concrete technology. (2001). (9th ed.). Khonkaen: Bangkok. Sampanpanit.

[8] Sastrachai Taitachang. The Study of Aluminum Dross Effect on Refractory Brick Properties. 2010. (Master's thesis, Faculty of Engineering, Rajamangala University of Technology Thanyaburi).

[9] Siriwat Chaichana. (1999). Concrete Technology Laboratory. (2th ed.). Bangkok. VJ Printing., Ltd.

[10] Sumate Santadwattana. (2011). Prediction on the Properties of Cellular Light Weight Concrete from Concrete Density. (Bachelor project, Faculty of Engineering, King Mongkut's Institute of Technology Ladkrabang).

[11] TIS. 409-2525. standard test method for compressive strength of cylindrical concrete specimens. Retrieved June 6, 2018, from http://research.rid.go.th/vijais/moa/

[12] TIS. 2601-2556. Industrial product standards for Cellular Lightweight Concrete. Retrieved June 6, 2018, from http://research.rid.go.th/vijais/moa/

[13] Jaruphan Phaiphuphim. (2014). Stress-Strain Curve of Cellular Lightweight Concrete. (Master's thesis, Faculty of Engineering, Ubon Ratchathani University)

[14] Phutthaporn Orathaiwan, Setsawat Chuaisrinual and Issara Sirikul. (2017). Study on Properties of Lightweight Concrete Mixing with Recycled Glass Bead after Configuration. (Bachelor project, Faculty of Engineering, King Mongkut's Institute of Technology Ladkrabang).

[15] Apichart Chalermwong. (2013). Study of Compressive Strength and Sound Absorption of Cellular Lightweight Concrete with Superplasticizer. (Master's thesis, Faculty of Engineering, Ubon Ratchathani University).

IOP Conf. Series: Materials Science and Engineering **639** (2019) 012035 doi:10.1088/1757-899X/639/1/012035

Effects of electrode configuration on chemical kinetics of Biodiesel production with electric field

Souksavath Phounsavath[1], Xayphone houngbounyaung[1], Vansay Maxsarik[1], Tanongkiat kiatsirirod[2]
Departments of Mechanical Engineering, Faculty of Engineering, National University of Laos.
Departments of Mechanical Engineering, Faculty of Engineering, Chiang Mai University (Thailand).

E-mail: phounsavath@live.com

Abstract: This paper presented chemical kinetics of biodiesel transesterification under electric field. A barrier discharge technique was carried out with 2 and 4 electrodes attached at outside and 1 electrode inside of a glass cylinder contained crude vegetable oil and methanol. The composition of the crude oil and methanol was 1:6 by mol. Crude Tung oil and crude palm oil were raw materials with an amount of 100 ml. The operating periods ware 50 s and 420 s for Palm oil and Tung oil, respectively. About 91 % and 88.5 % of biodiesel were found from the crude oils, respectively. With 3 and 5 electrodes, the chemical kinetics for the reaction rate (r) could be

Plan oil: $r = 0.081 C_{TG} e^{-27.106/RT}$ for three electrode,

$r = 0.088 C_{TG} e^{-31.081/RT}$ for five electrodes,

Tung oil: $r = 0.003 C_{TG} e^{-35.024/RT}$ for three electrode,

$r = 0.001 C_{TG} e^{-35.828/RT}$ for five electrodes.

1. Introduction

Fuel oil is an important issue in Laos, especially diesel fuel since each year the Country has to imports the fuel with a large number and the price of the oil increases continually. According to the government's renewable energy policy, bio diesel is one of the solutions.

Biodiesel mostly comes from long chain fatty acid molecules derived from vegetable oil or animal fat reacts with alcohol such as methanol or ethanol using catalyst. The reaction is called transesterification of which the yield is biodiesel and glycerin. Under normal heating process to generate the reaction, the operating period needs around 2 hours (Anucha Promwungkwa, 2549). Therefore, there are many research studies to find out the method that could shorten the reaction time. The heat source could come from microwave (Raweepat Singkham, 2553) of which the reaction time under microwave irradiation was between 10 to 50 second with a biodiesel yield of 95%. Noppadol Sirirat (2554) used electric field that is Electrolysis discharge to generate biodiesel of which the operating period was about 90 s for about 100 ml biodiesel.

In this paper, a use of electric field. A barrier discharge technique was carried out with 2 and 4 electrodes attached at outside and 1 electrode inside. Generate biodiesel from plan oil and Tung oil was considered. The chemical kinetics such as the reaction rate constant and the activation energy of the transesterification process were investigated.

Tung oil was considered. The chemical kinetics such as the reaction rate constant and the activation energy of the transesterification process were investigated.

2. Theory

Transesterification as shown in Fig. 1 is the process of exchanging the organic group R″ of an ester with the organic group R′ of an alcohol. These reactions are often catalyzed by the addition of an acid or base such as KOH of which the result was methyl ester and glycerin. (http://en.wikipedia.org/Transesterification).

Figure.1. Transesterification: methyl/ethyl alcohol + ester → glycerin + methyl/ethyl ester (http://en.wikipedia.org/wiki/Transesterification).

Diasakou et al. (1998) gave three-steps mechanisms of the methanolysis reaction of triglycerides to methyl esters as

Three steps:

$$TG + MeOH \xrightarrow{k} DG + ME$$
$$DG + MeOH \xrightarrow{k} MG + ME \quad (1)$$
$$TG + 3MeOH \xrightarrow{k} G + 3ME$$

Where TG = triglycerides, MeOH = methanol, G = glyceride, ME = methyl esters, DG = diglycerides, MG = monoglycerides, respectively.

Firstly, triglycerides were converted to diglycerides, then to monoglycerides, and finally converted to glycerides and methylester which was biodiesel.

Kusdiana and Saka (2001) simplified the three-step Diasakou's mechanisms to one step as:

$$\mathrm{TG} + 3\mathrm{MeOH} \xrightarrow{k} \mathrm{G} + 3\mathrm{ME}. \quad (2)$$

One mole of triglycerides reacts with three moles of methanol yield one of glycerin and three moles of methyl esters.

The reaction rate of the chemical reaction could be evaluated by

$$r = -\frac{dC_{TG}}{dt}$$

and

$$r = kC_{TG}^{m} C_{MeOH}^{n}. \quad (3)$$

Where r = reaction rate, k = constant of reaction, with temperature changes m = reaction order with respect to triglycerides concentration, n = reaction order with respect to methanol concentration. C_{TG} = triglycerides concentration and C_{MeOH} = methanol concentration.

IOP Conf. Series: Materials Science and Engineering **639** (2019) 012035 doi:10.1088/1757-899X/639/1/012035

The reaction rate (r) is normally simplified for m=1, n=0, then

$$r = kC_{TG}^{m} \text{ or}$$

$$r = -\frac{dC_{TG}}{dt} = kC_{TG}. \quad (4)$$

After integration, we obtained

$$\ln C_{TG,t} = -kt + \ln C_{TG,0}. \quad (5)$$

Where C_{TG} , t and C_{TG} , 0 are triglycerides concentration at time (t) and at the beginning (t = 0), respectively. The constant of reaction (k) is the slope of graph ln C_{TG}, v.s. t

The rate constant also depends on activation energy (Ea) and temperature of reaction (T) as

$$r = k_o C_{TG}\, e^{-Ea/RT}. \quad (6)$$

At any time, the temperature of reaction is changing then the instantaneous value of k is varying and the relation of the k value with time is given in a form of

$$\ln k = \ln k_o^{\,-Ea/RT.} \quad (7)$$

With the local values of T and k, then the value of Ea could be evaluated.

3. Experimental Setup

Fig.2. shows a test rig of our experiment. Methanol at 133 ml with KOH 1.1 mg was mixed with Plan oil and Tung oil at 100 ml in a cylinder. The molar ratio of methanol and Tung oil was 8:1. Under Electric field was DC generated by a 10 kV electrical supply.

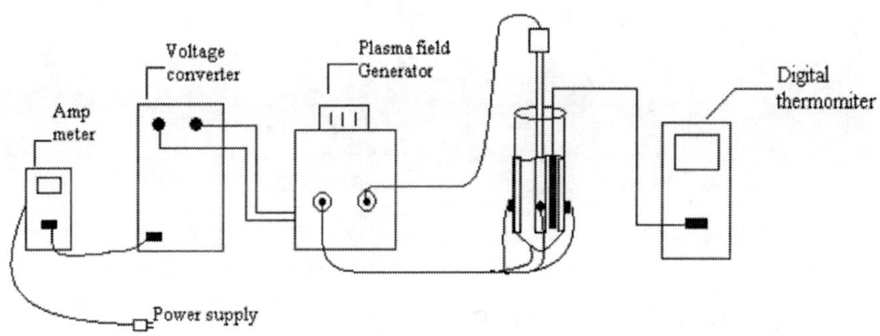

Figure. 2. Experimental setup of biodiesel production under Electric field.

4. Results and Discussions

4.1. Effect of electrode formatting

Test results of each characteristic of the electrode that affects the reaction Biodiesel oil generation and Glycerine continues to be tested in the Tung oil and palm oil. That uses three electrodes And five electrodes type It was found that the time that resulted in the occurrence of % yield of Biodiesel As shown in Figure 3 and Figure 4.

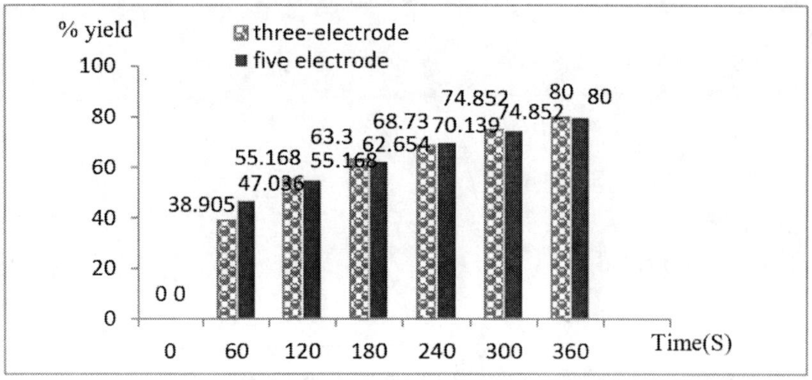

Figure 3. Comparison of reaction time with the characteristics of three electrode and five electrode bars using palm oil

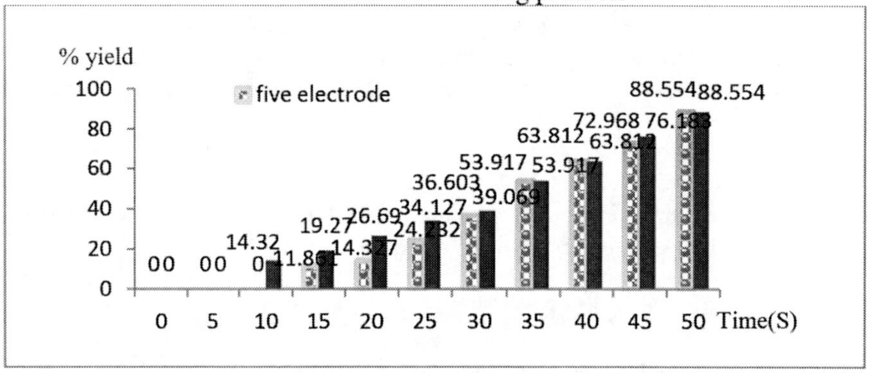

Figure 4. Comparison of reaction time with% yield of the characteristics of three electrode and five electrode using the Tung oil

4.2 Chemical kinetics of reaction rate

From the temperature relationship and Time of the reaction. Will give up the decline of $C_{TG,t}$ at any time shown in Figure 5 and 6 from the slope of the graph Shows the reaction constant (K) from the first reaction until the process ends.

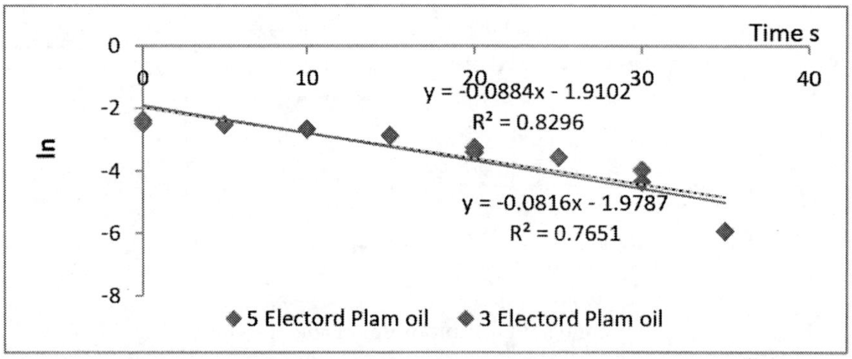

Figure 5. Reaction rate of biodiesel production process by formatting three electrodes and five electrode in palm oil

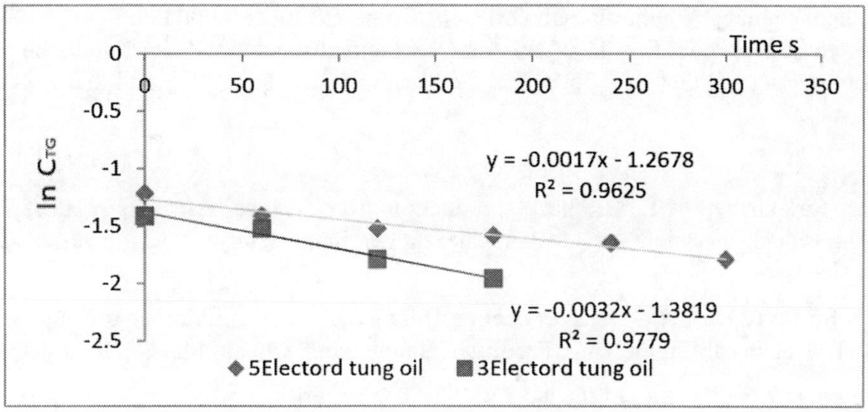

Figure 6. Reaction rate of biodiesel production process by formatting three electrodes and five electrode in the Tung oil.

From the slope of the graph showing E_a Activation, which has a high activation energy, indicating that this reaction is more difficult.

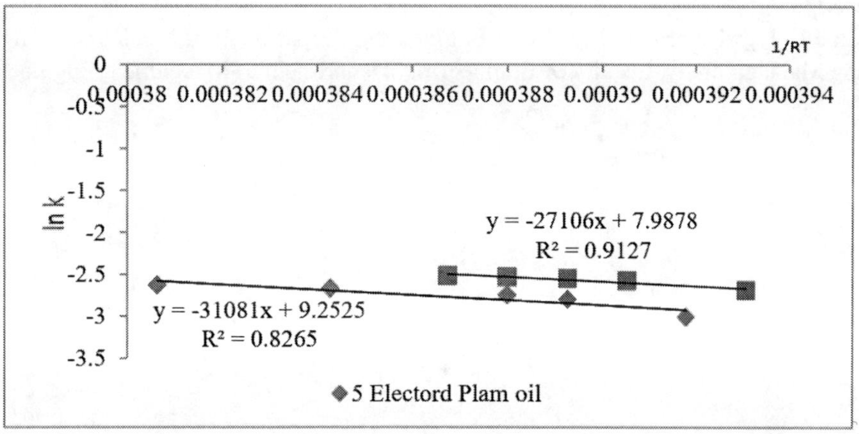

Figure 7. Graph showing the relationship between and 1 / RT Palm oil.

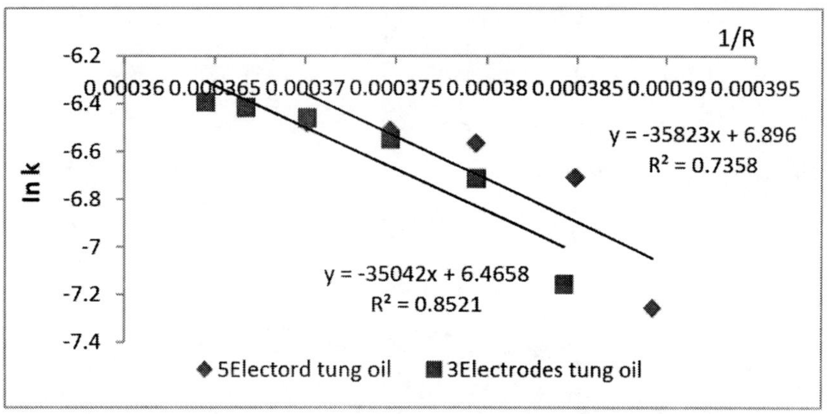

Figure 8. Graph showing the relationship between and 1 / RT of Tung oil.

IOP Conf. Series: Materials Science and Engineering **639** (2019) 012035 doi:10.1088/1757-899X/639/1/012035

In this research, biodiesel is produced from transesterification reaction of palm oil and Tung oil, and methanol by the form of electrodes in the electric field With a voltage of 10 kV which is determined by the reaction of chemical kinetics In palm oil using Three-electrode found that $E_a = 27.106$ J /mol and in the five electrode electrodes $E_a = 31.081$kJ / mol, respectively, and in the rocky oil in the stone Three-electrode electrodes found that $E_a = 35.042$ kJ / mol and in five electrode found that $E_a = 35.828$ kJ / mol, respectively.

5. References

[1] Raweepat Singkhan (2010), Biodiesel production from Transestérication of beef tallow with Microwave heating. Department of Mechanical Engineering, Faculty of Engineering, Chiang Mai University, (Thesis).

[2] Noppadol Sirirat (2011), Biodiesel production from Transesterication process using Electric field. Department of Mechanical Engineering, Faculty of Engineering, Chiang Mai University. (Thesis).

[3] Diasakou, M, (1998), Kinetics of the No-catalytic Transesterification of Soybean oil. Department of Mechanical Engineering, Faculty of Engineering, Chiang Mai University. (Thesis).

[4] Kusdiana, D., & Saka, S (2001), Kinetics of transesterificationin in Rapeseed oil to Biodiesel Fuel as Treated in Supercritical Methanol.

[5] Anucha Promwungkwa (2006). Biodiesel production from waste vegetable oil process in electronics industries. Department of Mechanical Engineering, Faculty of Engineering.

Acknowledgments

[1] National University of Lao, Faculty of Engineering, Department of Mechanical Engineering.
[2] Chiang Mai University, Faculty of Engineering, Department of Mechanical Engineering (Thailand).

IOP Conf. Series: Materials Science and Engineering **639** (2019) 012036 doi:10.1088/1757-899X/639/1/012036

Navigation of autonomous vehicle for rubber tree orchard

Worawut Kunghun*[1], Akapot Tantrapiwat[1], Phudit Chaidilokpattanakul[2]
[1, 2]Department of Mechanical Engineering, Faculty of Engineering, King Mongkut's Institute of Technology Ladkrabang, Bangkok, Thailand

59601321@kmitl.ac.th

Abstract. This research has studied the navigation of autonomous vehicles in rubber tree orchards. It was emphasised on the rubber tree plantation in Thailand. Due to the high labour costs in the rubber industry and the demand for natural rubber has continued to decline over the past several years. The farmer were struggled to reduce the labour costs in production. The development of autonomous vehicle technology in rubber plantation can be a way to ease the production costs, such as reducing the amount of labour in fresh latex harvesting. This work was aimed to develop algorithm of a vision system for replacement of expensive sensors used in typical autonomous vehicles. The vision system model was implemented using only one camera installed on the vehicle to search for calibrated targets which were put on the rubber tree trunk. There were two important parameters, tree row offset and target distance, which were concerned when determining the performance of the autonomous vehicle in the rubber tree orchards. In order to evaluate the accuracy of the system, a set of experiments was conducted at a distance of 1-5 meters between the target and the camera. The results were compared between the controlled environment in the laboratory and the uncontrollable environment in the real rubber plantation.

1. Introduction

Automatic vehicles in rough terrain are one of the major trends in control research. Such vehicles can be used widely in agriculture applications which lead to the solutions of modern agriculture development. Less labor and resource are required for the automation system in agriculture. This technology can also replace human limitations and reduce operating cost. This research interests on the application of the ground vehicle automatically for agriculture in the rubber plantation. The rubber industry in Thailand has been struggled due to low profit and high labour cost [1]. In the process of the production, harvesting latex sap is a tedious work and uses a lot of labour. To minimize the operating cost, automation in the rubber tree orchard is one of the most promising technologies to resolve the problem.

The autonomous vehicle and orchard navigation system have been studying for many applications. Z. Wang et al. developed Machine vision assessment of mango orchard in the current orchard and season, the correlation coefficient of determination between machine vision estimates of panicle area and multi-view panicle count and fruit yield per tree [2]. C. Pirat et al. had developed vision Based navigation for the scalability analysis of the ESA Autonomous Transfer Vehicle Guidance, Navigation & Control (GNC) performances and the Russian docking system. This paper used a single monocular camera on the chaser satellite and various sets of Light-Emitting Diodes (LEDs) on the target vehicle ensure the observability of the system throughout the approach trajectory [3]. G. Bayar et al. had developed a model-based control method for an autonomous agricultural vehicle that operates in tree fruit orchards. In the paper, they proposed an either state estimation or path following, relying instead only on data from a planar laser probability of a successful turn from one row into another, while respecting the maximum steering rate limits [4]. X. Zhang et al. did show an algorithm to the crop row detection for precision agriculture and automatic navigation. This research used a crop row detection method which was proposed for maize fields based on images acquired from a vision system. During the procedure of crop row detection, the position clustering algorithm and shortest path method were applied successively to confirm the final clustered feature point set [5]. W. Kunghun et al. had developed the vision model to use a single camera capturing calibrated targets based on the common pinhole model which were placed on the rubber tree trunks for study the errors and their uncertainties were monitored [6].

The prior works cited above did not focus on the operation between the trees rows. Most of the current researches of autonomous vehicle navigation systems had been using a laser scanner which is expensive and not practical for farming applications. In the rubber tree orchard, it is possible to permanently mount calibrated targets on the tree trunks and use them as the guidance for the autonomous vehicle system. The targets on the tree trunks for guidance the autonomous vehicle system will be easily searched by image processing technique. This method reduces the complication of terrain image processing and allows the mapping can be done with less performance computer platform. The same vision mapping technique can also applies to automatic navigation in other similar applications or other types of tree orchards.

This paper is aimed at finding ways to develop and implement vision system into the agriculture of natural rubber tree farming. The goal of this research is to develop an algorithm to autonomously guide a robotic vehicle platform along a rubber tree orchard row. There are two parameters important for navigation in the rubber tree orchard, target-camera distance parallel to the tree row and an offset distance between camera and tree row. This paper presents the test on real conditions in rubber plantations, which variations in the uncontrollable nature or the amount of light source that affects the parameters for navigation in the rubber tree row.

2. Orchard configuration and vision navigation to orchard geometry

2.1. Configuration of rubber tree orchard

The rubber tree orchard plantations usually have 3 meters row distance and 6-7 meters column distance. For the typical area of the plantation, it will use about 1600 square meters for 80 trees [1]. Rubber plantation farmers do harvest and rubber tree planting, etc. They work by walking along columns. Each column requires regular maintenance to facilitate daily work. The autonomous vehicle should therefore replicate the running path same as the work of the farmer.

The rubber tree orchard used in this research is well maintained and one of the rows was randomly selected as a test row. This experiment was conducted on a real rubber plantation located in Chachoengsao province, eastern of Thailand. The vision navigation was designed for 1-5 meter distance between row to row in the tree orchard. The sizes of rubber tree trunks were about 120-millimeter in diameter. In the actual orchard, it was found that there were several uncontrollable parameters which can affect the navigation performance. These include the straightness of rubber tree trunk, rough plantation areas of rubber trees and the amount of inconsistent light in each period. The effect of these parameters was compared with the system performance when working in controllable environment in laboratory.

2.2. Horizontal and vertical vision navigation to orchard geometry

The rubber tree orchard used in this research is well maintained and one of the rows was randomly selected as a test row. The vision model used for finding the rubber trees was developed based on the common pinhole model [7] as shown in Fig.1.

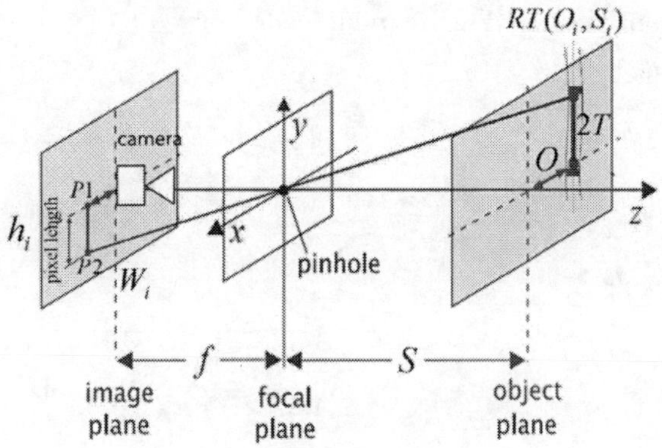

Figure 1. Pinhole model for the target position estimations [7]

By capturing a known size target the image position and pixel length between P1 and P2, which were measured by the number of pixels, can be used to determine the position of the separated red target on the image plane as shown in Fig.2.

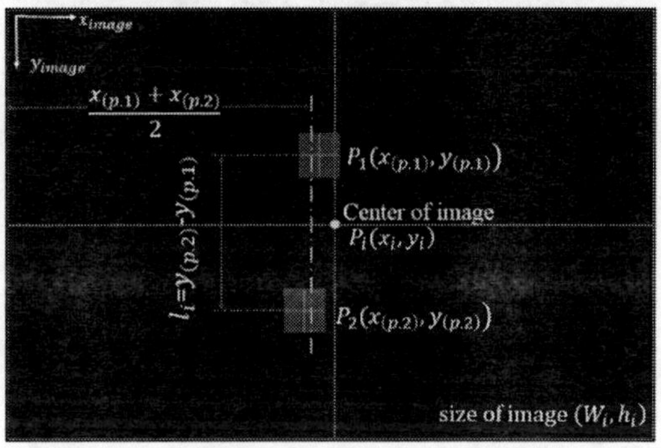

Figure 2. The position of binary image plane (x, y) coordinate systems

The target positions obtained from the pinhole vision model can only tell the positions of the trees in relation to the camera. To plan the camera path, these positioning data must be converted to the vertical vision navigation to orchard geometry model. Fig.1 shows the vertical plane of a rubber tree which is the target for the camera installed on an autonomous vehicle. The vertical vision navigation is very important because finding the separated red targets on the rubber tree to define as a region of interest (ROI) on an image plane. The image plane shown in Fig.2 is the picture retrieved from a webcam. Two red dots are detected as the targets. Then the center of both targets is identified as a position of rubber tree on image plane. Pixel position column on image plane $x_{(p.1)}$ and $x_{(p.2)}$ are the pixel coordinates in a column of image plane, when $y_{(p.1)}$ and $y_{(p.2)}$ are the pixel coordinate in the row of image plane. A pixel length distance l_i is calculated between pixel positions of $y_{(p.1)}$, $y_{(p.2)}$ and $P_i(x_i, y_i)$ is the center of image plane. This position data obtained from the pinhole model can be converted to the camera positioning on the orchard by the relationship of the vertical vision navigation on object plane.

The vertical model as shown in Fig.3 in the world coordinate is another significant angle for tracking separated Red target at install to rubber tree trunk. By estimate the optimum angles λ and distance H , it can be derived as shown in equation 1 and 2 where the parameter of $M = 0.1$ meters is the size of a single target. The view angle vertical field of view (VFOV) of the camera is $\psi = 43.30°$.

The distance between the centers of red target to center of separated target is $T = 0.35$ meters. The distance between centers of separated target to high edge of red target is $I = 0.4$ meters, and f is the focal length of the camera.

$$\lambda = 2\tan^{-1}\left(\frac{T + (\frac{M}{2})}{f \cdot \frac{2 \cdot T}{(l_i)}}\right) \tag{1}$$

$$H = \sqrt{\left(\frac{f \cdot \frac{2 \cdot T}{(l_i)}}{\cos(\frac{\psi}{2})}\right)^2 - \left(f \cdot \frac{2 \cdot T}{(l_i)}\right)^2} \tag{2}$$

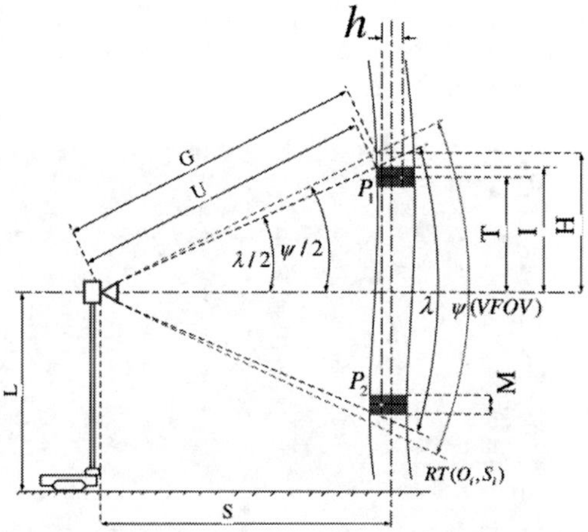

Figure.3 The vertical vision navigation object plane to orchard geometry model.

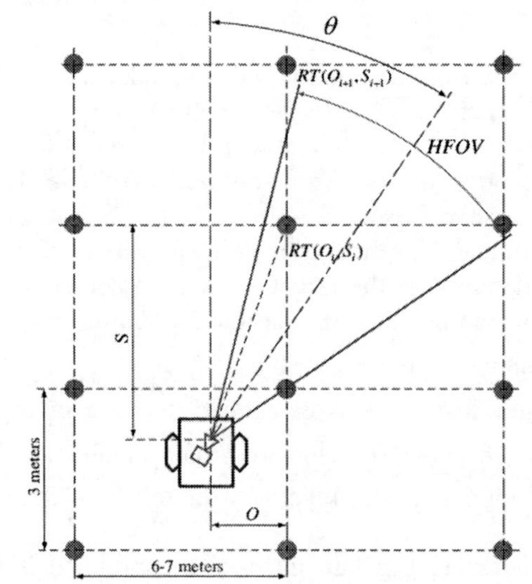

Figure.4 The horizontal vision navigation object plane to orchard geometry model.

From this position P1 and P2 on image plane as shown in Fig.2, the target positioning data on the image plane obtained from the pinhole model can be converted to the camera positioning on the real orchard by the relationship of two parameters. It can be seen that it can estimate target-camera distance parallel to the tree row (S), offset distance between camera and tree row (O).

$$S = \frac{\{2 \cdot T \cdot f \cdot [f \cdot \cos(\theta) + |\xi| \cdot \sin(\theta)]\}}{f \cdot (l_i)} \tag{3}$$

$$O = \frac{2\sqrt{2} \cdot T \cdot f \cdot \sqrt{f^2 + \xi^2 \cdot f^2 \cdot \cos(2\theta) + \xi^2 \cdot \cos(2\theta) - 2 \cdot f \cdot |\xi| \cdot \sin(2\theta)}}{2 \cdot f \cdot (l_i)} \tag{4}$$

When:

$$\xi = (\frac{x_{(p.1)}}{2} + \frac{x_{(p.2)}}{2} - \frac{W_i}{2}) \tag{5}$$

3. Experimentation Setup

From In this research, the experiment was made to compare between the controlled environment and the uncontrolled environment. By using actual rubber tree posts with 120 millimeters diameter, the camera used in this experiment has vertical field of view angle (VFOV) at 43.30° and a horizontal field of view angle (HFOV) at 70.42° for determining the region of Interest (ROI) on the image. The webcam Logitech C920 captured the images of the rubber tree with the size of 720x1280 pixels RGB image resolution on the world coordinate system as shown in figure 3. This dimension was based on the rubber tree trunk at the diameter of 120 millimetres. Each rubber tree was mounted with two red color bands with a vertical distance defined by the target length 0.7 meters. The target identifying method was made by capturing all frame rate the image frame in RGB format. Then, the image input was pre-processed by the kernel filter procedure this method is to eliminate the high frequency noise that is in the image or reduce the variance in the image to make the image more detailed. Because the Red target colour was selected to be distinctive from the common rubber tree orchard background. After that, the target areas in image were classified by a low level gain and high level gain on HSV format. The Red value from the image was then sorted out and the image was set to binary image that the target was presented in white on black background. The target distance is 0.7 meters, to assess the accuracy and efficiency of parameter for navigation.

Figure.5 Detection of separate targets on the rubber tree on the uncontrolled environment of actual tree orchard

4. Results

Fig.6 has shown the magnitude errors of O obtained from the navigation experiments by comparison between the controlled environment and uncontrolled environment. It can be seen that magnitude errors target row offset (O) in the controlled environment were minimal when comparing to uncontrolled environment and they were small throughout the test ranges. Because the area of the rubber plantation is different from the lighting conditions from the controlled environment, the rubber tree is not straightened the surface of the rubber plantation not flat terrain.

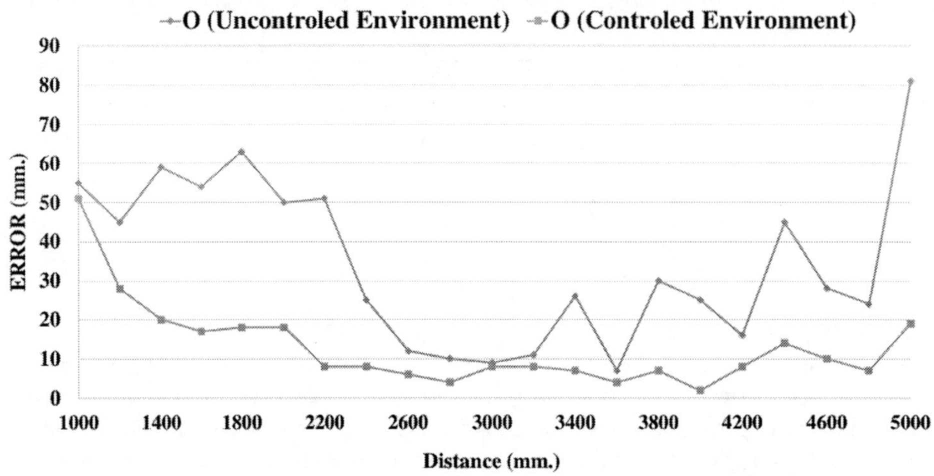

Figure.6 The magnitude error of the target row offset (O) for uncontrolled and controlled environment.

The magnitude errors of the distance parallel to the target (S) Fig.7 when the camera locates in the longest distance from the target of the test is highly error compared the camera near to the target. In a controlled environment have the result of a constant magnitude error throughout the distance for testing. But the experiment for the uncontrolled environment throughout the test distance by the average is considered to have an error that is still acceptable.

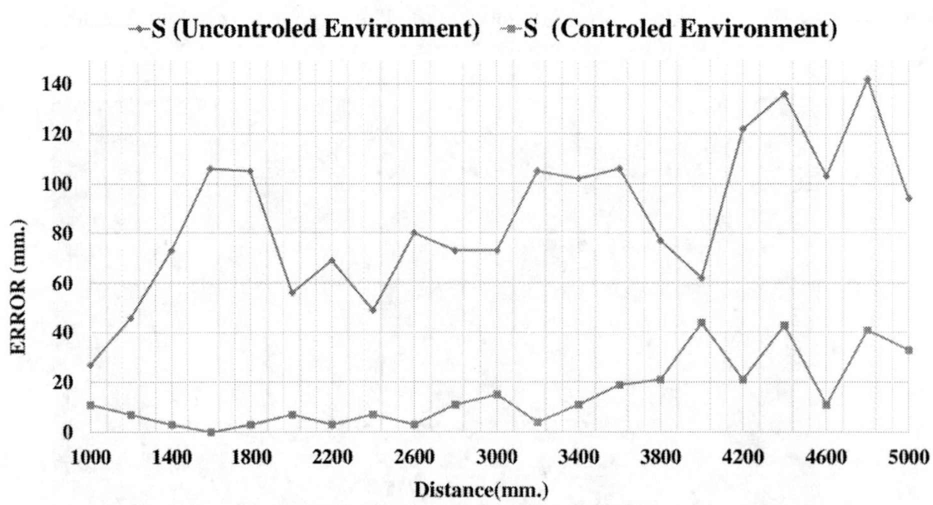

Figure.7 The magnitude error of the distance parallel to the target (S) for uncontrolled and controlled environment.

5. Conclusions

In this research, the navigation of autonomous vehicle for rubber tree orchard was successfully evaluated for the magnitude of errors. It has shown that the methodology was effectively applied to estimate the distance parallel to the tree row (S) and offset distance between camera and tree row (O) using a two separated targets as the reference for estimating parameters in the actual rubber tree orchard. The experiment showed that different environment results in twice the accuracy of S and O parameters. By the maximum errors of O and S can be converted to the location errors in the tree orchard geometry, and it has shown the accuracy of 8.1 centimetres in the O - direction, 14.6 centimetres in the S-direction for uncontrolled environment and 5.1 centimetres in the O - direction, 4.1 centimetres in the S-direction for controlled environments. Though O-direction and S-direction of error seem to be large, the navigation of autonomous vehicle for rubber tree orchard method is still feasible. The next step will be to design an automatic car using a camera system for navigation in the rubber plantation.

References

[1] Rubber Authority of Thailand, "Rubber Research Institution of Thailand," Thai rubber statistics, Retrieved from http:// http://www.rubber.co.th.

[2] Z. Wang, J. Underwood and K. B.Walsh, "Machine vision assessment of mango orchard flowering," Journal Computers and Electronics in Agriculture, Vol.151, Centre for Intelligent Systems, Central Queensland University, Australia, 2018, pp 501-511.

[3] C. Pirat , F. Ankersen , R. Walker and V. Gass, "Vision Based Navigation for Autonomous Cooperative Docking of CubeSats," Acta Astronautica, Vol.146, European Space Agency and Swiss Federal Institute of Technology in Lausanne (EPFL), Station 13, 1015, Lausanne, Switzerland, 2018, pp 418-434.

[4] G. Bayar, M. Bergerman, A. B. Koku and E. i. Konukseven, "Localization and control of an autonomous orchard vehicle," Computers and Electronics in Agriculture, Vol.115, Mechanical Engineering Department, Middle East Technical University, Dumlupinar Bulvari, Cankaya, 06800 Ankara, Turkey, 2015, pp 118-128.

[5] X. Zhang, X. Li, B. Zhang, J. Zhou, G. Tian, Y. Xiong and B. Gu,"Automated robust crop-row detection in maize fields based on position clustering algorithm and shortest path method," Computers and Electronics in Agriculture, Vol.154, College of Engineering, Nanjing Agricultural University, Nanjing 210031, Jiangsu, PR China, 2018, pp 165-175.

[6] W. Kunghun and A. Tantrapiwat," Development of a Vision Based Mapping in Rubber Tree Orchard," International Conference on Engineering Applied Sciences and Technology (IEEE), Mechanical Engineering Department Faculty of Engineering, King Mongkut's Institute of Technology Ladkrabang Bangkok, Thailand, 2018.

[7] R. Hartley, A. Zisserman, Multiple View Geometry in Computer Vision ,2rd ed., Cambridge University, 2003, pp.153-177.

IOP Conf. Series: Materials Science and Engineering **639** (2019) 012037 doi:10.1088/1757-899X/639/1/012037

A Study of Vibrations on Monopole Billboard Construction Induced by Earthquake Excitation

Nuthnapa Triepaischajonsak[1], Chak Chantalakhana[2] and Khemapat Tontiwattanakul [2]

[1] Electrical and Electronic Standards Laboratory Thailand Institute of Scientific and Technological Research Bangkok, Thailand
[2] Department of Mechanical & Aerospace Engineering King Mongkut's University of Technology North Bangkok Bangkok, Thailand

Corresponding author. E-mail: nuthnapa@tistr.or.th

Abstract. Effects of earthquake on large billboards become significant issue after occurrence of disasters recently in Thailand. Two set of measurements were carried out, finding natural frequencies of actual large billboard. First set of measurements was made using accelerometers and the last one was made using non-contact laser vibration measurement. The test results show that, at low frequency, these natural frequencies correspond to transient vibrations of earthquake motion. The distinguished vibration occur at frequencies between 0.09 to 85 Hz. A simple billboard model is also presented here. The solutions to the problem are present especially for suture zones in Thailand.

Keyword: Monopole Billboard, earthquake excitation, vibration, Laser measurement

1. Introduction

Thailand has increasingly been affected from earthquake especially in suture zone as shown in figure 1 [1]. The vibration of significance occur at frequencies between 0.2 to 20 Hz [2]. Although many papers [3 to 6] studied strength of billboard structure and large structure for avoiding fatigue or reducing its probability, those concerned the effects in term of wind load as its regulation in Thailand [7]. Wind-induced structural vibration, can lead to damage due to fatigue, however mostly its damage can be found either on its advertisement board or the whole billboard falling down. This is specific only for monopole billboard construction as shown in Figure 2 [8 to 11]. This paper studied the effects of large billboard due to earthquake excitation, which recently becomes a significant issue as earthquakes increase in magnitude especially in the northern part of Thailand. Two sets of measurement were carried out at Bangpoo, Samutprakarn province.

2. Billboard Description

Two sided outdoor advertising billboard monopole, as shown in Figure 3, situates at Bangpoo, Samutprakarn province, where sea breeze blow during measurements taken. Billboard considered is of dimensions 7m wide and 4m height. The total height of pole is 7.5 m with its circumference of 1.84 m. At the top of the billboard, there are six spotlights hanging with cantilever (three spotlights for both sides).

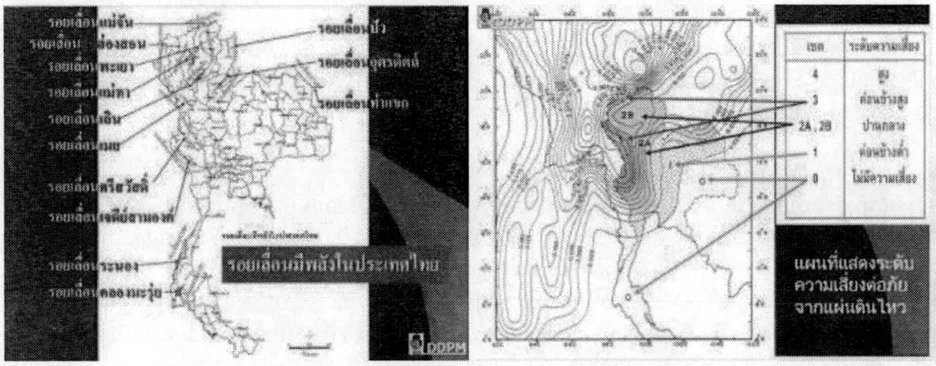

Figure 1. Suture zones in Thailand [1].

Figure 2. Billboards were blown over by strong winds [8 to 11].

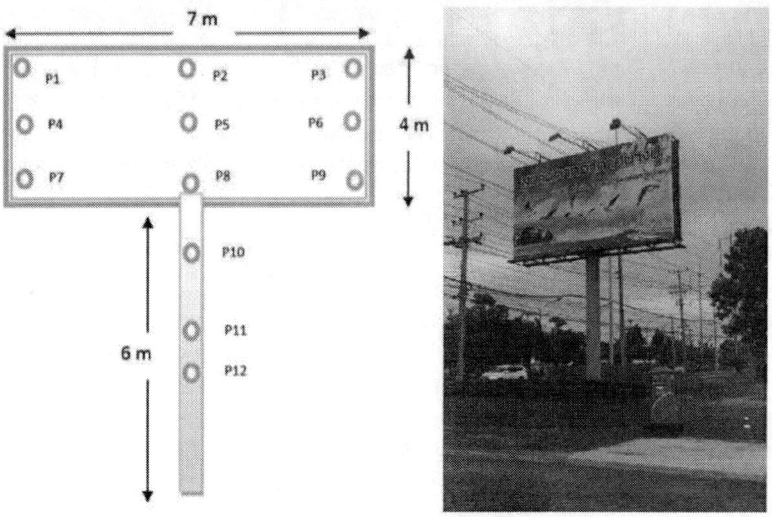

Figure 3. Specifications of the billboard and its measurement points.

IOP Conf. Series: Materials Science and Engineering **639** (2019) 012037 doi:10.1088/1757-899X/639/1/012037

3. Measurements

Measurements were separated into two parts: measuring vibration with accelerometers and non-contact laser vibration measurement. Such measurements were carried out to find natural frequency corresponding to vibration of earthquake motion.

3.1 Measuring vibration with accelerometers

For the first measurement, accelerometers were mounted 2m from the bottom as shown in Figure 4. The following equipment was used: accelerometers B&K type 4507B001 with frequency range 0.1 Hz to 6 kHz. The vibration response were recorded at a sampling frequency of 2 kHz for 120 seconds. The measurement were taken at the time of trucks passing by. Because of sea breeze blowing, this was also one source of those loads taken. In or der to excite thin-walled vibration mode, small impact was applied. Although, this mode should be excited using large impact. Because of not allowing any damage to the billboard, small impact was done.

Figure 4. First measurement and its posistion, 2m from the base.

3.2 Non-contact laser vibration measurement

For second measurement, remote sensing vibrometer RSV-150 was used with controller RSV-E-150B, at a sampling frequency of 48 kHz for about 20s each point. The vibrometer was set up about 100m perpendicular to the billboard. The position of 14 measurements is shown in figure 3. At P12, this is the same point as in the first measurement. The vibration response using accelerometer was also acquired to compare those data at the same point.

4. Conclusion

Results show separately for both measurements as shown in Figure 6 and Figure 7. The natural frequency found at P8 might be the combination mode between board and pole as shown in Figure 8. The simple model was constructed using SOLIDWORKS program with the properties of ASTM A36 steel.

4.1 Results: Measuring vibration with accelerometers

Figure 5. Second measurement: non-contact laser vibration measurement.

The natural frequencies found are 69.1 Hz, 82.1 Hz, 150.4 Hz, 183 Hz and 200 Hz. Due to the small impact, these frequencies correspond to thin-walled vibration

At very low frequency, such frequencies found are 1.628 Hz, 2.116 Hz and 2.686 Hz. These frequencies correspond to spotlights on the top of the billboard.

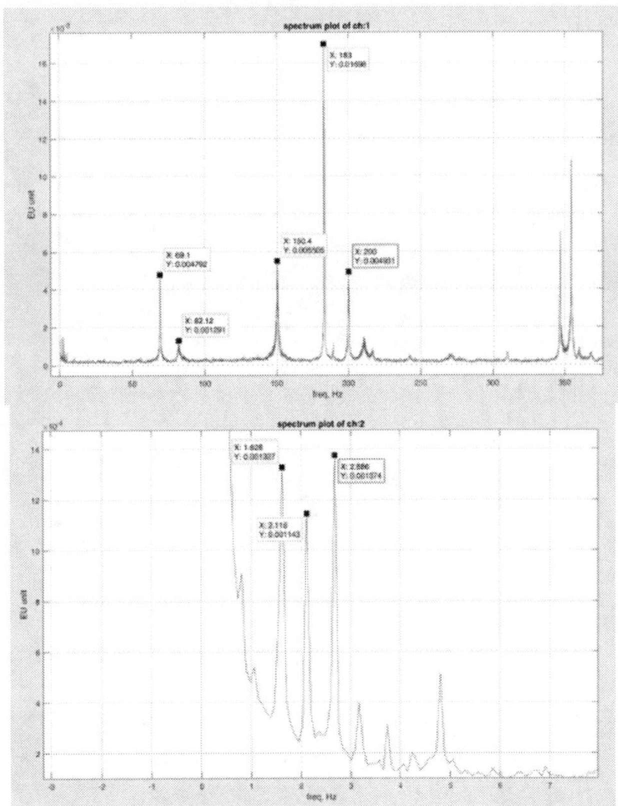

Figure 6. Vibration response from first measurement.

4.2 Results: Non-contact laser vibration measurement

The natural frequencies found, for laser vibration measurement, are shown in Table I.

Table1. All Natural Frequencies Found From Laser Vibration Measurement

Position.	Natural frequencies found (Hz)
P1	0.14, 0.32, 0.41, 0.69, 2.61, 7.14, 11.05
P2	0.09, 0.23, 0.6, 2.33
P3	0.18, 2.52, 3.36
P4	0.09, 0.32, 2.61, 3.39, 11.08, 53.1, 74.84
P5	0.92, 1.6, 2.61
P6	0.09, 0.27, 0.87, 3.39
P7	0.09, 0.14, 2.38, 3.43, 10.44, 11.26
P8	0.14, 2.49, 10.44, 45.14, 62.48, 74.34, 107.12
P9	0.09, 0.14, 2.71, 3.39
P10	0.09, 0.27, 2.5, 10.39
P11	0.08, 0.14, 2.43, 10.07, 11.03
P12	0.09, 2.56, 10.3

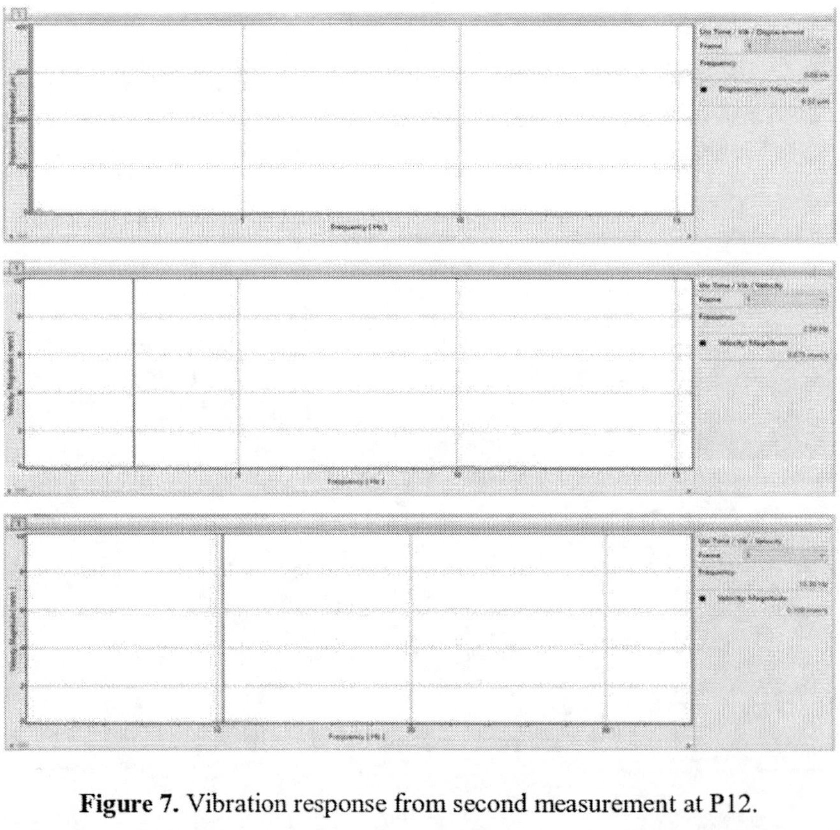

Figure 7. Vibration response from second measurement at P12.

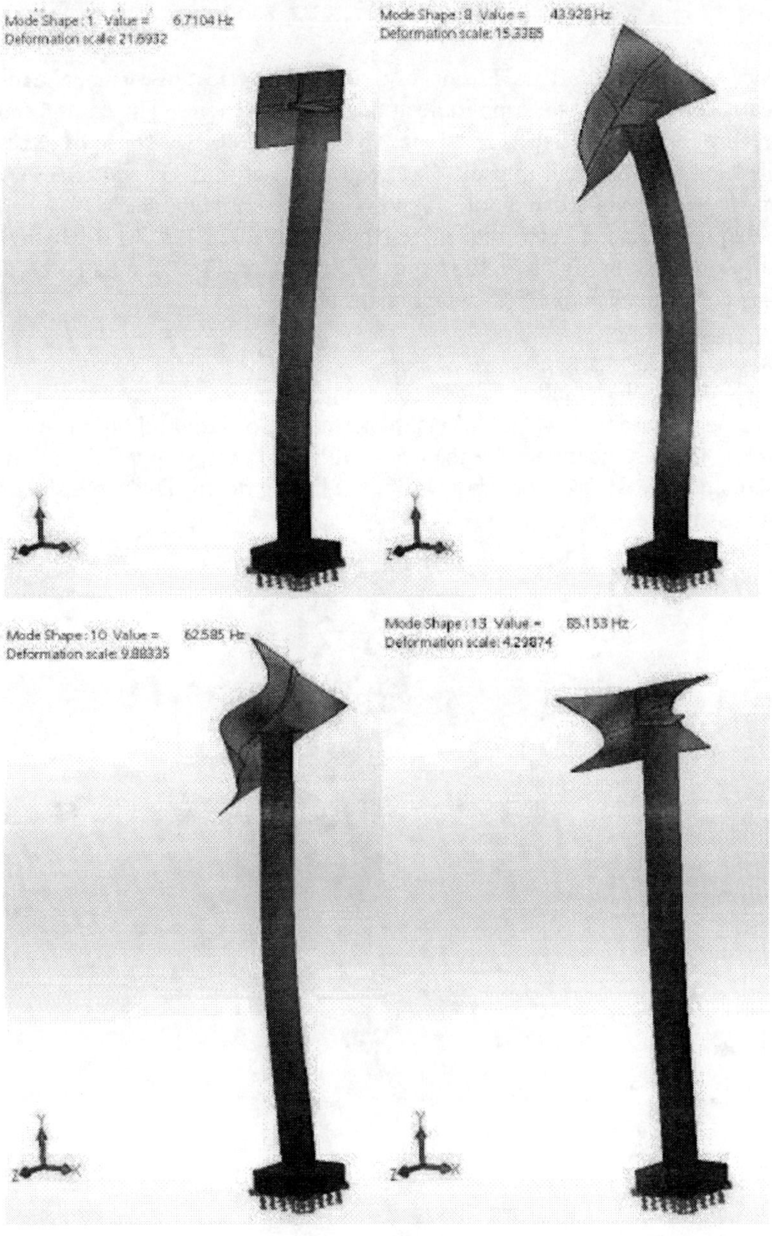

Figure 8. Vibration response from second measurement at P12.

It was found that the whole body of billboard vibrate at frequency of 0.09 Hz. This might transmit though ground borne vibration as its ground property of soft soil. Four types of the ground were classified [12]. However, this paper cannot identify the vibration source at frequency of 0.09 Hz.

As a sensitivity of remote sensing vibrometer RSV-150 is better than accelerometer B&K 4507B001, the frequency at 0.09 Hz can be found only from non-contact laser vibration measurement. At frequencies of 1.628 Hz, 2.116 Hz and 2.686 Hz, they corresponded to the spotlights hanging with cantilever on the top of Billboard. Such vibrations transmitted through the pole. Therefore they were found from accelerometer measurement at P12. These frequencies were also found on the board and point P12 from non-contact laser vibration measurement.

5. Conclusions And Suggestions

5.1 Conclusions

Thailand has increasingly been affected from earthquake. The effect of earthquake might cause damage on large billboard depending on the amplitude at natural frequencies. The natural frequencies, of large billboard, found are 69.1 Hz, 82.1 Hz, 150.4 Hz, 183 Hz and 200 Hz. These frequencies correspond to thin-walled vibration, due to small impact. The frequencies of 69.1 Hz and 82.1 Hz, this might be the combination mode between board and pole as shown in the simple model.

At low frequency range, such frequencies found are 1.628 Hz, 2.116 Hz and 2.686 Hz. These frequencies correspond to spotlights on the top of the billboard. For the middle frequency range, about 10 to 11 Hz, this corresponds to mode shape of board only.

5.2 Suggestions

Earthquake vibration should be taken in to account as occurrence of disasters is increased. In order to prevent the damage, Figure 9 shows additional construction of large billboard. As the damage mostly found either on its advertisement board or the whole billboard falling down, the solution to this problem might reduce damage at least from the whole billboard falling down. This issue should be considered as a regulation, if possible.

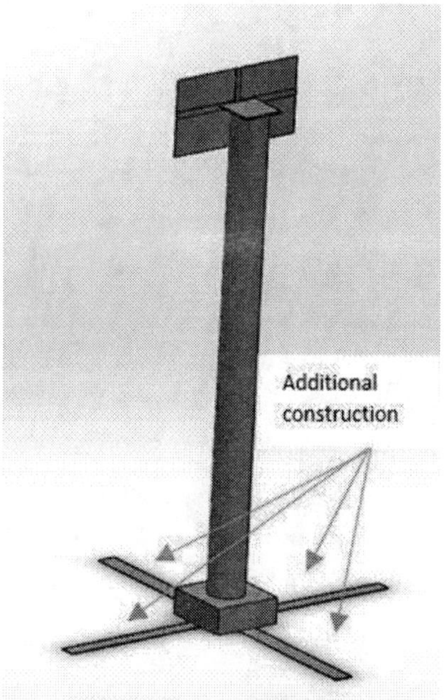

Figure 9. Solution to the problem specific only on monopole billboard construction..

Acknowledgment

The first measurement using accelerometers was supported by Asst.Prof. Dr. Chak Chantalakhana and Asst.Prof. Dr. Khemapat Tontiwattanakul of Department of Mechanical & Aerospace Engineering, King Mongkut's University of Technology North Bangkok. The author would like to thank the following offices and persons for their supports in equipment, technical skill and laboring: Mr. Kittiphong Nimarkorn, Ms. Chalothorn Bhamornsut and their colleagues of Performance and Safety Inspection Laboratory, Dr. Anat Hasap of Railway Transportation System Testing Centre, Dr. Wadee Wichaidit of

IOP Conf. Series: Materials Science and Engineering **639** (2019) 012037 doi:10.1088/1757-899X/639/1/012037

Industrial Metrology and Testing Service Centre, Thailand Institute of Scientific and Technological Research; Bangpu Recreation Center.

References

[1] Department of Disaster Prevention and Mitigation, Ministry of Interior, Thailand.
[2] Retrieved as of https://www.src.com.au/earthquakes/seismology-101/what-is-an-earthquake, Earthquake Monitoring since 1976, Seismology Research Centre.
[3] P. Warnitchai, S. Sinthuwong, and K. Poemsantitham, "Wind Tunnel Model Tests of Large Billboards," Advances in Structural Engineering, Vol.12 No.1, pp.103-112, 2009.
[4] V. Boonyapinyo, "Wind-Related Disaster Risk Reduction Activities in Thailand," Department of Civil Engineering, Thammasat University.
[5] Retrieved as of https://blog.gooshared.com/view/52#sthash.DMpyFxSz.dpuf, C. Jitjaujun.
[6] S. Neridu, V. D. Kumar, and A. Dongre, "Impact on Structural Behavior due to Installation of Billboard," Journal Teknologi (Sciences & Engineering), pp.89-98, 2017.
[7] Department of Works and Town & Country Planning (DPT).
[8] Retrieved as of http://users.ece.utexas.edu/~kwasinski/ikegrid.html
[9] Retrieved as of http://www.ci.texarkana.tx.us/gallery.aspx?PID=87
[10] Retrieved as of https://www.dailymail.co.uk/news/article-2499851/Typhoon-Haiyan-Full-horror-destruction-Philippines-revealed.html
[11] Retrieved as of https://www.google.com/imgres?imgurl=https%3A%2F%2Fi.ytimg.com%2Fvi%2FB3u9InHoplw%2Fmaxresdefault.jpg&imgrefurl=https%3A%2F%2Fwww.youtube.com%2Fwatch%3Fv%3DB3u9InHoplw&docid=e8mid5eV9ve53M&tbnid=eHSLDZg54xSJVM%3A&vet=10ahUKEwit3PXBytLgAhUHq48KHTFYDJcQMwhLKBMwEw..i&w=1280&h=720&itg=1&bih=568&biw=1114&q=billboard%20damage%20from%20wind&ved=0ahUKEwit3PXBytLgAhUHq48KHTFYDJcQMwhLKBMwEw&iact=mrc&uact=8
[12] N. Triepaischajonsak, D.J. Thompson, C.J.C. Jones and J. Ryue, "Track-based control measures for ground vibration – the influence of quasi-static loads and dynamic excitation," 10th International Workshop on Railway Noise, Nagahama, Japan, October 2010, pp.237-244.

IOP Conf. Series: Materials Science and Engineering **639** (2019) 012038 doi:10.1088/1757-899X/639/1/012038

Qualitative analysis for sweetness classification of longan by near infrared hyperspectral imaging

W Sahachairungrueng[1] and S Teerachaichayut[1,2]

[1] Department of Food Process Engineering, Faculty of Agro-Industry King Mongkut's Institute of Technology Ladkrabang, Bangkok 10520, Thailand
[2] sontisuk.te@kmitl.ac.th

Abstract. Near infrared analysis is a nondestructive technique used for determining the quality of various materials including fruit and other food. The objective of this study was to test whether near infrared hyperspectral imaging could be used for classifying sweetness of longan. One hundred and twenty samples were divided into a calibration set (n= 80) and a prediction set (n= 40). The average absorbance spectra from samples in the wavelength range of 935-1720 nm were used in this study. The sweetness of longan was represented by total soluble solids (TSS) which was used to separate fruit into a low sweet (TSS≤ 21.30^{O}Bx) and high sweet fruit (TSS> 21.30^{O}Bx). A classification model was developed in order to classify groups of longan based on sweetness, where 0 = low sweet and 1 = high sweet, by partial least squares discriminant analysis (PLS-DA). Spectra were preprocessed using a Savitzky-Golay smoothing method in order to obtain the optimal performance of the classification model. The results showed an accuracy of the classification model in the calibration set of 85% and the accuracy was 77.5% in the prediction set. Therefore, it was concluded that near infrared hyperspectral imaging has a potential for classifying longans nondestructive based on sweetness.

1. Introduction

Longan (*Dimacorpus longan* Lour) is a tropical fruit in the sapindaceae family similar to lychee and rambutan. It is grown in southern China and countries in Southeast Asia including Thailand. The flesh of longan is sweet and juicy and is normally eaten fresh, but can be processed to many kinds of product such as frozen longan, dried longan, canned longan and longan juice [1]. Thailand is one of the largest producers and exporters of longan products in the world [2]. In 2015, Thailand exported 565,559 tonnes of fresh, frozen and dried longan with a value of 401.49 million USD. Since then the value has progressively increased to 451.80 in 2016, 494.17 in 2017 and 501.69 million USD in 2018 [3]. Sweetness is a major quality factor of longan that can affects its acceptability to consumers. It would therefore be an asset to the export and processing industries to have a reliable, accurate, fast, simple nondestructive method of assessing sweetness on each individual fruit before it was exported or processed. Therefore fresh longans were tested qualitative in order to separate them into low sweet and high sweet longans.

Sweetness of longan cannot be sorted by visual inspection and there is no report about a nondestructive technique that uses for classifying the sweetness of longan. One method that has been successfully used to grade the quality of fruit nondestructively are near infrared techniques. Fruit that have been tested include Marian plum [4], mangosteen [5]; [6], pineapple [7], sweet corn [8] and Sweet Tamarind [9]. Therefore, near infrared hyperspectral imaging was tested on longans, cv. "Edo" in order to classify them into low sweet and high sweet groups.

2. Meterials and Methods

2.1 Longan samples

Samples of longan were purchased from a local fruit market in Bangkok, Thailand were selected for good appearance and size uniformity (1.5 cm) and stored in an air conditioning room at about 25°C for about 24 hr before measurement. Each fruit was scanned by the near infrared hyperspectral imaging unit, peel and seed were removed and the juice was extracted and TSS was measured using a digital refractometer (PR101, Palette Series, Atago Co., Ltd., Tokyo, Japan).

2.2 Near infrared hyperspectral imaging unit

Near infrared hyperspectral imaging, was a hyperspectral camera (Specim FX17e, Spectral Imaging Ltd., Oulu, Finland) in reflectance mode at a scanning speed of 5.5 mm/s and the scanning positioning speed was 20 mm/s was used to acquire spectral data in the range 935-1720 nm of each sample.

2.3 Data analysis

The acquired spectral data from each sample gave the spectral data including the background. Therefore the region of interest (ROI) was defined with the same dimensions (30x30 pixels), in only that the area of each sample so the background would be excluded. Average absorbance spectrum from the ROI of each sample was used for analysis. The spectra feature of samples was considered in order to eliminate outliers (abnormal spectra) from this analysis. Low sweet samples with a TSS $\leq$ 21.30°Bx were defined as "0" and high sweet samples with a TSS > 21.30°Bx were defined as "1". Samples (N= 120) were divided into a calibration set (N= 80) and a prediction set (N= 40). In the calibration set there were 38 low sweet samples and 42 high sweet samples and in the prediction set there were 19 low sweet samples and 21 high sweet samples. From the calibration set, the classification model was established using partial least squares discriminant analysis (PLS-DA). Spectral pretreatment methods, smoothing (Savitzky-Golay), first derivative differentiation, second derivative differentiation, multiplicative scatter correction (MSC) and standard normal variate transformation (SNV) were investigated in order to obtain the highest accuracy of classification for establishing the model. The cut off value of 0.5, mid-way between low sweet longans (0) and high sweet longans (1), was used to evaluate correct or incorrect prediction. Then the acquired classification model was tested for accuracy by samples in the prediction set. Analysis was carried out using the Unscrambler software (CAMO, Osla, Norway), Microsoft Office Excel 2007 and the UmBio Evince hyperspectral image analysis software (Prediktera Evince, version 2.7.5, Sweden).

3. Results and Discussions

The spectra feature of longan (N= 120) showed the main peaks at around 1200 and 1450 nm (figure 1) which were previously reported to be the absorbance peaks of water [10], which, as with other fresh fruit, are mainly composed of water. As described above, the samples were separated based on TSS with TSS= 19.89-21.30°Bx for the low sweet group and TSS= 21.35-22.90°Bx and the average absorbance spectra both groups in the wavelength range of 935-1720 nm are shown in 'figure 2', which clearly shows a difference over most of the range.

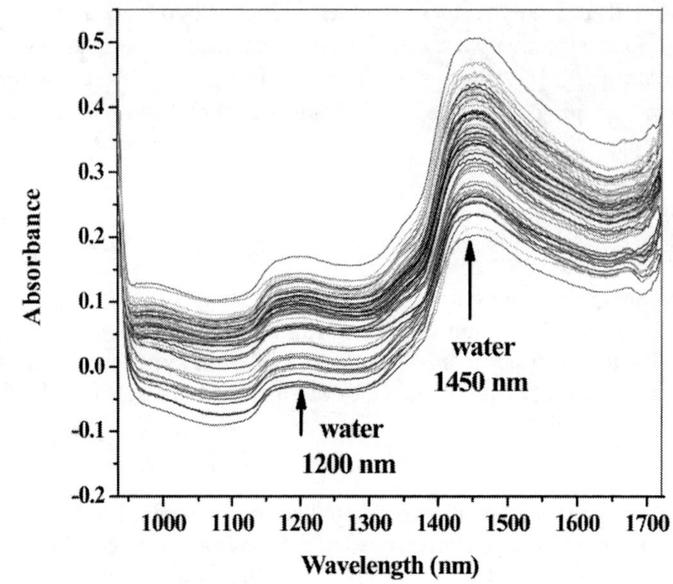

Figure 1. The original spectra of longan in the wavelength range of 935-1720 nm.

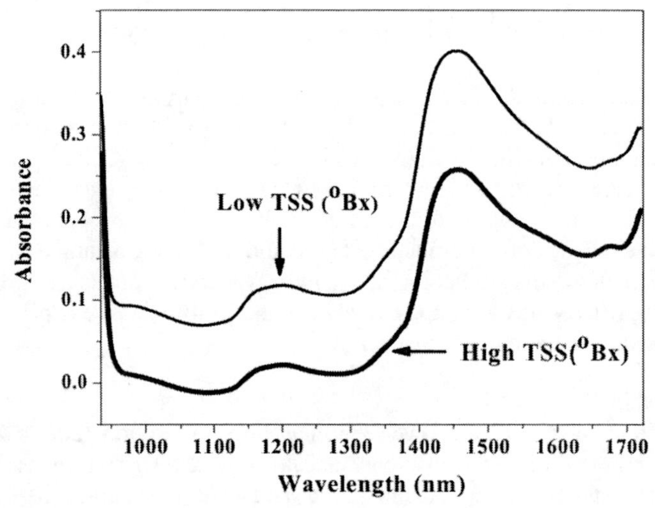

Figure 2. Average absorbance spectra of low sweet longans and high sweet longans.

The spectral pretreatments were carried out in order to obtain the best condition for developing the classification model (table 1), which shows that Savitzky-Golay smoothing methods gave the best results with a total accuracy of 82.5%. The accuracy of the model for the low sweet group was 79% and for high sweet longans was 85.7%. This indicates that the Savitzky-Golay smoothing spectral pretreatment gave the best results and was therefore used to establish the classification model in this research. The accuracy of the classification model in the calibration set gave an accuracy of 85% (81.6% for low sweet longans and 88.1% for high sweet longans) while the classification model in the prediction set had an accuracy of 77.5% (68.4% for low sweet longans and 85.7% for high sweet longans) (table 2).

Table 1. Spectral pretreatments for the classification model in the calibration set.

Spectral pretreatment	N[b]	F[a]	Low Sweet		High Sweet		%Total accuracy
			Correct	Incorrect	Correct	Incorrect	
Original	80	1	29/38	9/38	36/42	6/42	81.2
Smoothing[c]	**80**	**1**	**30/38**	**8/38**	**36/42**	**6/42**	**82.5**
1st Derivative[d]	80	4	30/38	8/38	32/42	10/42	77.5
2nd Derivative[e]	80	2	28/38	10/38	33/42	9/42	76.2
MSC[f]	80	2	32/38	6/38	30/42	12/42	77.5
SNV[g]	80	2	32/38	6/38	31/42	11/42	78.8

[a.]Factors, [b.]Number of samples, [c.]Savitzky-Golay smoothing, [d.]Savitzky-Golay first derivative differentiation, [e.]Savitzky-Golay second derivative differentiation, [f.]Multiplicative scatter correction, [g.]Standard normal variate transformation

Table 2. Results of the classification model for the low sweet longans and the high sweet longans in the calibration set and in the prediction set.

Spectral pretreatment	F[a]	N[b]	Calibration				
			Low Sweet		High Sweet		% total accuracy
			Correct	Incorrect	Correct	Incorrect	
Smoothing[c]	1	80	31/38	7/38	37/42	5/42	85
Spectral pretreatment	F[a]	N[b]	Prediction				
			Low Sweet		High Sweet		% total accuracy
			Correct	Incorrect	Correct	Incorrect	
Smoothing[c]	1	40	13/19	6/19	18/21	3/21	77.5

[a.]Factors, [b.]number of samples, [c.]Savitzky-Golay smoothing

The scatter plot of actual value versus predicted value in both the calibration set (figure 3a) and the prediction set (figure 3b) of the classification model for low (0) and high sweet longans (1) using the cut off value of 0.5 show the accuracy of the classification model.

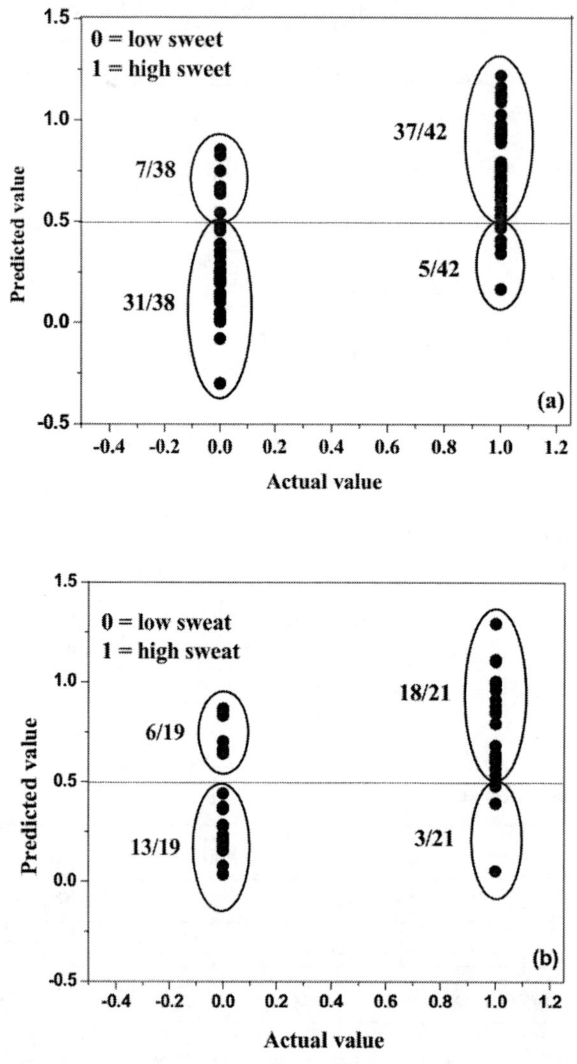

Figure 3. Scatter plots of actual and predicted value by the classification model: (a) in the calibration set and (b) in the prediction set.

The results showed that the classification model for prediction obtained the accuracy of 77.5%. Therefore it can be used for rough screening for the sweetness discrimination of intact longans [12].

IOP Conf. Series: Materials Science and Engineering **639** (2019) 012038 doi:10.1088/1757-899X/639/1/012038

4. Conclusions

The average absorbance spectra from ROI of samples by the near infrared hyperspectral imaging technique were preprocessed by Savitzky-Golay smoothing spectral pretreatment in order to establish the classification model for low sweet longans and high sweet longans. The results of prediction were appropriate for classification with 85% in the calibration set and 77.5% in the prediction set. Therefore the near infrared hyperspectral imaging technique has a potential to use to classify between low sweet and high sweet longans.

5. References

[1] Tokusoglu O and Swansan B G 2015 *Biometrics* 42-1

[2] Office of Agricultural Economics Import-Export Statistic [Online] Available: http://www.oae.go.th. [Accessed: Dec. 20 2018]

[3] Postharvest Technology Innovation Center Logistics of longan in the eastern part of Thailand [Online] Available: http://www.phtne.org/2018/01/737/ [Accessed: Dec. 20, 2018]

[4] Teerachaichayut S, Phonmakham S and Suktanarak S 2017 *Acta Hort. (ISHS)* **1179** 25-30

[5] Teerachaichayut S, Kil K Y, Terdwongworakul A, Thanapase W and Nakanishi Y 2007 *Postharvest Biol Technol* **43** 202-206

[6] Teerachaichayut S, Terdwongworakul A, Thanapase W and Kiji K 2011 *J. Food Eng* **106** 206-211

[7] Sukwani S and Teerachaichayut S 2013 *Acta Hort. (ISHS)* **989** 395-399

[8] Suktanarak S, Supprung P and Teerachaichayut S 2017 *Acta Hort. (ISHS)* **1152** 47-54

[9] Teerachaichayut S, Suktanarak S and Kasemsumram S 2014 *Acta Hort. (ISHS)* **1053** 113-119

[10] Workman J and Weyer L 2012 *Practical guide and spectral atlas for interpretive near-infrared spectroscopy* (2nd ed Taylor & Francis Group LLC USA)

[11] Williams P 2007 *Near-infrared Technology-Getting the Best out of Light* (PDK Grain Nanaimo British Columbia and Winnipeg Manitoba Canada)

Acknowledgment

This research was funded by the Faculty of Agro-Industry, King Mongkut's Institute of Technology Ladkrabang, Thailand (2562-01-07002). The authors are grateful to Assoc. Prof. Panmanas Sirisomboon for technical help and Prof. A K Thompson for editing English language in this paper.

Feasibility of a photoelectric sensor technique for nondestructive prediction of granulation disorder in tangerines

S Teerachaichayut[1,2], T Kijpadung[1] and V Cheevathumrat[1]

[1] Department of Food Process Engineering, Faculty of Agro-Industry King Mongkut's Institute of Technology Ladkrabang, Bangkok 10520, Thailand
[2] sontisuk.te@kmitl.ac.th

Abstract. Granulation is a physiological disorder in Tangerine (*Citrus reticulata*) that can cause an internal disorder which cannot be determined by visual inspection. A nondestructive technique, which is simple and low cost, that could determine whether an individual fruit suffered from granulation would be of help to the citrus industry. The intensity signal of photoelectric sensor, when light passed through tangerines 'Keaw Dumnuan' was investigated and showed that the signal was significantly larger ($p < 0.05$) for normal tangerines and those suffering from granulation. In order to determine a suitable non-destructive method for detect granulation disorder in tangerines that could be applied commercially, a simple and low cost prototype of this photoelectric sensor technique was developed. A set of 52 fruit were used and their average intensity signal from four measurements was used for discriminant analysis. The cut off value of 61.7 mV was used for classification. If a fruit had the signal intensity equal to or less than 61.7 mV, it had granulation, while fruit that had the signal intensity of more than 61.7 mV was normal. Using this signal intensity discrimination the accuracy of classification was 90.4%. It was also shown that the accuracy of prediction was related to the severity of granulation in each fruit. Therefore, the photoelectric sensor technique was shown to be feasible for use in nondestructively detect of granulation in tangerines and be possibly used in future grading systems.

1. Introduction

Tangerines can be grown in every part of Thailand and are an important commercial favorite fruit. Granulation is a physiological disorder which affects to the quality of citrus fruits, not only in Thailand, but throughout the world. Granulation causes the juice sacs to become enlarged, hard, dry, assume a grayish color, have little free juice tasteless and are unsatisfactory for consuming [1]. Granulation is an internal defect of tangerines and cannot be seen by visual inspection, therefore a reliable, fast and nondestructive technique for their detection would be of benefit to both producers and consumers. Various nondestructive techniques have been used to predict internal defects in fruits. X-ray imaging were used to predict core rot in apples [2] and translucency in pineapple [3], color and firmness measuring to predict pitting damage in peaches [4] magnetic resonance sensor to detect internal browning in apples [5] proton magnetic resonance to determine internal browning and watercore in apples [6], NMR to study on internal browning in pears [7] and MRI and X-ray CT to detect core breakdown in pears [8]. For quality detection of citrus, machine vision and ultraviolet fluorescence was used to detect freeze damage in oranges [9], acoustic impulse response to determine the firmness of mandarins [10], ultrasonics to determine the hydration and firmness of the oranges

[11], magnetic resonance imaging to detect freeze damage of orange [12] and [13], electronic nose techniques to evaluate mandarin maturity [14] and gas sensors to detect freeze damage in oranges [15]. There are some reports which use various techniques to evaluate granulation disorder of intact tangerines including X-rays [16] and the use of capacitance based techniques [17].

Basic principles of materials describe that materials are made up of molecules and molecules are composed of atoms that are joined together by bonds. The most common bonds in agricultural and food materials are joined between carbon (C), oxygen (O), hydrogen (H), nitrogen (N) and phosphorus (P) atoms therefore molecules such as cellulose, oil, protein, starch and water contain many atomic groupings such as C-H, N-H and O-H. Whenever materials are irradiated by energy from light, this energy is absorbed by molecules in the materials. The molecules are in different forms such as protein, water, starch, oil, fiber and others, which will absorb different energy levels based on their characteristics [18]. By this fundamental, when normal tangerines and granulation tangerines are irradiated by light energy, they shall absorb different level of energy. In this case, it can be a measure of the difference of energy absorbance by the quantity of light intensity signal that passed through tangerines.

Light technology techniques, such as near infrared spectroscopy (NIRS), have been used to predict the internal qualities of agricultural and food products such as sweet corn [19], tomato [20], marian plum [21] and [22], sweet tamarind [23], mangosteen [24] and [25], pineapple [26] and [27], limes [28] and [29], guava [30], hen's eggs [31] and [32], pork meatball [33] and jelly [34]. Good results for prediction were obtained from the NIRS technique but it is quite expensive. A photoelectric sensor is a lower cost and is simple to use, therefore this technique was evaluated for the detection of granulation. Photoelectric sensors, such as phototransistors, photocells, light dependent resistors (LDR), photodiodes, respond to light in the visible and infrared radiation range [35]. This technique is fast, simple and nondestructive and therefore can be possibly applied to an on-line grading system. In a recent report, the photoelectric sensor was combined with density measurements for predicting the freshness of hen's eggs [36]. Therefore, we decided to study the photoelectric sensor technique for nondestructively prediction of the granulation disorder in tangerines.

2. Materials and Methods

Tangerines cultivar 'Keaw Dumnuan' were purchased from a local fruit market in Thailand. Fifty two fruit were selected that were similar in size and without blemishes. Their maximum diameter and minimum diameters were measured with a vernier caliper (Auto-lock vernier caliper, Long Jer, Taiwan) and their mass with an electronic digital balance (WANT, Jiangsu, China). A low cost prototype photoelectric sensor was designed and developed in order to measure the intensity signal of each fruit as shown in figure 1. The transformer (220V-6V), light source (50W halogen lamp), photoelectric sensor (KODENSHI, silicon phototransistor, ST- 325, spectral sensitivity between 500-1,050 nm and the peak wavelength at 880 nm) and a fruit holder were assembled in a special box, which was designed with a door for controlling the light. Each sample was placed on the fruit holder and the light source was switched on so that transmitted light passed through the fruit to the photoelectric sensor. Four measurements were taken on each fruit (four points in the horizontal position at the equator in every 90o rotation). The average signal was calculated and used for analysis. After measurement, each sample was peeled and inspected and any internal granulation disorder recorded by visual inspection and divided to a set of sound samples and a set of granulated samples. The set of granulated samples were then divided into two groups based on the severity of granulation by measuring the percentage of granulated area on the surface of each segment. This measurement was referred to as the index of severity. On this basis each fruit was classified into one of three groups: 0% granulation, 1-50% granulation and 51-100% granulation. Statistical analysis was carried out with R Statistical Software using ANOVA and Duncan's Multiple Range Test where appropriate at $p < 0.05$.

IOP Conf. Series: Materials Science and Engineering **639** (2019) 012039 doi:10.1088/1757-899X/639/1/012039

Figure 1. Instrument for photoelectric sensor measurement and sample presentation.

3. Results and Discussions

Sound samples and granulated samples were separated by visual inspection as shown in figure 2. The characteristic of the fruit used and intensity signal are shown in table 1. There were twelve sound fruit and forty granulated fruit. The signals from the fruit in sound group were significantly higher ($p < 0.05$) than those in the granulation group (figure 3). This difference was probably due to the acquired intensity signal that was measured from transmitted light that had passed through samples and implies that the tangerines with granulation absorbed higher light energy than those of sound tangerines.

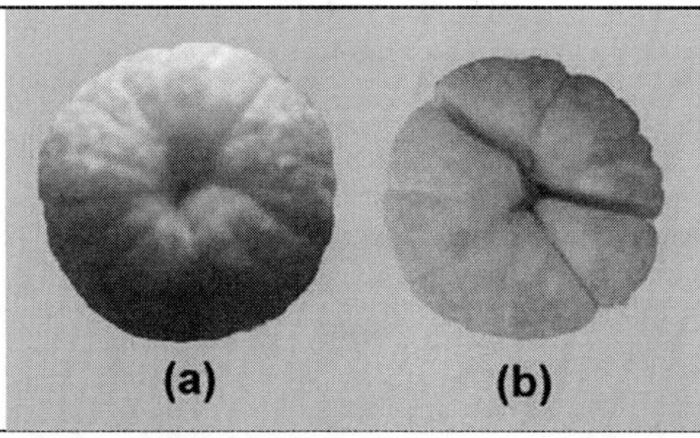

Figure 2. Samples of tangerine with different internal quality (a) sound (b) granulation.

Table 1. Characteristics of the sample (n = 52) of tangerines used for testing the photoelectric sensor.

	Range	Average
Mass (g)	73.5-204.9	134.2
Diameter (mm)	56.9-77.0	65.3
Signal intensity (mV)	25.0-173.0	50.6

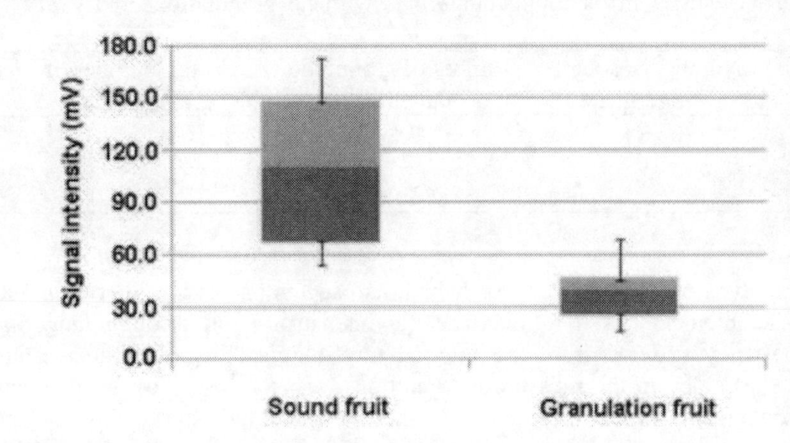

Figure 3. Signal intensity of tangerines with or without granulation, where sound (0% granulation) and granulation (1-100% granulation).

The intensity signal of tangerines was found to be related to the severity of granulation (figure 4) where the signals of sound fruit were significantly higher ($p < 0.05$) than the signals from fruit with 1-50% granulation, which, in turn, were significantly higher ($p < 0.05$) than the signals of 51-100% fruit.

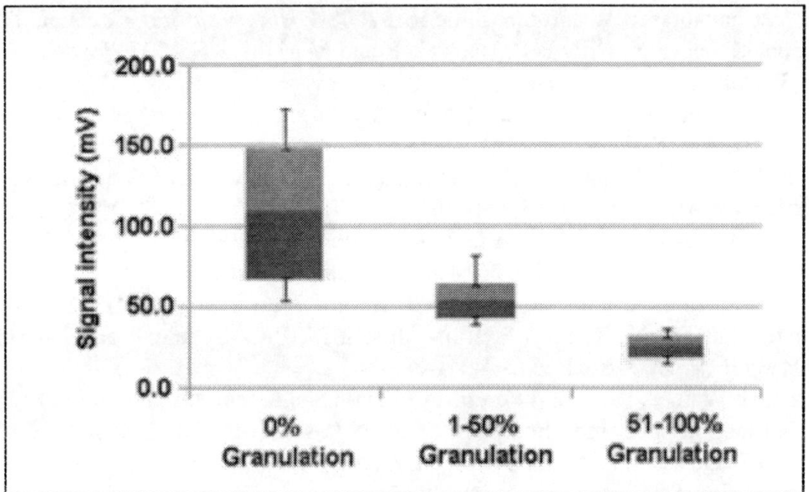

Figure 4. Box plot of signal intensity of tangerines with different levels of granulation.

A box plot of the intensity signal in the group of sound compared to the group of fruit suffering granulation were investigated and the cut off value of 61.7 mV was defined, which was the mean value between a value at 90% percentile of normal group and a value at 10% percentile of granulation group. Therefore, for discrimination analysis, if the intensity signal of fruit was equal or less than 61.7 mV, those fruit was assumed to be the granulation fruit. If the intensity signal of fruit was higher than 61.7 mV, those fruit was assumed to be the sound fruit. The results of discrimination analysis showed that the accuracy for prediction in sound group was 83.3% the accuracy for prediction in granulation group was 92.5%, with an overall accuracy of 90.4% (table 2). It had higher accuracy when compared to the use of capacitance based techniques with the accuracy of 75.6% in the previous study.

Table 2. Signal intensity classification for tangerines, with or without granulation, using a threshold value of 61.7 mV (>61.7 mV= sound tangerines with no granulation; <61.7 mV= tangerines with granulation).

Acquisition	Number of sound tangerines	Number of granulation tangerines	Total Number of tangerines	Classification accuracy (%)
Total	12	40	52	
correct	10	37	47	90.4
Incorrect	2	3	5	

4. Conclusion

A low cost prototype photoelectric sensor, which had spectral sensitivity between 500-1,050 nm, was developed, assembled and used to measure the intensity signal through tangerines to determine whether or not they suffered from the internal physiological disorder called granulation. Results showed 90.4% accuracy in the measurement and also some measure of the intensity of granulation within individual fruit, which indicates that there is potential for developing this prototype for commercial use in commercial online grading of citrus fruit.

5. References

[1] Sharma R R, Singh R and Saxena S K 2006 *Sci. Hortic. (Amsterdam)* **111(1)** 91–6
[2] Schatzki T F, Haff R P, Can I, Le L C and Toyofuku N 1997 *Trans. ASAE* **40** 1407–15
[3] Haff R P, Slaughter D C, Sarig Y and Kader A 2006 *J. Food Process Preserv.* **30** 527–33
[4] Crisosto C H, Valero C and Slaughter D C 2007 *Appl. Eng. Agric.* **23 (2)** 189–94
[5] Chayaprasert W and Stroshine R 2005 *Postharvest Biol. Technol.* **36 (3)** 291–301
[6] Cho B K, Chayaprasert W and Stroshine R L 2008 *Postharvest Biol. Technol.* **47 (1)** 81–9
[7] Hernandez-Sanchez N, Hills B P, Barreiro P and Marigheto N 2007 *Postharvest Biol. Technol.* **44 (3)** 260–70
[8] Lammertyn J, Dresselaers T, Van Hecke P, Jancsok P, Wevers M and Nicolai B M 2003 *Postharvest Biol. Technol.* **29 (1)** 19–28
[9] Slaughter D C, Obenland D M, Thompson J F, Arpaia M L and Margosan D A 2008 *Postharvest Biol. Technol.* **48** 341–46
[10] Wang J A, Gomez H and Pereira A G 2006 *J. Food Qual.* **29** 392–404
[11] Morrison D S and Abeyratne U R 2014 *J. Food Eng.* **141** 107–12
[12] Kim S M, Milczarek R and McCarthy M 2008 *Mod. Phys. Lett. B* **22 (11)** 941–46
[13] Hernandez-Sanchez N, Barreiro P, Ruiz-Altisent M, Ruiz-Cabello J and Fernandez-Valle M E 2004 *Appl. Magn. Reson.* **26 (3)** 431–45
[14] Gomez A H, Wang J, Hu G and Pereira A G 2006 *Sens. Actuators B. Chem.* **113 (1)** 347–53
[15] Tan E S, Slaughter D C and Thompson J F 2005 *Postharvest Biol. Technol.* **35 (2)** 177–82
[16] Yantarasri T and J Sornsrivichai 1998 *Acta Hort. (ISHS)* **464** 494
[17] Teerachaichayut S, Terdwongworakul A, Keawsumnuk K, Rangsi M and Seangkeaw K 2012 *Proc. Int. Conf. of the CIGR-AgEng. (Spain)* p 0122
[18] Williams P C and Norris K H 1987 *Near-infrared technology in the agricultural and food industries* (Minnesota: American Association of Cereal Chemists)
[19] Suktanarak S, Supprung P and Teerachaichayut S 2017 *Acta Hort. (ISHS)* **1152** 47–54
[20] Jannok P, Petcharaporn K and Teerachaichayut S 2014 *Acta Hort. (ISHS)* **1053** 93–9
[21] Teerachaichayut S, Phonmakham S and Suktanarak S 2017 *Acta Hort. (ISHS)* **1179** 25–30
[22] Phonmakham S, Suttivijitpukdee N and Teerachaichayut S 2018 *Acta Hort. (ISHS)* **1213** 537–42
[23] Teerachaichayut S, Suktanarak S and Kasemsumram S 2014 *Acta Hort. (ISHS)* **1053** 113–9
[24] Teerachaichayut S, Kil K Y, Terdwongworakul A, Thanapase W and Nakanishi Y 2007 *Postharvest Biol. Technol.* **43** 202–6
[25] Teerachaichayut S, Terdwongworakul A, Thanapase W and Kiji K 2011 *J. Food Eng.* **106** 206–11
[26] Srivichien S, Terdwongworakul A and Teerachaichayut S 2015 *J. Food Eng.* **150** 29–34

[27] Sukwani S and Teerachaichayut S 2013 *Acta Hort. (ISHS)* **989** 395–9
[28] Huong H T and Teerachaichayut S 2017 *Acta Hort. (ISHS)* **1179** 31–36
[29] Teerachaichayut S and Ho H T 2017 *Postharvest Biol. Technol.* **133** 20–5
[30] Klinbumrung N and Teerachaichayut S 2018 *AIP Conf. Proc.* **2030** 020209-1–5
[31] Suktanarak S, Teerachaichayut S, Jannok P and Supprung P 2018 *Acta Hort. (ISHS)* **1213** 637–42
[32] Suktanarak S and Teerachaichayut S 2017 *J. Food Eng.* **215** 97–103
[33] Sukthanaruk S, Boonpiam S and Teerachaichayut S 2018 *AIP Conf. Proc.* **2030** 020210-1–5
[34] Onnom P and Teerachaichayut S 2018 *AIP Conf. Proc.* **2030** 020211-1–5
[35] Genn R C 1983 *Practical Solid State Circuits* (New Jersey: Prentice-Hall, Business and Professional Division)
[36] Teerachaichayut S, Pansiri J, Nguanprasert P and Thepwapee W 2018 *AIP Conf. Proc.* **2030** 020264-1–5

Acknowledgement
The authors would like to thanks Department of Food Process Engineering, Faculty of Agro-Industry, King Mongkut's Institute of Technology Ladkrabang for all supports. The authors are grateful to Prof. A K Thompson for editing English language in this paper.

Phenolic Compound Extraction from Industrial Tomato Waste by Ultrasound-Assisted Extraction

N Sengkhamparn[1,*] and N Phonkerd[1]

[1]Faculty of Applied Science and Engineering Khon Kaen University, Nong Khai Campus Nong Khai, Thailand, 43000

nipaporn@kku.ac.th

Abstract. The tomato fruits have been reported that it contained high nutrient and bioactive compounds. The various Ultrasound-Assisted Extraction (UAE), including extraction temperature of 30, 50 and 70oC and extraction time of 10, 30 and 50 min, were performed in order to extract phenolic compound from tomato waste from industry. The total phenolic compound, total flavonoid content and also retained carotenoid in obtained extract were determined. Moreover, the DPPH radical scavenging activity of extract were studied. The results showed that the extraction temperature was a main factor which affected the amount of extracted bioactive compound. Moreover, the antioxidant activity of the extract was related to the total flavonoid and carotenoid content which can be draw the regression as follows; Y = 55.81 +0.75X1 + 0.96X2 when radical scavenging activity, total flavonoid and carotenoid content was Y, X1 and X2, respectively. This regression analysis showed high significant (R2 = 0.74). Moreover, the UAE condition of 50oC for 50 min was a best condition which gave the highest antioxidant activity and in accordance with the high amount of total flavonoid and carotenoid content.

1. Introduction

Tomato (*Solanum lycopersicum* L.) is one of the health plants which provide important nutrients such as dietary fiber, protein sugars, mineral (potassium and folate), vitamin (A, E, C) and contain bioactive compound like carotenoids and phenolic compound [1,2]. They are consumed as a fresh vegetable and also processed into many food products like ketchup, paste, sauce, juice. Due to their health properties, tomatoes and also tomato products were increasingly demand for years [2]. During commercial tomato manufacturing, it generally produces a lot of wastes. The component of waste obtained from tomato product industry are varied depending on product processing. These wastes can be used as raw material for bioactive compound extraction such as carotenoids [3,4], pectin [5]. The wastes provided from ketchup industry were consisted of seed, skin and some pulps [3]. Many researches have been studied about the carotenoid contents in tomato by-product however the phenolic contents in by-product have not been limited studied.

Phenolic compounds are found in plants tissue of fruits, vegetables and grains which produced for against microorganism infections and insect injuries [6]. Moreover, they also may reduce risk of prostate, heart diseases and cancer and also have biological effect like anti-inflammatory, anti-allergic and antibacterial [2,6]. There have been reported that the flovanones such as naringenin and flavonols such as quercetin, rutin and kaempferol were the main phenolic compound in tomato [7]. Even, these phenolic compound was be stable during industry processing,but the thermal processing can be also affected to the phenolic content. Crozeir et al [8] reported that the quercetin content in tomatoes was decreased during microwaving, boiling and frying. Moreover, Chang and Liu [9] found that the total flavonoid and phenolic content was be declined during hot air drying. In addition, Vallverdú-Queralt et al. [7] found that during step of paste processing, the phenolic and antioxidant activity was differently changed. Kelebek et al [10] reported that the processing of paste by using low temperature (cold break) can increased the phenolic acids more than that using high temperature (hot break).

Ultrasound-assisted extraction (UAE) has been reported that it was an effective method for the extraction of bioactive compounds due to low extraction time and solvent consumption [11,12]. This technique is used the ultrasonic wave that produce a cavitation phenomenon and caused the damage of solid thus the solvent penetration and mass diffusion will be occurred [12]. However, the extraction temperature and time were affected to the yield. Espada-Bellido et al. [12] found that the extraction temperature was the importance factor for anthocyanin and phenolic compound from mulberry (Morus nigra) pulp.

Therefore, this study was aimed to use UAE for phenolic compound extraction from industrial tomato waste. The antioxidant activity of obtained extract was also determined.

2. Material and Methods

2.1 Material
Tomato waste from the tomato paste industry was dried at 60°C for 24 h, then was ground and kept at -18°C before further experimentation.

2.2 Physicochemical of tomato waste powder
The chemical composition including protein, fat, crude fiber, ash, and carbohydrate content of tomato waste powder was determined according to AOAC 2000 [13]. The results are expressed as g per 100 g dry tomato waste power. The Color of powder was also observed and expressed as L*(lightness), a* (redness) and b* (yellowness).

2.3 Ultrasound-assisted extraction
The tomato waste powder (5 g) was added to 95 % ethyl alcohol at a solid/solvent ratio of 1/20 w/v and sonicated at 30, 50, and 70°C for 10, 30, and 50 min each. Then, the separation was performed by using Hexane containing 0.1 % BHT for fractionation of hydrophobic and hydrophilic part. The hydrophilic part was collected and concentrated by using rotary evaporator. The crude extracted was stored at -18°C for further analysis.

2.4 Total phenolic content
The total phenolic content of the crude extract was determined using Folin-ciocaltau method. The 0.4 mL of crude extract was mixed with 4 mL of 2% Na_2CO_3 and 0.2 mL of Folin-ciocaltau reagent. Then, the mixed solution was left for 30 min and the absorbance of the solutions was measured at 750 nm. Results are expressed as micrograms of gallic acid equivalent per g of dry weight of tomato waste powder.

2.5 Total flavonoid content
The total flavonoid content of the crude extract was determined. The 500 μL of extract was mixed with $AlCl_3$ and left for 10 min. the absorbance of the solution was measured at 430 nm and the results were expressed as micrograms of catechin equivalent per g of dry weight of tomato waste powder.

2.6 Carotenoid content
The carotenoid content of crude extract was determined using spectrophotometer at 503 nm. The results were expressed as mg beta-carotene equivalent per g of dry weight of tomato waste powder.

2.7 DPPH radical scavenging activity

The 2 mL of crude extract was mixed with 2 mL of 0.2 mM DPPH in ethanolic solution. The solution was left in the dark for 10 min. the absorbance of the solution was measured at 515 nm. The ethanol was used as a control solution instead of the extract. The DPPH radical scavenging activity was calculated with the following equation: Radical scavenging activity (%) = (Acontrol-Asample)x100 / Acontrol

2.8 Statistical analysis

All results were done in triplicate and expressed as the mean with standard deviation and the treatment comparison was done by Duncan's New Multiple Range Test at significance level of $p<0.05$. Moreover, the correlation and regression analysis were performed at significance level of 0.05.

3. Material and Methods

3.1 Physicochemical of tomato waste powder

The tomato waste from industry was dried and chemical composition and physical properties were determined. The results showed that the tomato waste powder contained of protein, crude fat, crude fiber and carbohydrate of 22.81$\pm$0.24, 11.89$\pm$0.56, 54.61$\pm$0.78, 2.75$\pm$0.02 and 7.88$\pm$1.12 % (% dry basis), respectively. Moreover, it exhibited the lightness, redness and yellowness as 51.30$\pm$0.03, 22.45$\pm$0.18 and 39.87$\pm$0.56, respectively.

3.2 Total phenolic content

The total phenolic content of crude extract is showed in figure 1. The results showed that the extraction time was not significantly affected to the total phenolic content, while the extraction temperature was significantly affected the total phenolic yield. The extraction temperature of 50°C gave the lowest yield, while that of 70°C and 30°C were comparable.

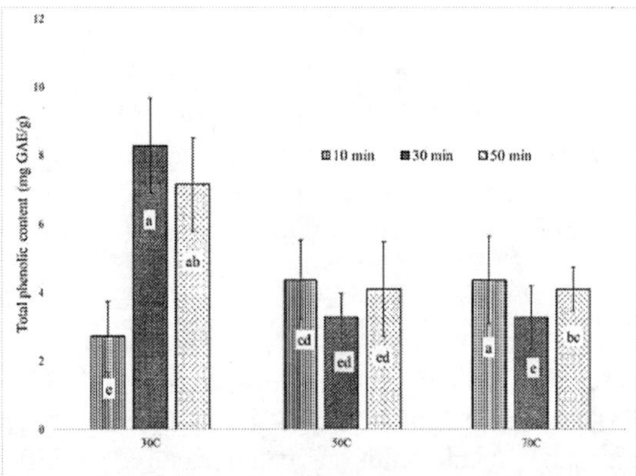

Figure 1. The phenolic content of crude extract. (Different letter in each bar shows the significant difference ($p<0.05$) of the values)

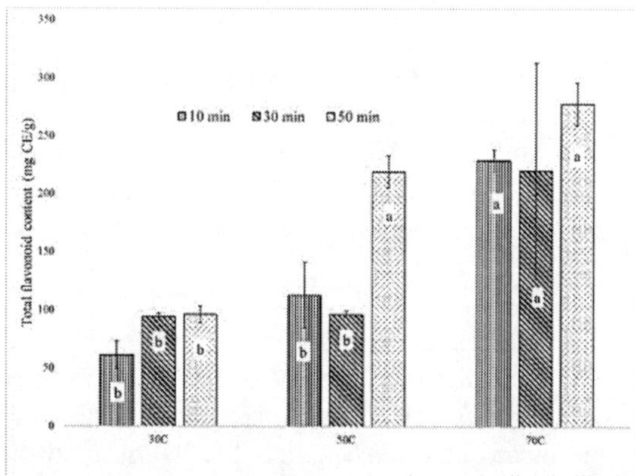

Figure 2. The flavonoid content of crude extract. (Different letter in each bar shows the significant difference ($p<0.05$) of the values)

3.3 Total flavonoid content

The total flavonoid content of crude extract is showed in figure 2. The results showed that the higher extraction temperature, the higher flavonoid yield. Moreover, the extraction time of 30 min and 50 min were comparable.

3.4 Carotenoid content

Due to the extract showed an orange solution, therefore the carotenoid content was determined and compared to beta-carotene. The results are showed in table 1. These results showed that even though the fractionation was performed, still carotenoid was found to retain in the extract. Moreover, the higher extraction time, the higher carotenoid content was found.

3.5 DPPH radical scavenging activity

The phenolic compound has been reported to be an antioxidant compound which found in natural plant. Therefore, the DPPH radical scavenging activity of extract were determined as shown in table 1. The results exhibited that the higher extraction temperature, the higher activity was observed.

Table 1. Carotenoid content and radical scavenging activity of the extract.

Extraction condition	Carotenoid content (mg ß- carotene/g DM)	Radical scavenging activity (%)
30 °C/10 min	52.35 ± 15.93^{b}	61.84 ± 8.17^{c}
30 °C/30 min	53.01 ± 10.85^{b}	67.73 ± 2.08^{c}
30 °C/50 min	48.37 ± 14.05^{b}	66.69 ± 5.58^{c}
50 °C/10 min	78.19 ± 9.37^{ab}	71.52 ± 8.25^{bc}
50 °C/30 min	100.06 ± 68.41^{ab}	72.59 ± 12.41^{ab}
50 °C/50 min	134.30 ± 29.32^{a}	85.87 ± 1.37^{a}
70 °C/10 min	117.95 ± 12.78^{a}	81.73 ± 2.51^{a}
70 °C/30 min	58.31 ± 28.11^{b}	79.84 ± 9.52^{a}
70 °C/50 min	84.50 ± 14.03^{ab}	86.19 ± 0.53^{a}

* Different letter in each column shows the significant difference ($p<0.05$) of the values.

3.6 Regression analysis

The multiple regression coefficients analysis for radical scavenging activity found that both total flavonoid and carotenoid content was found to be significant ($p<0.05$). Moreover, the multiple regression coefficient analysis and the ANOVA for this regression are showed in table 2 and 3 with the correlation coefficient of 0.74. The relation of residual and radical scavenging activity observed was plotted (figure 3) and showed that no irregular values was found.

Table 2. Coded regression coefficients, standard error and p values for building the model.

	Coefficients	*Standard Error*	*t Stat*	*P-value*
Intercept	55.81	2.47	22.59	1.11×10^{-17}
flavonoid	0.75	0.17	4.50	0.000148
carotene	0.96	0.39	2.46	0.02159

Table 3. ANOVA for the regression

	df	*SS*	*MS*	*F*	*Significance*
Regression	2	2081.81	1040.907	39.06762	2.83×10^{-8}
Residual	24	639.45	26.64373		
Total	26	2721.26			

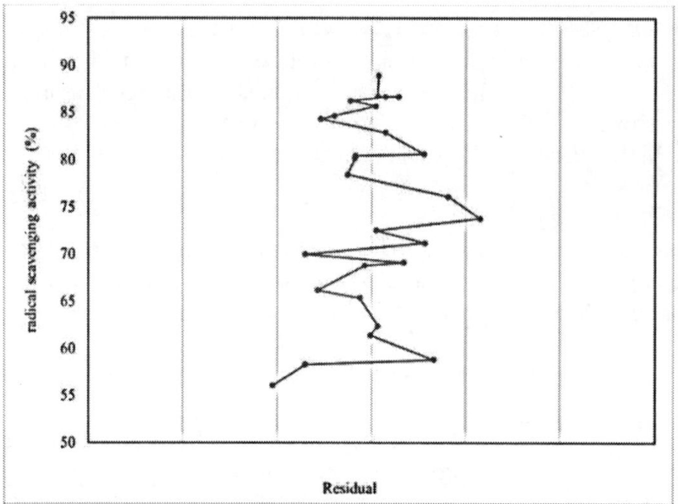

Figure 3. Plot of residual versus measured radical scavenging activity of the crude extracts.

4. Discussion

During the commercial processing of tomato paste, the large amount of waste was produced. This waste has been used as a raw material for bioactive compound extraction [3-5]. The tomato waste was obtained from tomato industry in Nongkhai, Thailand. The chemical composition of tomato waste powder was determined. The results showed that the tomato waste powder contained mainly crude fiber which was 54.61 % which was lower than that of reported by Kanyakam and Uriyapongson [14]. This probably due to the method, this study has used the hot acid and alkali solution to digest while Kanyakam and Uriyapongson [14] used the enzyme which more specific for determine the fiber content. The tomato waste powder showed higher redness and yellowness pointed to lycopene content in the sample. However, the phenolic content has been also reported that it found in tomato and the research in this point of view was limited.

The UAE method has been reported that it was an efficiency method due to lower time and solvent consumption. Therefore, the phenolic compound was extracted from industrial waste by UAE. The various UAE condition were performed and the results showed that the extraction temperature was more important factor than extraction time for total phenolic content (figure 1). The extraction temperature of 50°C yielded the lowest phenolic compound this was probably due to the activity of polyphenol oxidase (PPO). The activity of PPO in tomato has been reported that it has highest activity at 55°C [15], hence the phenolic compound was degraded higher than others. Manieddine et al [16] reported that some phenolic compound in tomato has been degraded by heat treatment for example hydroxycinnamic acid, while some phenolic such as caffeic acid and chlorogenic acid has be found to be increased. Moreover, the extraction temperature of 70°C was comparable to that of 30°C. This may be explained by the cavitation effect during extraction. The cavitation effect was a phenomenal concerning the collapse of bubbles near by the surface of material, hence caused the damage of cell wall. This effect was interrupted during higher temperature due to the high-pressure vapor of solvent. Therefore, the cavitation effect at temperature of 70°C was lower than that of 30°C so the phenolic yield could be declinable extracted at temperature of 70°C.

Similar to phenolic yield, the extraction temperature was more important factor. The results showed that the higher extraction temperature, higher flavonoid was extracted (figure 2). This can be explained by the increasing of flavonoid solubility to the solvent at the higher extraction temperature. Moreover, the cavitation effect seems to less affected to the flavonoid yield than phenolic yield. In addition, the flavonoid content in the extract was higher than phenolic content in accordance to Vallverdú-Queralt et al [7] which reported that the main phenolic compound was flovanones and flavonols. However, the results were variated due to the tomato waste powder was consisted of seed, skin and pulp.

The obtained extract showed red-yellow liquid, this may be because of the retain of carotenoid compound. The carotenoid compound in the extract was determined and compared to the beta-carotene standard. The results (table 1) showed that the extraction temperature of 50 and 70°C was comparable and the extraction temperature of 50°C for 50 min gave the highest carotenoid content. This can be explained by the cavitation effect during sonication. This effect enhanced the temperature of solvent and also the micro-jet on material surface which caused the higher solubility of bioactive compound. At higher extraction temperature (70°C) the cavitation phenomenon was affected by vapor of solvent; hence the lower yield of carotenoid was found. Moreover, this results also pointed that even though the fractionation of the crude extract by Hexane, still some carotenoid retained in the hydrophilic part. This carotenoid compound can be affected to antioxidant activity of the extract.

Antioxidant activity of extract was performed using DPPH radical scavenging assay. The results (table 1) showed the extraction temperature still was an important factor. Moreover, the extraction condition at 50°C for 50 min and at 70°C for all extraction gave the high radical scavenging activity. This probably due to at this condition, the bioactive compound especially flavonoid and carotenoid was extracted. Moreover, the correlation coefficient of each bioactive compound and antioxidant activity was 0.83 and 0.75 for flavonoid and carotenoid content, respectively.

The Multiple regression coefficients for antioxidant activity showed that only the total phenolic compound was not significant at a significance level of 0.05. Therefore, both the total flavonoid and carotenoid content were used for regression analysis and the coefficient was showed in table 2 and the regression was significant, by the significant F was lower than 0.05 (table 3). It can be drawn the regression equation ($R^2 = 0.74$) as follow;

$$Y = 55.81 + 0.75X_1 + 0.96X_2$$

when Y was radical scavenging activity, X_1 and X_2 was total flavonoid and carotenoid content, respectively. The residual plot versus radical scavenging measured exhibited that that the values are low and does not appear to show any regular trend. It can be concluded that extraction condition at 50°C for 50 min was the best condition due to it gave high flavonoid content and also carotenoid content which showed the highest antioxidant activity.

5. Conclusion

The tomato waste from industry was used as raw material for phenolic compound extraction. The various UAE condition were performed. The results showed that the extraction temperature was the main factor that affect to the yield of bioactive compound. The higher extraction temperature, the higher flavonoid and carotenoid content was found. The total phenolic content was found lesser than flavonoid content. Moreover, radical scavenging activity of extract was related to the amount of flavonoid and carotenoid in the extract. This can be concluded that the extraction condition of 50°C for 50 min assist with ultrasonication was the best condition.

6. Acknowledgement

The researchers gratefully thank the Research and Technology Transfer Affairs of Nong Khai Campus, Khon Kaen University (KKU-NKC60-009), for their financial support. Moreover, we are also thankful to Supalak Prasertsar and Pimchanok Haputon for their help in experimentation.

References

[1] Coyago-Cruz E, Corell M, Moriana A, Mapelli-Brahm P, Hernanz D, Stinco C M, Beltrán-Sinchiguano E and Meléndez-Martínez A J 2019 *Food Chem.* **277** 480–9
[2] Katırc N, Işık N, Güpür Ç, Ozge Guler H, Gursoy O and Yilmaz Y *Journal of the Saudi Society of Agricultural Sciences*, https://doi.org/10.1016/j.jssas.2018.11.003
[3] Kaur D, Wani A A, Oberoi D P S and Sogi D 2008 *Food Chem.* **108** 711–8
[4] Poojary M M and Passamonti P 2005 *Food Chem.* **188** 84–91
[5] Grassino A N, Brnčić M, Vikić-Topić D, Roca S, Dent M and Brnčić S R 2016 *Food Chem.* **198** 93–100
[6] Hongyan L, Zeyuan D, Tao W, Ronghua L, Steven L and Rong Tsao H 2012 *Food Chem.* **130** 928–36
[7] Vallverdú-Queralt A, Medina-Remón A, Casals-Ribes I, Andres-Lacueva C, Waterhouse A L and Lamuela-Raventos R M 2012 *LWT - Food Sci. Technol.* **47** 154–60
[8] Crozier A, Lean M E J, McDonald M S and Black C 1997 *J. Agr. Food Chem.* **45** 590–5
[9] Chang C H and Liu Y C 2007 *J. Food Sci.* **72** E532–40
[10] Kelebek H, Selli S, Kadiroğlu P, Kola O, Kesen S, Uçar B and Çetiner B 2017 *Food Chem.* **220** 31–41
[11] Ma Y-Q, Chen J-C, Liu D-H and Ye X-Q 2009 *Ultrason. Sonochem.* **16** 57–62
[12] Espada-Bellido E, Ferreiro-González M, Carrera C, Palma M, Barroso C G and Barbero G F *Food Chem.* **219** 23–32
[13] *AOAC.17th ed.*2000 (USA: Washington DC)
[14] Kanyakam K and Uriyapongson J 2010 *Agr. Sci. J.* **41(3/1)(Supplement)** 289–92
[15] Shahriar saeidian and Bahaaldin Rashidzad 2013 *IRJABS.* **4 (11)** 3306–11
[16] Mahieddine B, Amina B, Faouzi S M, Sana B and Wided D 2018 *AOAS.* **63** 135–9

Biological activities and GABA content of Ma-Lod (*Elaeagnus latifolia* Linn.)

M Meeploy and C Niwaspragrit

Expert Centre of Innovative Agriculture, Thailand Institute of Scientific and Technological Research (TISTR), 35 Mu 3, Tambon Khlong Ha, Amphoe Khlong Luang, Changwat Pathum Thani 12120, Thailand

Corresponding author: maneerat@tistr.or.th

Abstract. *Elaeagnus latifolia* Linn. is a type of wild edible fruit plant belonging to the family Elaeagnaceae locally known as Ma-Lod in the Upper North of Thailand. In this study, the fruit of twelve Ma-Lod trees, namely ML-1, ML-2, ML-3, ML-4, ML-5, ML-6, ML-7, ML-12, ML-16, ML-17, ML-18, and ML-19, collected from Sangkhom District, Nong Khai Province were investigated for tyrosinase inhibitory activity, antioxidant activity, free radical scavenging activity, total phenolic content and GABA content. The results show that the reducing power of Ma-Lod flesh extracts highly correlated with their phenolic contents indicating phenolic compounds could be one of the main contributors to oxidant reducing power. Among twelve tested Ma-Lod trees, fruit flesh extracts of ML-1, ML-4, ML-5, and ML-17 were considered to be a promising source of biologically active compounds due to their high total phenolic contents, antioxidant activity and outstanding of tyrosinase inhibitory activity. GABA was also found in Ma-Lod flesh extract and the high recovery of GABA after passing through the Dowex 50W-X8 column was obtained. This wild and underutilized fruit exhibits many valuable characteristics which deserve conservation and utilization. It could be a potential source of natural biologically active compounds in the development of value-added products such as antioxidant-rich products and cosmetics. In addition, new investigation of the presence of GABA in Ma-Lod fruit might lead to the development of GABA food supplement in the future.

Keywords: γ-aminobutyric acid, biological activity, *Elaeagnus*, Ma-Lod, HPLC-ELSD, tyrosinase inhibition

1. Introduction

The genus *Elaeagnus* (silverberry or oleaster) contains about 50 species of flowering plants belonging to the family Elaeagnaceae. They are found around the world which are mainly distributed in subtropical regions of Asia, Europe and some parts of North America. *Elaeagnus latifolia* Linn. is a type of wild edible fruit distributed in Norhteast India, Vietnam and Thailand. In Thailand, *Eleagnus latifolia* Linn. is the commonly found species in the Upper North of Thailand. It is known with the Thai common name Ma-Lod. This deciduous shrub is mostly grown in the back yard garden of many families of the region and harvested during March and April in 3-4 picking. Fruit could be utilized for making jam, jelly, pickle and refreshing drinks. The fruit of many members of this genus including

Elaeagnus latifolia Linn., is a very rich source of vitamins, especially vitamins A, C and E, essential fatty acids, minerals, flavonoids, carotenoids and other bioactive compounds. Different parts of Ma-Lod plant, especially the fruits and flowers, have been used traditionally as an astringent, analgesic, antipyretic and laxative herbal medicine. The objective of this study was to determine the presence of γ-aminobutyric acid (GABA), total phenolics and biological activities, including the *in vitro* mushroom tyrosinase inhibitory activity, antioxidant activity and free radical scavenging activity in Ma-Lod flesh extracts.

2. Materials and methods

2.1. Materials

The fruits of twelve Ma-Lod trees, namely ML-1, ML-2, ML-3, ML-4, ML-5, ML-6, ML-7, ML-12, ML-16, ML-17, ML-18, and ML-19, were observed and collected from Sangkhom District, Nong Khai Province in March 2015. They were transported to the laboratory in cold conditions (4°C) and cleaned by rinsing under running water and pat them dry with paper towels. Maturation stages of Ma-Lod fruit were visually classified based on the color of the exocarp as follows: stage 1 greenish-yellow, stage 2 yellow, stage 3 yellowish-orange, stage 4 orange, stage 5 orangish-red, stage 6 red and stage 7 dark red. Stages 6 and 7 were chosen for this study. Pure Ma-Lod flesh samples were obtained by removing the seeds and the flesh was cut into small pieces, frozen in liquid nitrogen and kept at -80°C until use. GABA, tyrosinase from mushroom, L-3,4-dihydroxyphenylalanine (L-DOPA), 2,2-diphenyl-l-picrylhydrazyl (DPPH), 2,4,6-tris(2-pyridyl)-s-triazine (TPTZ), 6-hydroxy-2,5,7,8-tetramethylchroman-2-carboxylic acid (Trolox), iron (III) chloride hexahydrate, gallic acid and kojic acid were purchased from Sigma-Aldrich (St. Louis, USA). HPLC-grade acetonitrile was purchased from BDH (Lutterworth, UK). Water was purified by a Milli-Q plus system from Millipore (Bedford, USA). Dowex 50W-X8 was purchased from Merck (Darmstadt, Germany). Folin-Ciocalteu's reagent was purchased from Carlo Erba (France).

2.2. Preparation of Ma-Lod flesh extract

Frozen pieces of Ma-Lod flesh were ground into a fine powder using a mortar and pestle and liquid nitrogen. A 10 g sample of the finely ground Ma-Lod flesh was weighed accurately and placed into a 100-mL Erlenmeyer flask. Using a pipette, 100 mM potassium phosphate buffer (pH 7.8) was added to the flask at a ratio of 1 : 1 (w/v). The flask was then covered with aluminium foil and sealed with parafilm. The sample was shaken at 40°C at 150 rpm for 1 h in an orbital shaking incubator. After shaking, the sample was then sonicated in an ultrasonic bath for 30 min in order to enhance extraction efficiency. The sample was centrifuged at 8,000 x g for 15 min using a refrigerated centrifuge. The actual volume of supernatant was recorded. Triplicate 1-mL aliquots of the supernatant were evaporated using a centrifugal vacuum concentrator at 35°C. The extraction yield was 9.94±1.15% (w/w). Ma-Lod flesh extract was stored at -20°C until use.

2.3. 2,2-diphenyl-1-picrylhydrazyl (DPPH) radical scavenging assay

The DPPH radical scavenging assay was carried out according to the method of Yamasaki *et al.* [1]. The ability to scavenge DPPH radical (scavenging activity, %) was calculated by the following equation. All determinations were performed in triplicate.

$$\text{Scavenging activity (\%)} = [(A_{control} - A_{sample})/A_{control}] \times 100 \quad (1)$$

where $A_{control}$ is the absorbance of DPPH radical in ethanol at 517 nm; A_{sample} is the absorbance of DPPH radical solution mixed with Ma-Lod flesh extract solution at 517 nm. The free radical scavenging activity of the flesh extract was expressed as IC_{50}. The half-maximal inhibitory concentration (IC_{50}) value, which is the sample concentration that inhibits the formation of DPPH radicals by 50%, was obtained by probit analysis [2].

2.4. Ferric reducing antioxidant power (FRAP) assay
The FRAP assay was carried out according to the method of Benzie and Strain [3] with minor modifications. The FRAP values were calculated and expressed as millimole Trolox equivalents (TE) per gram dry weight of sample based on the calibration curve plotted using Trolox as a standard at concentrations ranging from 0.01 to 0.15 mM. All determinations were performed in triplicate.

2.5. Tyrosinase inhibitory activity assay
The inhibitory activity of Ma-Lod flesh extract on mushroom tyrosinase was determined using the dopachrome method with L-DOPA as a substrate, according to the method of Jiménez *et al.* [4] with minor modifications. Kojic acid, a standard tyrosinase inhibitor, was used as a positive control in the assay. The percentage inhibition of the tyrosinase activity was calculated by the following equation. All determinations were performed in triplicate.

$$\text{Tyrosinase inhibition (\%)} = \{[(A - B) - (C - D)] / (A - B)\} \times 100 \quad (2)$$

where A is the absorbance at 475 nm with L-DOPA without sample, B is the absorbance at 475 nm without L-DOPA and sample, C is the absorbance at 475 nm with sample and L-DOPA, and D is the absorbance at 475 with sample and without L-DOPA.

2.6. Determination of total phenolic content
The amount of total phenolic content of Ma-Lod flesh extract was determined according to the Folin-Ciocalteu micro method [5]. The total phenolic contents were calculated based on the calibration curve plotted using gallic acid as a standard at concentrations ranging from 0.05 to 0.50 mg/mL and expressed as milligram gallic acid equivalents (GAE) per gram dry weight of sample. All determinations were performed in triplicate.

2.7. Determination of the GABA content in aqueous extract of Ma-Lod flesh
The fruit of ML-1 was chosen for evaluation of GABA contents at three different stages of maturation including stages 3, 5 and 7. The extraction and the analysis of GABA by high-performance liquid chromatography coupled with an evaporative light scattering detector (HPLC-ELSD) were carried out according to the method of Meeploy and Deewatthanawong [6].

2.8. Separation of GABA from aqueous extract of Ma-Lod flesh
The separation of GABA was carried out on the Dowex 50W-X8 column according to the method of Carmona *et al.* [7].

2.9. Statistical analysis
Samples were analysed in triplicate and the results were given as mean ± standard deviation (SD).

3. Results and discussion

3.1. Tyrosinase inhibitory activity of Ma-Lod flesh extracts
Tyrosinase is a copper monooxygenase widely distributed in animals, plants and microorganisms. It is responsible for the synthesis of melanin in animals and browning in plants. Therefore, finding an effective tyrosinase inhibitor is beneficial in antibrowning of foods, medicinal and cosmetic applications in relation to hyperpigmentation. The tyrosinase inhibitory activity of Ma-Lod flesh extracts at the final concentration of 6.00 mg/mL were determined using the dopachrome method. It was found that all the flesh extracts exhibited inhibitory activity against mushroom tyrosinase ranging from 72.41±2.42% to 15.99±3.29%. The fruit flesh extract of ML-17 possessed the highest inhibition of tyrosinase activity (figure 1). Meanwhile, the visual color change of a reaction mixture was also

observed, color alteration from colorless to reddish brown colored dopachrome was obviously reduced in the reaction mixture containing the fruit flesh extracts of ML-1, ML-4, ML-5, and ML-17. This finding was related to their high total phenolic contents (figure 2 and figure 5).

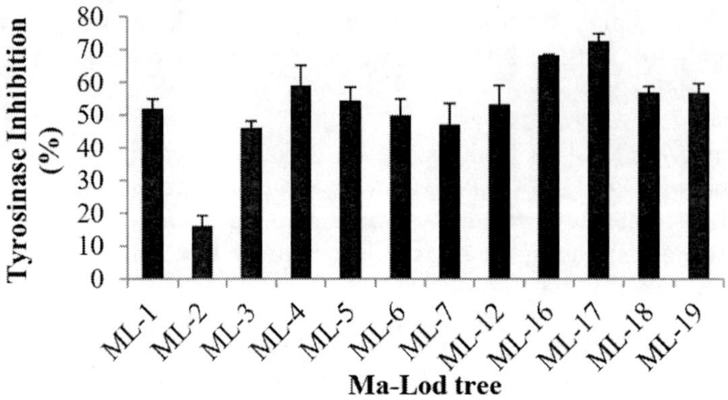

Figure 1. The tyrosinase inhibition of the fruit flesh extract of twelve Ma-Lod trees (*Elaeagnus latifolia* Linn.). Results represent the means ± SD of three replications.

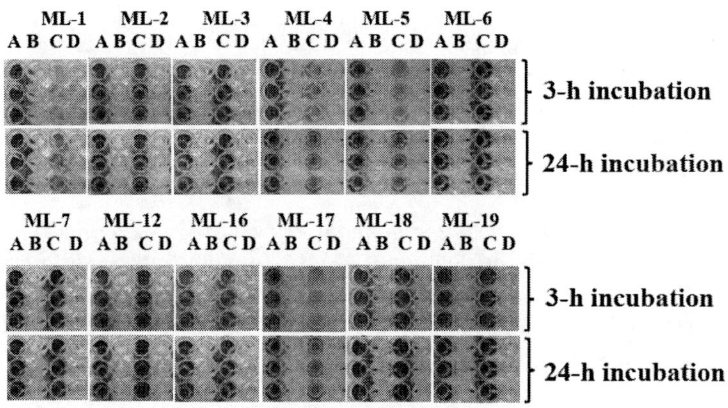

Figure 2. The tyrosinase inhibition of the fruit flesh extract of twelve Ma-Lod trees (*Elaeagnus latifolia* Linn.) was determined using the dopachrome method with L-DOPA as a substrate. The assay was conducted in a 96-well microtiter plate in a total volume of 250 μl. The final concentration of Ma-Lod flesh extracts in the reaction mixture was 6.0 mg/mL. Color of dopachrome formation was observed after incubation in the dark at room temperature for up to 24 h. (A) control, (B) blank of control, (C) sample, and (D) blank of sample.

The inhibition of tyrosinase activity of Ma-Lod flesh extract could be caused by one of the following: 1) reducing agents such as ascorbic acid, which is used as a melanogenesis inhibitor because of its capacity to reduce back *o*-dopaquinone to DOPA, 2) alternative enzyme substrates such as some phenolic compounds, which have a good affinity for the enzyme prevent dopachrome formation, 3) specific tyrosinase inactivators, these inhibitors inhibit tyrosinase activity by inducing the enzyme catalyzing and 4) specific tyrosinase inhibitors, they reversibly bind to tyrosinase and reduce its catalytic capacity. *Elaeagnus latifolia* fruit contains at least three classes of compounds including ascorbic acid, carotenoids, and phenolic compounds, that can inhibit tyrosinase activity [8].

3.2. Ferric reducing antioxidant power (FRAP) of Ma-Lod flesh extracts

The antioxidant capacity of Ma-Lod flesh extracts was determined using FRAP assay. In this assay, ferric ions are reduced to ferrous ions in the presence of a water-soluble antioxidant (or a reducing agent) which form a blue-colored ferrous tripyridyltriazine complex (Fe^{2+}-TPTZ) at pH 3.6. The increase in absorbance at 593 nm is proportional to the total ferric reducing power of the tested sample. Among the Ma-Lod flesh extracts tested in this study, ML-1 possessed the highest antioxidant capacity followed by ML-17, ML-5, and ML-4 with the FRAP values of 169.45±1.16, 113.57±1.80, 96.29±5.92, and 79.64±3.49 mmole TE/g of dried Ma-Lod flesh extract, respectively. The fruit flesh extracts of ML-12 showed the lowest antioxidant capacity with the FRAP value of 12.79±0.77 mmole TE/g of dried Ma-Lod flesh extract (figure 3). The reducing power of Ma-Lod flesh extracts is probably due to the action of hydroxyl group of the phenolic compounds which might act as electron donors.

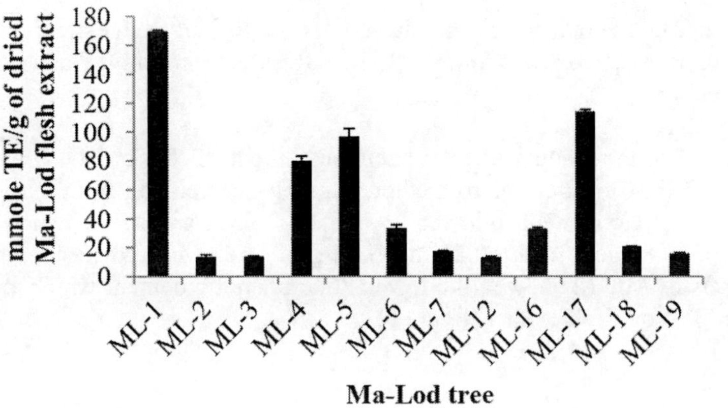

Figure 3. Ferric reducing antioxidant power (FRAP) values of the fruit flesh extract of twelve Ma-Lod trees (*Elaeagnus latifolia* Linn.). Results represent the means ± SD of three replications.

3.3. DPPH radical scavenging activity of Ma-Lod flesh extracts

In this study, all Ma-Lod flesh extracts exhibited the moderate activity on scavenging DPPH radicals. The fruit flesh extracts of ML-1, ML-2, ML-3, ML-4, ML-5, ML-6, and ML-7 had IC_{50} values below 5.00 mg/mL, while the IC_{50} values of ML-12, ML-16, ML-17, ML-18, and ML-19 ranged from 6.48 to 24.74 mg/mL (figure 4). The lower IC_{50} value means the higher antioxidant activity. The DPPH radical scavenging of Ma-Lod flesh extracts is probably due to the action of phenolic hydroxyl groups that are prone to donate a hydrogen atom or an electron to free radical . In addition, the structure of phenolic compounds have extended conjugated aromatic system to delocalize an unpaired electron. Total antioxidant capacity from both DPPH and FRAP assays does not show similar trend (figure 3 and figure 4), this could be due to different mechanism of assay method, structure of different phenolic compounds, the antioxidant protection mechanism exhibited by compounds and also the synergistic effects of different compounds [9].

3.4. Total phenolic content of Ma-Lod flesh extracts

Phenolics are widely distributed in the plant kingdom and are the most abundant secondary metabolites of plants with more than 8,000 phenolic structures currently known. Flavonoids are mainly present in fruits, their phenolic hydroxyl groups are able to chelate metals, reduce lipid peroxidation and have shown a high antioxidant and free radical scavenging activities [10].

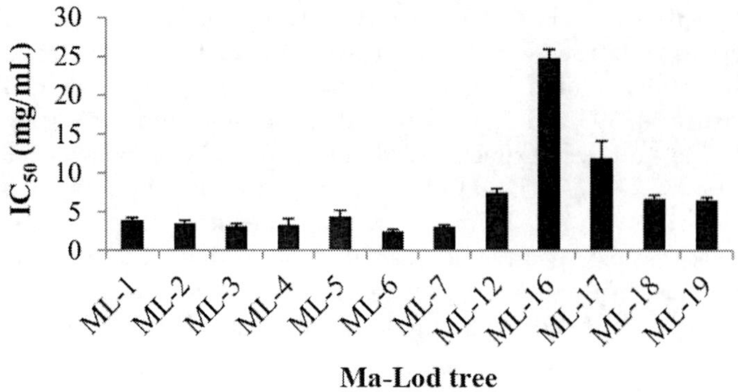

Figure 4. DPPH radical scavenging activity of the fruit flesh extract of twelve Ma-Lod trees (*Elaeagnus latifolia* Linn.). Results represent the means ± SD of three replications.

In the present study, the amount of total phenolic content of Ma-Lod flesh extract was determined according to the Folin-Ciocalteu micro method. Among the Ma-Lod flesh extracts tested, ML-1 had the highest total phenolic content followed by ML-17, ML-5, and ML-4 which value of 31.49±0.34, 19.75±0.74, 16.66±0.80, and 14.99±0.21 mg GAE/g of dried Ma-Lod flesh extract respectively. The fruit flesh extracts of ML-12 showed the lowest total phenolic content which value of 3.56±0.41 mg GAE/g of dried Ma-Lod flesh extract (figure 5).

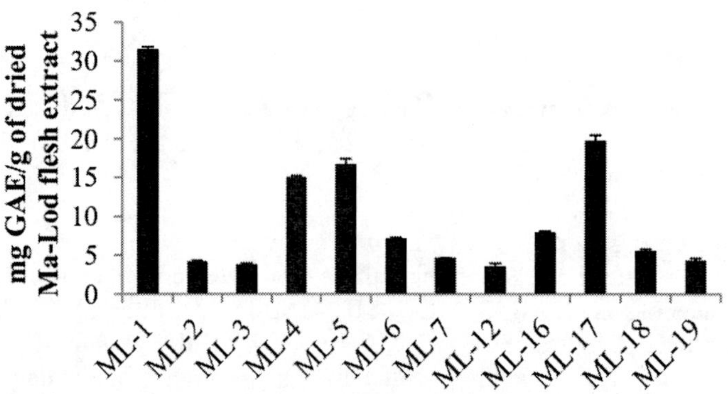

Figure 5. Total phenolic content of the fruit flesh extract of twelve Ma-Lod trees (*Elaeagnus latifolia* Linn.). Results represent the means ± SD of three replications.

The correlation analysis between antioxidant capacity and total phenolic content were also determined. The results show a high correlation between FRAP and total phenolic content (r = 0.9965), which suggested that total phenolics in Ma-Lod flesh extracts were mostly responsible for the ferric reducing power (figure 6). Poor correlation was obtained between the DPPH and total phenolic content (data not shown), indicating phenolic compounds do not strongly scavenge free radicals. Antioxidant activity could also be contributed by synergistic action of other phytochemicals which are not determined in this study such as carotenoids, ascorbic acid, polysaccharides and amino acids. In addition, we found that the inhibitory effects of the flesh extracts of ML-1, ML-4, ML-5, and ML-17, on mushroom tyrosinase are related to their total phenolic contents (figure 2 and figure 5) and FRAP values (figure 2 and figure 3). These results support the fact that polyphenols are highly potent antioxidants and several polyphenols are accepted as tyrosinase inhibitors. Phenolic compounds such as catechin, epicatechin, kaempherol, quercetin, coumarines, caffeic acid and stilbenes are found in *Elaeagnus angustifolia* Linn. fruit extract. These phenolic compounds showed monophenolase and/or diphenolase inhibitory activity toward mushroom tyrosinase [11].

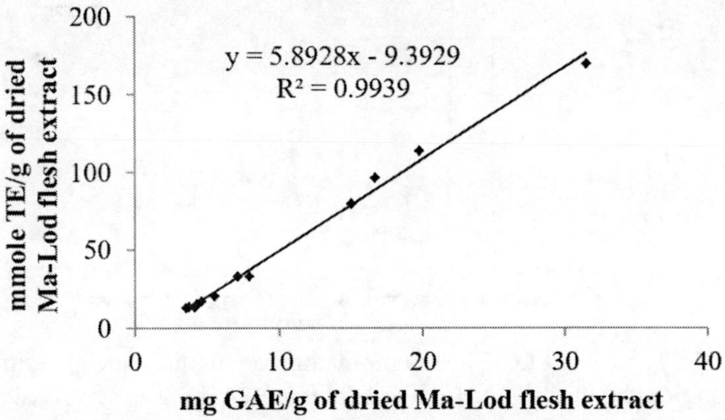

Figure 6. The positive correlation between the antioxidant capacities measured by the ferric reducing antioxidant power (FRAP) assay and total phenolic content of Ma-Lod flesh extracts.

3.5. GABA content

In this study, the fruit of ML-1 was chosen for evaluation of GABA contents at three different stages of maturation including stages 3, 5, and 7. Chromatographic separation was carried out on the Inertsil®NH_2 HPLC column. GABA was separated with the retention time of 54.6500 min. It was found in stages 3, 5, and 7 with the amounts of GABA content at 0.1301±0.0054, 0.1640±0.0032, and 0.0913±0.0080 mg/g fresh weight, respectively (figure 7). The results show that the highest level of GABA content was detected in stage 5 and decline level of GABA content started from stage 5 to stage 7, indicating the beginning of fruit senescence. It decreased almost 2-fold for stage 7. Previous research studies have reported that GABA content in the fruit of cherry tomato peaked before starting maturation and then declined during ripening. A similar pattern was observed during fruit development in Micro-Tom. GABA accumulation in fruits at the early development stage might protect immature seed and the GABA levels decline during the ripening stage when seeds have already matured. In addition, the contents of GABA found in bitter melons, tomatoes, apples, and rambutans were 0.41-2.06 mg/g dry weight, 0.50-0.67, 0.007, and 0.48-0.94 mg/g fresh weight, respectively [6,12].

After extraction of Ma-Lod flesh with water, 5-mL of a filtered sample extract was passé through the Dowex 50W-X8 column. GABA was eluted with 2 N NH_4OH at a flow rate of 0.8 mL/min. Of the 3 mL effluent and eluate fractions were collected. The HPLC-ELSD chromatogram of GABA in the 22nd fraction resulted in high GABA recovery of 98.08% obtained from the column (figure 8). From figure 7 and figure 8, the results show that numerous impurities in the aqueous extract of Ma-Lod flesh were removed after passing through the Dowex 50W-X8 column.

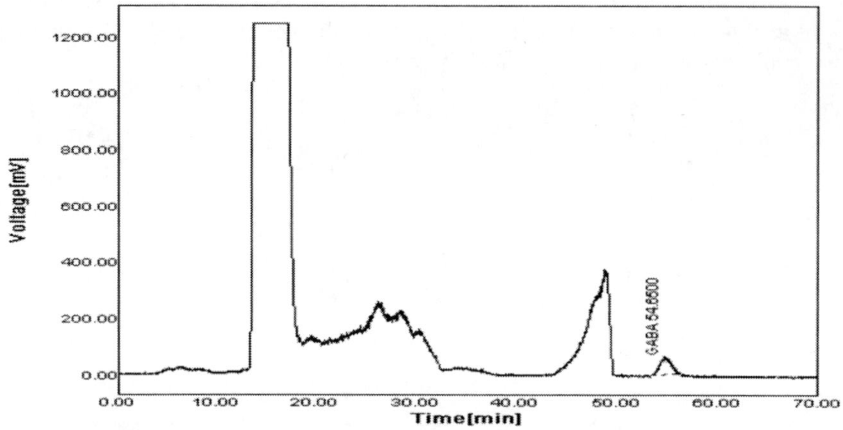

Figure 7. Typical HPLC-ELSD chromatogram of the aqueous extract of Ma-Lod flesh separating on the Inertsil® NH_2 HPLC column. GABA was separated with retention time of 54.6500 min.

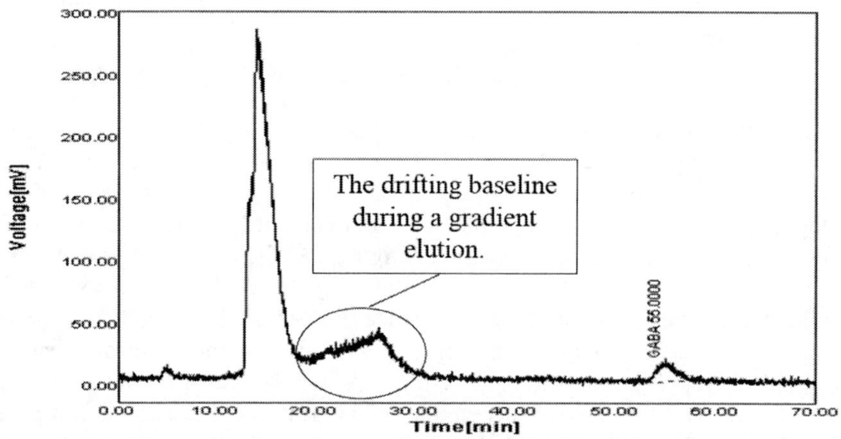

Figure 8. HPLC-ELSD chromatogram of the eluate (the fraction 22) from the Dowex 50W-X8 column.

4. Conclusions

According to the results of this study, there was a high correlation between the FRAP value and total phenolic content, indicating phenolic compounds could be one of the main contributors to oxidant reducing power of Ma-Lod flesh extracts. The fruit flesh extracts of ML-1, ML-4, ML-5, and ML-17 could be the promising sources of biologically active compounds due to their high total phenolic contents, antioxidant activity and outstanding of tyrosinase inhibitory activity among twelve tested Ma-Lod trees. In addition, GABA was also found in Ma-Lod flesh extract and the high recovery of GABA was obtained after separation it by ion exchange chromatography. This study shows that this wild edible fruit could be considered as a potential source of natural biologically active compounds for application in food supplement, medicine, and cosmetics in the future.

Acknowledgements

The authors thank Thailand Institute of Scientific and Technological Research (TISTR) for facilities and necessary equipments.

IOP Conf. Series: Materials Science and Engineering **639** (2019) 012043 doi:10.1088/1757-899X/639/1/012043

References

[1] Yamasaki K, Hashimoto A, Kokusenya Y, Miyamoto T and Sato T 1994 *Chem. Pharm. Bull. (Tokyo)* **42** 1663-65
[2] Finney D J 1971 *Probit Analysis* (Cambridge: Cambridge University Press) p 333
[3] Benzie I F and Strain J J 1999 *Methods Enzymol.* **299** 15-27
[4] Jiménez M, Chazarra S, Escribano J, Cabanes J and Carmona F G 2001 *J. Agric. Food Chem.* **49** 4060-63
[5] Singleton V L, Orthofer R and Lamuela-Raventós R M 1999 *Methods Enzymol.* **299** 152-178
[6] Meeploy M and Deewatthanawong R 2016 *J. Chromatogr. Sci.* **54** 445-452
[7] Carmona E, Gomes C and Trolin G 1980 *Acta Pharmacol. Toxicol. (Copenh)* **46** 235-240
[8] Chang T S 2009 *Int. J. Mol. Sci.* **10** 2440-75
[9] Bóhm F, Edge R, McGarvey D J and Truscott T G 1998 *FEBS Lett.* **436** 387-9
[10] Khattak K F 2012 *J. Med. Plants Res.* **6** 5196-5203
[11] Hamidpour R, Hamidpour S, Hamidpour M, Shahlari M, Sohraby M, Shahlari N and Hamidpour R 2016 *J. Tradit. Complement Med.* **7** 24-29
[12] Saito T, Matsukura C, Sugiyama M, Watahiki A, Ohshima I, Iijima Y, Konishi C, Fujii T, Inai S, Fukuda N, Nishimura S and Ezura H 2008 *J. Japan Soc. Hort. Sci.* **77** 242-250

IOP Conf. Series: Materials Science and Engineering **639** (2019) 012044 doi:10.1088/1757-899X/639/1/012044

The optimization for decreasing ruptured billets when harvesting sugarcane with chopper harvester

K Kosum[1], and S Bun-art[1]

[1]School of Agricultural Engineering, Suranaree University of Technology, Muang, Nakhon Ratchasima, Thailand

E-mail: kk_changnoi@yahoo.com

Abstract. The purpose of this study was to analyze the level of factors that are optimized for the amount of ruptured billets when harvesting sugarcane with the sugarcane harvester. A factorial experimental design was used to study the effect of three factors: sugarcane type, planting and soil type. The experiments showed that the three factors affected the quantity of ruptures billets. Also, the variance analysis was done at a significant level of 0.05, and it was found that the sugarcane type, soil type and the planting technique had an influence on the amount of ruptures billets. And ratoon-sugarcane double-row and sand is the level of suitable factors that can reduce the amount of ruptures billets by 25%.

1. Introduction

Sugarcane is an economic crop that the Thai government encourages farmers to plant in order to expand the area planted as articulated under the Agricultural Crops Strategy [1]. In 2014, Thailand produced 103.7 million tons of sugarcane, and up to 11.3 million tons of sugar. About 65 percent of the output was exported to major customer countries including Indonesia, Japan, China and South Korea. The revenue brought into the country by sugar and sugarcane exports was 102,103 million baht [2].

In the present day, agricultural machinery is widely used to facilitate all steps of sugarcane production, especially in harvesting which takes place during November and May of each year. During these harvest times, farmers require a lot of labor to cut sugarcane so that it can reach the sugar mill before the closure of the season. But, at the present time, agricultural laborers have moved away from the industry [3] because agricultural work is difficult and produces a low income. As a result, the newest generation of workers have moved away from working in agriculture. In addition, the increase in the minimum wage increased the cost of production for farmers. One solution to the labor shortage was to use sugarcane harvesters which could replace the workers who once harvested sugarcane during the harvest season.

Various harvesting practices are used in sugarcane production, each requiring different harvesting technologies to ensure effective and efficient field operations. Based on how the sugarcane is presented to the harvester, harvesting practices are classified into two groups: 1) harvesting sugarcane with pre-harvest burning (burnt sugarcane harvesting) and 2) harvesting sugarcane without burning (green sugarcane harvesting). Based on the form of the harvested materials, two classes of harvesters have been developed, whole stalk harvesters and chopper harvesters [4].

A typical whole stalk harvester system consists of a topper, a base cutter, a feeding mechanism, and a discharging mechanism. The topper is designed to sever sugarcane tops and then discharge the severed tops to the side of the harvesting rows. Topped sugarcane stalks are then cut by the base cutters at about

30 mm above the ground level [5]. The feeding mechanism includes a set of rollers to convey the sugarcane stalks to the discharging mechanism. The discharging mechanism then delivers harvested stalks to either a wagon or onto the field.

Chopper harvesters include the components of whole stalk harvesters but have extra components including choppers and extractors. The functionality of the chopper and extractor are to chop whole stalk into billets and to separate leafy materials. During the harvesting process of the chopper harvester, the discharging mechanism is used to deliver the billets to a wagon or a truck [4].

Sugarcane harvesters used for sugarcane harvesting are both imported from overseas and produced by local manufacturers. In the operation of both the domestic and imported sugarcane harvesters, there are problems that cause a loss of productivity. Linedale [6] identified four major types of harvesting loss which includes: 1) extractor loss; sugarcane thrown from the harvester during the cleaning processes, 2) boot and elevator loss; billets falling from the harvester during its operation, 3) pick-up loss; sugarcane broken off or run-over and not taken into the harvester, and 4) spillage loss; billets spilt during transfer to the haul out equipment and en route to, or at, the tram siding or road transport systems.

Loss caused by the work process of chopper harvester can be divided into 1) Quantitative losses (weight), when the sugarcane is cut into billets, there will be an open surface on both ends of the sugarcane. Some billets have a rupture which increases the surface area without peeling when exposed to the dry weather of winter in Thailand. This makes the water in the sugarcane billets to evaporate into the atmosphere more easily. 2) Quality losses, when the sugarcane has an open surface area that is exposed to the environment, microorganisms can enter inside the sugarcane. These microorganisms can multiply rapidly until the sugarcane is spoiled. Consequently, the sugarcane will be rejected for purchase due to its condition because it will negatively affect the quality of sugar and machinery in the production process.

From the problems discussed above, the cost of using a sugarcane harvester has increased every year. Therefore, this study aims at determining the amount of ruptured billets and the factors affecting the amount of ruptured billets to provide information for decision making in controlling the operation of sugarcane harvester and to reduce such losses. The hypothesis of this research is that harvesting with sugarcane harvester affects the amount of ruptured billets.

2. Materials and methods

The research method established after a preliminary survey in Khonkhen and Chaiyaphum province in Thailand was selected for the analysis of the problem and to determine the value of the factors used in the experiment. In Khon Kaen, 3 sugarcane varieties were selected for analysis. The control factors included sugarcane without burning and the use of a chopper harvester (Austoft 8000 sugarcane harvester series, FPT Cursor 9 engine 4 valves 9 liters - 353hp (260kW) @ 2100 rpm). The sugarcane type, soil type and planting technique were the initial factors used to determine the amount of impurities and the statistical analysis from sugarcane harvested with sugarcane harvesters.

2.1. Experiment to find factors affecting the amount of ruptured billets with sugarcane harvesting by chopper harvesters.

The research was conducted using the 2k factorial experiment design because this experimental method leads to a high performance while monitoring the influence of many factors [7].

Table 1. Shows the factors and levels of each factor used in the experiment.

Factor	Level	
	-1	1
Sugarcane type	Plant	Ratoon
Type of planting	Single-row	Double-row
Soil type	Sand	Clay

2.1.1. Determining the level of factors used in the experiment. The experiment used two levels of factors: low (-1) and high (1) as shown in Table 1 with the three factors: sugarcane type, type of planting, and soil type. Usually, sugarcane is a plant that can be harvested many times. There are 2 types of planting: single-row and double-row. The soil type is in accordance with the soil conditions in the study area.

2.1.2. Experimental design. In this experiment, we used a statistical program (Minitab V. 18) to determine the order of experiments and the experimental design (DOE). For the study of three replicates, there are 24 units in the experiment.

2.1.3. Experiment conducted to determine the amount of ruptured billets found in sugarcane harvested by chopper harvester. The operator adjusted the speed of the harvester within the limits of the machine under the given working conditions in order to achieve the requirements for raw material quality and process productivity [8]. In preparation for sugarcane testing, the sugarcane was sampled on the trucks in the location where the sugarcane was harvested [9] by placing sample containers over the sugarcane on the front, center and rear of the truck being used with the sugarcane harvester. Then, the sugarcane from all 3 sample containers were collected to determine the total weight of the ruptured billets.

2.2. Statistical analysis of results

The results were interpreted using a statistical analysis to determine the relationship between the initial factors and the response factors. The statistical analysis was divided into 2 parts as follows:

2.2.1. Residual analysis. Using the residuals derived from the model chosen in Section 2.1, the residuals were normalized by distributing the variance of the residuals of various factors. The constant variance and decay needs to be random.

2.2.2. Analysis of variance. The variability source of the model was examined by considering the P-values of the various terms in the table for the analysis of variance compared with the statistical significance. ($p = 0.05$)

3. Results and discussion

3.1. The amount of ruptured billets found in sugarcane harvesting trial

Minitab V.18 was used to help plan the experiments using a 2k factorial design with three factors: sugarcane type, type of planting and soil type. There were 24 experimental units. It was found out that the average of the ruptured billets is 0.48 kg per 1 experiment.

3.2. Statistical analysis

3.2.1. Residual analysis. When the residuals were analyzed, it was found out that the model of each response was not abnormal because the distribution of residual is normal. The variance of the factors is uniform and there were no out-of-control signals for the data in the control charts. The results show that the responses are normal.

3.2.2. Analysis of Variance. Based on the comparisons of table 2, the statistical analysis shows that the sugarcane type, soil type and planting technique had a statistically significant influence on the amount of ruptured billets.

3.2.3. Analysis of the impact of factors influencing the response. The co-factors influencing the amount of ruptured billets are shown in figures 1. It was found out that the ruptured billets of the sugarcane is caused by the influence of sugarcane type and soil type. There are intersections on the line that connects the average of the results of both factors. Therefore, the increase or decrease of the ruptured billets will be changed by both factors accordingly. The ratoon-sugarcane has a wider period of ruptured billets between the sand and clay than plant-sugarcane. This implies that the ratoon-sugarcane planted in clay soil has the most ruptured billets.

Table 2. Shows results of analysis of variance of ruptured billets.

Source	P-Value
Model	0.005
Linear	0.616
sugarcane type	0.919
Planting	0.368
soil type	0.339
2-Way Interactions	0.001
sugarcane type*planting	0.806
sugarcane type*soil type	0
planting*soil type	0.229
3-Way Interactions	0.049
sugarcane type*planting*soil type	0.049

The double-row planting technique has a period of ruptured billets between the sand and clay that is wider than single-row. This implies that the most ruptured billets are planted in clay soil using the double-row planting technique.

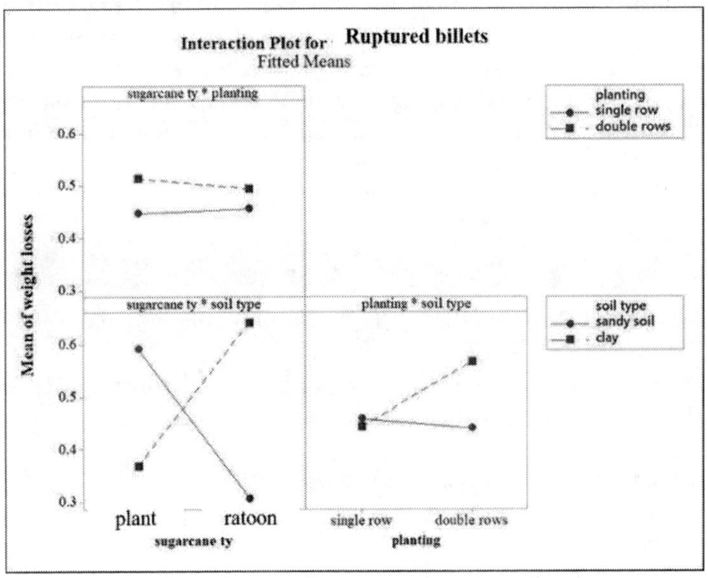

Figure 1. shows the effect of the co-factors influencing the amount of ruptured billets

3.2.4. Determination of optimum values of factors affecting the ruptured billets when harvesting sugarcane of chopper harvester. The result of the analysis of the most optimal value of factors affecting the ruptured billets from the operations process of chopper harvester is shown in table 3.

Table 3. Shows the optimal value of factors affecting the ruptured billets when harvested sugarcane with chopper harvester.

1. Parameters

Response	Goal	Lower	Target	Upper	Weight	Importance
weight losses	Minimum		0.14	0.93	1	1

2. Solution

Solution	sugarcane type	Planting	soil type	weight losses Fit	composite desirability
1	ratoon sugarcane	double-row	sand	0.23	0.88607

3. Multiple Response Prediction

Variable	Setting
sugarcane type	ratoon sugarcane
Planting	double rows
soil type	sand

The prediction response of the rupture of the billets equals to 0.230, and the overall composite desirability level is equal to 0.8861. It can be concluded that ratoon-sugarcane, sandy soil and double-row planting technique are the appropriate factors that have led to the 25 percent decrease of ruptured billets harvested by chopper harvester as compared to the average ruptured billets from the past tests.

4. Conclusion

This research was designed using a 2k factorials to determine the amount of ruptured billets when sugarcane is harvested using the chopper harvester considering 3 factors: sugarcane type, soil type and planting technique. Based on the analysis of variance at the level of significance 0.05, it was found out that the sugarcane type, soil type and planting technique had an influence on the amount of ruptured billets. And from the optimal level of factors analysis, it was also found out that the type of sugarcane is ratoon-sugarcane, the soil type is sand and the planting technique is double-row. These are the suitable factors that reduce the amount of ruptured billets when harvesting sugarcane using the chopper harvester.

References

[1] Office of Agricultural Economics 2013 *Agricultural economics information* (Bangkok: Ministry of Agriculture and Cooperatives)
[2] Office of Agricultural Economics 2014 *Agricultural statistics of Thailand crop year 2014* (Bangkok: Ministry of Agriculture and Cooperatives)
[3] Office of Agricultural Economics Region 4 2014 *Study on the use of agricultural technology* (Kaen: Ministry of Agriculture and Cooperatives)
[4] Shaochum, M, Patrict, A S Manoj, K, and Qin, Z 2014 Sugarcane harvester technology: a critical overview *Appl. Eng. Agric.*, *30*(**5**) 727
[5] Esquivel, B M, Marrero, S M, Ponce, E, Guerrero, A C, Stainlay, T, Villaruz, J, Wood, A, and Dibella, L 2008 Evaluation of the automatic base-cutter control system in the Australian sugarcane industry *Proc. 2008 Conf. Australian Soc. Sugar Cane Tech.* pp 322-327
[6] Linedale, A L 1997 *Final report srdc project bs82s* (BSES Publication SRDC final report sd97009)
[7] Montgomery, D C 2009 *Design and analysis of experiments* (N Y: John Wiley & Sons, Inc.)
[8] Peloial, P R, Milan, M and Romanelli, T L 2010 Capacity of the mechanical harvesting process of sugar cane billets. *Sci. Agric. (Piracicaba, Braz.)* ***67*(6)** 619

IOP Conf. Series: Materials Science and Engineering **639** (2019) 012044 doi:10.1088/1757-899X/639/1/012044

[9] Opanukul, W, Nakwattananukul, S, Jongsukwai, K, Tunhaw, M, Tongdaeng, B, Sarntoonpituk, D, Sookniyom, S 2012 Study of sugarcane harvester used in Thailand. Department of Agriculture

Acknowledgement
The researcher would like to thank Khun Anupong Niamcharoen, Khun Songsak Phachada and Khun Yothin Kaewsriprom for their advice and assistance. I would also like to thank Suranaree University of Technology for providing financial support for this research.

Real-Time Monitoring of Dissolved Oxygen at a Vicinity of an Aquatic plant

Kansuk Patthamawan, Tamura Keisuke and Xing-Zheng Wu

Life, Environment and Applied Chemistry, Fukuoka Institute of Technology, Fukuoka, Japan

mbm18202@bene.fit.ac.jp, tamy138045c@i.softbank.jp and wu@fit.ac.jp

Abstract. Real-time monitoring of dissolved oxygen (DO) concentration at a vicinity of an aquatic plant in water by the fluorescent quenching method is discussed. Ru(II) complex (Tris(2,2'-bipyridyl) ruthenium(II) chloride) was used a fluorescent probe, whose fluorescence was quenched by DO. The fluorescence of the Ru(II) complex at a vicinity of the aquatic plant was excited by a semiconductor laser of 405 nm, and detected by a PMT. Experiment results showed that the DO concentration at the vicinity of *Egeria Densa* decreased with time in the respiration processes, while increased with time in the photosynthetic processes. It was further applied to monitor DO at vicinities of the aquatic plant in water environments containing heavy metal ions such as Cu^{2+}.

1. Introduction

Dissolved oxygen (DO) is an important parameter to measure water quality as the DO level either too high or low can harm aquatic life and affect water quality [1]. The DO in water varies with temperature, pressure, and saltiness. It is well known that DO decreases as the temperature increases.

Aquatic plants play important roles in the water quality and water environment. They absorb or degrade toxic substances such as heavy metal irons, salicylic acid, carbonic acid, etc. [2]. Many countries use some aquatic plants to treat water pollution problem [3].

Normally, to monitor the physiological activities of plants, oxygen or carbon dioxide concentration in certain container holding the plants is monitored by an oxygen or carbon dioxide sensor. However, the monitored oxygen or carbon dioxide is spatially averaged, which reflects the average concentration of oxygen or carbon dioxide released or absorbed by the whole plants in the container. It is difficult to monitor and distinguish the oxygen or carbon dioxide released or absorbed by different organs (e.g., leaves, stems, or roots) of the plants. Therefore, it is nearly impossible to distinguish the physiological activities of each part of the plants in the conventional method.

Recently, we have developed the optical detection system that could real-time in-situ simultaneous monitoring of the DO at a vicinity of micrometres from an aquatic plant surface [3-5]. Moreover, the system could monitor in both the respiration and photosynthetic process [5]. This real-time method allows to monitor DO change in physiological activity at each part of a plant.

Here, we used Excel macro to proceed the calculation of this work. Here, we apply this method to monitor DO change at vicinities of aquatic plant in different water environment.

2. Experimental

2.1. The optical detection system for plant monitoring

The optical detection system for plants monitoring was almost same as reported before [3-5]. This system was placed in the darkroom. A wavelength of about 660 and 450 nm red-blue LED and electric power supply were placed outside the darkroom to illuminate the aquatic plant for the photosynthetic process. Light from a semiconductor laser with a wavelength of 405 nm was reflected by the dichroic mirror and then was focused to a vicinity of the aquatic plants in the culture dish with a culture solution by the objective lens. The culture dish was filled with 20 ml of 10^{-6} M Ru(II) complex (Tris(2,2'-bipyridyl) ruthenium(II) chloride) solution either without or with Cu^{2+} (10^{-4} M). Fluorescence of the Ru(II) complex passed through the dichroic mirror and was detected by a photomultiplier (PMT). Also, an interference filter with a wavelength of 589 ± 25 nm was placed in front of the PMT. In addition, a commercial temperature sensor and DO sensor were inserted into the culture dish. Temperature, DO, and fluorescence signal were simultaneously monitored.

A piece of slide glass was covered on the aquatic plants to prevent possible movement. *Egeria Densa* was used as a model aquatic plant.

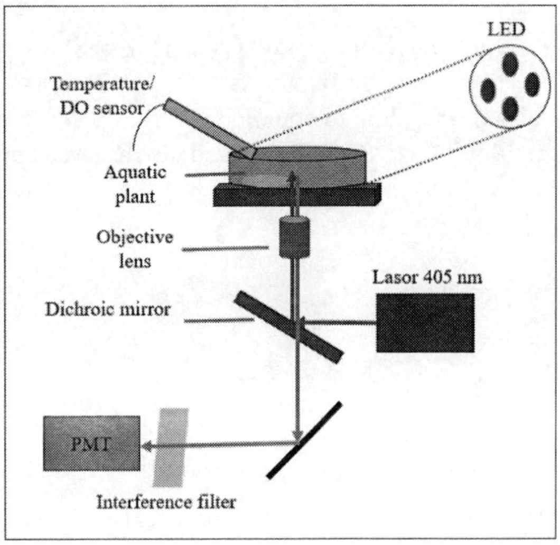

Figure 1. The optical detection system for plant monitoring

2.2. Preparation of sample solutions

Stock solution of the 10^{-4} M Ru(II) complex was prepared by dissolving 0.0075 g of the ruthenium complex in 100 ml water. The stock solution was diluted by 100-fold to prepare a 10^{-6} M Ru(II) complex solution. Preparation of a mixed solution of 10^{-6} M Ru(II) complex and 0.008 M sodium sulfite (Na_2SO_3) was prepared by adding 1 ml of the Ru(II) complex stock solution and 0.0504 g of Na_2SO_3 to 100 ml volumetric flask, then the water was added to 100 ml. Stock solution of 0.02 M Cu^{2+} solution was prepared by dissolving 0.3929 g of the $CuSO_4 \cdot 5H_2O$ in 100 ml water. Then, the stock solution of Cu^{2+} was diluted to desired concentration when adding to the culture dish.

3. Results and Discussions

Quenching of the Ru(II) complex fluorescence by DO is described by the Stern-Volmer equation:

$$\frac{F_0}{F} = 1 + K_{sv}C_{DO} \quad (1)$$

Where, F_0 is the fluorescence intensity of the Ruthenium complex solution without DO, F is the fluorescence intensity of the Ruthenium complex solution with DO. K_{sv} is the Stern-Volmer-constant and C_{DO} is the DO concentration. This equation is transformed as follows:

$$C_{DO} = (\frac{F_0}{F} - 1)K_{sv}^{-1} \quad (2)$$

As shown in equation (2), C_{DO} can be calculated from F_0 and F of a sample solution as long as K_{sv} is known beforehand.

Values of K_{sv} were obtained as follows. Firstly, F and C_{DO} of the 10^{-6} M Ru(II) complex solution without the aquatic plant were measured, and their relationships with temperature T were approximated as cubic equations of (3) and (4), respectively.

$$F = 0.04837T^3 - 3.74319T^2 + 96.55830T - 828.82074 \quad (3)$$

$$C_{DO} = 0.23984T^3 - 18.43152T^2 + 471.95482T - 4018.19559 \quad (4)$$

Secondly, F_0 of the 10^{-6} Ru(II) complex solution containing 0.008 M Na_2SO_3 were monitored, and its relationships with T was also approximated as a cubic equation of (5).

$$F_0 = 0.00133919T^3 - 0.06924T^2 + 0.89438T + 2.18662 \quad (5)$$

Thirdly, Ksv was calculated according to equation (1) from above F, C_{DO}, and F_0. Figure 2 showed calculated results of K_{sv} at 24.9 °C - 25.8 °C. The calculated K_{sv} was approximated as a cubic function too, as shown in Fig. 2.

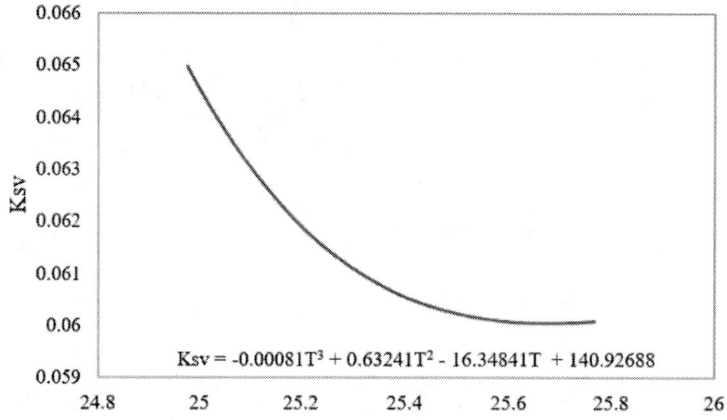

Figure 2. Relationship of K_{sv} to temperature

In order to verify validity of the above K_{sv}, fluorescence intensity F of another 10^{-6} M Ru(II) complex solution without the aquatic plants were monitored for 4 hours, and C_{DO} was calculated according to equation (2) from the monitored F and K_{sv} in Figure 2. The calculation results were further compared with the monitored DO concentration by the DO sensor, and results were shown in figure 3. The calculated C_{DO} agreed well with the one obtained by the DO sensor, although the calculated ones were with large noise. Therefore, the method can be used for monitoring DO.

Next, the method was applied to monitor DO at vicinity of the aquatic plant. Fluorescence intensity F_v at vicinities of the aquatic plants were monitored, and C_{DO} at the vicinities was calculated according to equation (2) from F_v and K_{sv} in figure 2. Figure 4A showed an example of the calculated C_{DO} at vicinities of the aquatic plant in a respiration process (The red-blue LED was off). Both the calculated C_{DO} at the vicinities (0 µm) and DO concentration obtained from the sensor at 1 cm away from the aquatic plants surface decreased with time. However, the C_{DO} at 0 µm decreased faster with time than that at 1 cm. Figure 4B showed another example of DO monitored in a photosynthetic process (The red-blue LED was on). The DO concentration at the vicinity of the aquatic plants of 0 µm increased with time faster than that at 1 cm in the photosynthetic process. The photosynthesis process of aquatic plants produced oxygen, and then spread to the culture solution [3,5]. While the respiration process of aquatic plants absorbed DO firstly at vicinities of the plants. Therefore, DO at vicinities of the plants changed with time faster than that at 1 cm away from the plants.

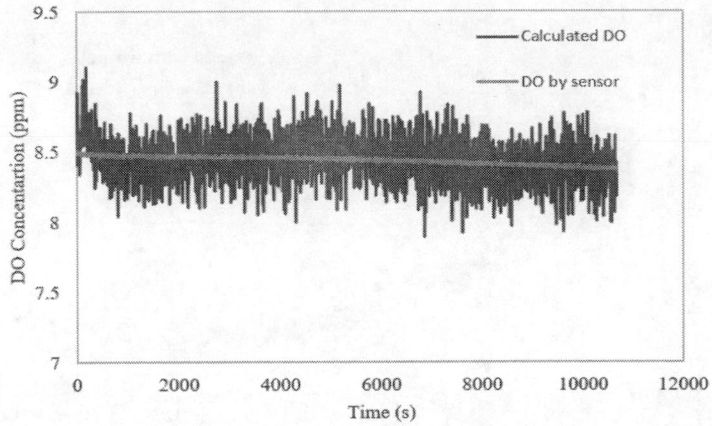

Figure 3. A comparison of DO concentration change with time obtained by the present method and commercial DO sensor.

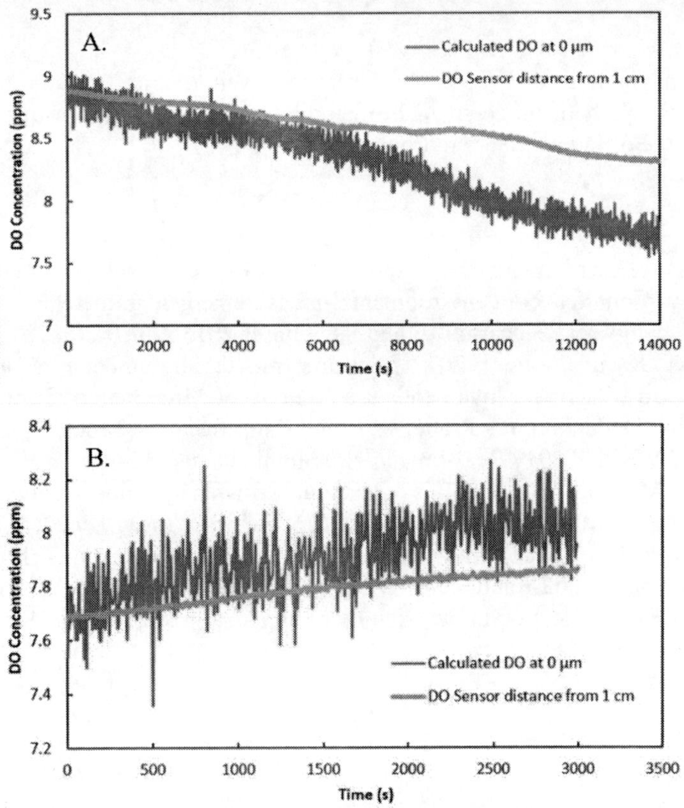

Figure 4. DO concentration changes with time in a respiration (A). and photosynthetic process (B).

IOP Conf. Series: Materials Science and Engineering **639** (2019) 012045 doi:10.1088/1757-899X/639/1/012045

DO change at vicinities of the aquatic plants in water environment containing heavy metal irons such as Cu^{2+} is being monitored too. Figure 5 showed an example of preliminary experimental results of DO monitoring at vicinities of the aquatic plant in water containing of 10^{-4} M Cu^{2+} in a respiration process. It is clear that DO changes at vicinities with time is different from that in Fig. 4A. This suggested that existence of Cu^{2+} might have affected absorption of DO in respiration process of the aquatic plants. Its details and mechanisms are being studied and will be reported later.

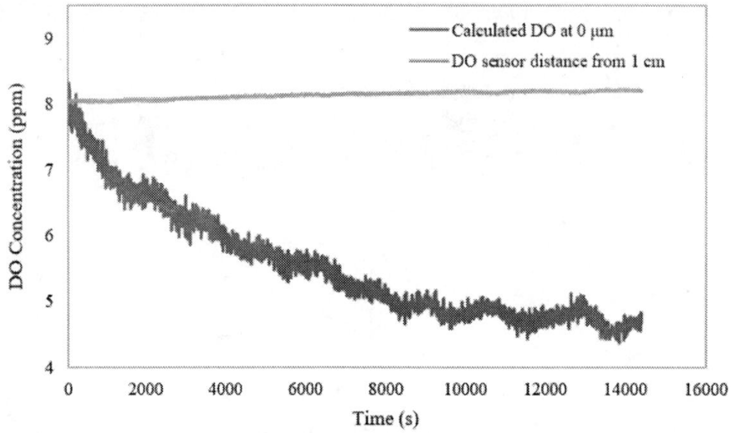

Figure 5. DO concentration changes with time in a respiration process containing Cu^{2+} of 10^{-4} M.

4. Conclusion

DO concentration at vicinities of an aquatic plant surface were monitored real time and in-situ in both a respiration and photosynthetic processes. The method is useful in monitoring of DO change at vicinities of aquatic plant in the different water environment.

References

[1] Fondriest Environmental. Inc. 2013 Dissolved Oxygen Online: http://www.fondriest.com/environmental-measurements/parameters/

[2] Manzatua C, Nagya B, Ceccarinib A and Iannellic R 2015 *Mar. Pollut. Bull.* **101** 605

[3] Wu X-Z Wu X and Inoue T 2017 Real-time in-situ Simultaneous Monitoring of Dissolved Oxygen and Materials Movements at a Vicinity of Micrometers from an Aquatic Plants by Combining Deflection of a Probe Beam and Fluorescence Quenching *Anal. Sci.* **33** 351-355

[4] Huang L and Wu X-Y 2017 Real-time in-situ Simultaneous Monitoring of Dissolved Oxygen and Materials Movement at Vicinities of an Aquatic Plants by Fluorescence Quenching/Deflection with an improved calculation method *SDRP Journal of Plant Sci.*, **2**. 1-7

[5] Wu X.-Z and Huang L 2018 Improvements on the Fluorescence Quenching/Deflection Method for Real-time in situ Simultaneous Monitoring of Dissolved Oxygen and Material Movement induced Beam Deflection in the Vicinity of an Aquatic Plants *Anal. Sci.* **34** 1335-1337

Physical properties and calorific value of briquettes produced from Palmyra palm waste with molasses binder

K Utchariyajit, V Panprasert, L Chayawat, W Jungthanasombat, P Janprom and M Choatchuang

Department of Science and Technology, Kanchanaburi Rajabhat University, Kanchanaburi, Thailand

Corresponding author: k.utchariyajit@kru.ac.th

Abstract. This work aims to produce briquette made from Palmyra palm waste using molasses as a binder. The amount of binder to charcoal ratios was studied. The physical properties i.e., density, moisture content, ash content, volatile matters, fixed carbon, calorific value and cooking efficiency were investigated. It was found that the amount of binder had a significant effect on the quality and properties of the briquette. The results showed that the amount of binder decreased with moisture content and ash content decreased while density and calorific value increase. All briquettes produced with molasses binder pass the criteria of the Thai Community Product Standard (tcps 238/2547). These results provide an alternative energy for household use.

Keywords: Briquette, Palmyra palm, molasses, physical properties, Calorific value

1. Introduction

The demand for fuel energy has been increasing over the years in developing countries. However, non-renewable source of energies are depleted rapidly. So, a decreasing availability source has been concerned. A number of researches tend to study the utilization of agricultural waste and residues. Energy in the form of charcoal has been the most conventional source of renewable energy in developing countries. Large amount of agricultural waste and residues are generated annually. The utilization of them instead of firewood as alternative energy source could help due to its environmental friendliness and preventing deforestation.

However, charcoal has many drawbacks which can be overcome by way of compression into a uniform solid with binder and water either by hand or press machine for briquette production to improve fuel quality giving more practical and economical. This is to increase strength, durability, long time burning and reduce cost of transportation, handling and storage. Moreover, it is a desirable fuel because of non-sparking and smokeless fire. Briquettes have been developed as an option for value addition of biomass wastes being sustainable energy development such as rice husk [1], sawdust [2-3], banana tree waste [4-5], bagasse [6], organic waste [7] and coconut shell [8]. Binding agents such as cassava starch [1], molasses [3,9] and clay [5-6] have been employed. Briquette quality and properties were influenced by the proportion of binders [5]. A study by Canage, et. al. [10] reported that higher amount of molasses increase fixed carbon, volatile matters and moisture content while ash content decrease.

Palmyra palm is one of the sources of biomass available in Thailand. It generated large amount of trash. The strategy is to minimize a plenty of waste achieving zero waste by value addition. Moreover, it could help enhance environmental protection and conservation. So, Palmyra palm waste can be utilized as a raw material for the production of charcoal briquette as a fuel in a household use to reduce the wood charcoal consumption. Molasses, by-product obtained from the sugar manufacturing process, is highly viscous, inexpensive and easily available which are the characteristic required to use as an excellent binder.

This research work intent to produce briquettes developed from Palmyra palm waste with molasses as binding agent including study their quality and properties to determine the compatibility and effectiveness of these briquettes.

2. Methodology

2.1. Raw materials

Palmyra palm waste which is the agriculture waste for this study was collected from cultivated areas in Kanchanaburi province, Thailand. Sample was dried on sunlight to reduce moisture before being carbonized in a 200 liter drum pyrolyzer to obtain charcoal. For the binder ingredient, three conditions of molasses contents were prepared at the ratios of 5%, 10% and 15% of the weight of charcoal.

2.2. Methods of briquette production

After carbonization, the charcoal prepared was milled and sieved to obtain fine powder. The fine powder charcoal was then homogeneously mixed manually with molasses, as binding agent, at the different mixing ratio of 5%, 10% and 15% w/w, followed by adding water as required. Thereafter, the mixtures were briquetted via the cold extrusion process with a designed shape. The resulting briquettes were dried in an oven at a temperature of 100 °C for 48 hours to reduce moisture content.

2.3. Characterization

The quality and properties of briquettes produced i.e., density, moisture content, ash content, volatile matters, fixed carbon and calorific value were determined according to the ASTM standard. Data analysis was done with SPSS statistical software to test the difference for the parameters in briquettes produced. The results were assessed to be statistically significant when $P<0.05$.

3. Results and discussion

All briquette mixtures obtained from Palmyra palm waste and molasses binder at the proportion of 5, 10 and 15% w/w could keep well the cylinder shape along the extruder machine showing that these briquettes have good forming ability. The briquettes obtained have a uniform black color, cylindrical shape of 100 mm length, 40 mm diameter and 10 mm with hollow holes in the middle in order to let air penetrate inside enhancing combustion process as shown in Figure 1. They have smooth surface without crack indicating that 5% w molasses contents mixed with Palmyra palm waste charcoal was enough to coat the surface. While some of briquettes produced from others agriculture waste required a large proportion of binder, such as tamarind peel using tapioca starch and molasses as a binder at a ratio of 33-100% w/w [11], water hyacinths using molasses at the ratio of 166-233% w/w [9]. In this study, briquettes can be formed cylindrical shape without adding molasses. However, it is brittle and easily broken. The addition of molasses as binder improved the endurance of the briquette to breakage. Therefore, the briquette production is necessary to use the binding agent in order to store and transport to users and use as fuel.

The effects of the Palmyra palm waste charcoal and molasses binder proportions on the briquettes at various properties are discussed in detail according to the values obtained in Table 1.

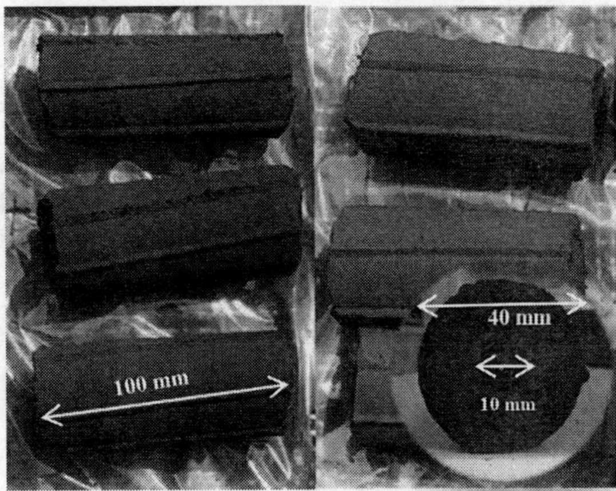

Figure 1. Briquette produced from Palmyra palm waste and molasses binder.

Table 1. Mean values of the properties of briquettes according to the different binder contents

Briquette properties	Molasses (% w/w)			p-value
	5%	10%	15%	
Density (g/mL)	1.02±0.08	1.00±0.10	0.99±0.11	0.681
Moisture content (%)	5.84±0.36 [a]	6.27±0.44 [b]	6.32±0.46 [b]	0.000
Ash content(%)	5.69±0.64 [a]	5.70±0.61 [b]	6.44±0.73 [b]	0.011
Volatile matters (%)	44.58±2.00 [b]	42.96±1.82 [a]	46.95±1.37 [c]	0.000
Fixed carbon (%)	43.64±0.46 [b]	45.06±1.51 [c]	40.16±1.02 [a]	0.000
Calorific value (cal/g)	6,915.08±74.06 [b]	6,738.26±35.39 [a]	6,689.80±37.56 [a]	0.001

Data presented in the table are mean values of three replicates. Values with same alphabet (a,b,c) in the same row are not significantly different at α=0.05. Using Duncan multiple range test

3.1. Densities

Densities of all briquettes obtained are displayed in Table I. The densities were in the range of 0.99-1.02 g/cm^3. Comparison between each briquette indicated that briquette mixed 5 %w/w molasses had the highest density followed by 10 % and 15 %w/w, respectively. So, density increased as molasses proportion decreased. However, there were no significant differences ($p>0.05$) on density of briquettes prepared from the different molasses contents. Density is an important property of briquette. High density is desirable because the higher value of density, the easier for transportation, handling and storage.

3.2. Moisture content

The moisture contents were in the range of 5.84 - 6.32%. Sample prepared using 5% w/w molasses gave the lowest moisture content. As the molasses proportion increased, the moisture content of the briquette also increased. Due to the molasses has a hygroscopic property, moisture can be trapped inside the briquette. This result corresponds to the study by Carnage et al. [10]. The different molasses proportions significantly influenced on moisture content of the briquettes ($p<0.001$). However, samples prepared at the proportions of 10% and 15% w/w had no significant different. High moisture content causing ignition and combustion are difficult [12]. In addition, it results in a lower calorific value [13]. However, the moisture content of all briquettes met the standard of the Thai Community Product Standard (tcps 238/2547) because briquette should have the moisture content not exceeding 8 %.

3.3. Ash content

Ash content is one of the factors for prediction of burning efficiency. High ash content causes incomplete combustion. Sample prepared at 5%w/w molasses content gave the lowest ash content. As increasing molasses proportion, the higher ash content was found because it is known to mainly contain inorganic particles. This was also observed in the study by Tanui et al. [14]. The different molasses proportions were significantly impacted on ash content of the briquettes ($p<0.05$). However, samples prepared at the ratios of 10% and 15% w/w were no significant differences. It is useful when briquette has small amount of ash content because it is easy to eliminate after use.

3.4. Volatile matters

Table 1 shows the effect of molasses proportion on the volatile matters content. The volatile matters of briquettes produced using molasses for 5%, 10% and 15%w/w were 44.58, 42.97 and 46.95% respectively. The volatile matters varied significantly with different molasses proportions used in the briquette production ($p<0.001$). Briquette produced from 10% molasses content had the least volatile matters while 15% molasses content briquette had the highest volatile matters. High volatile matters mean they are ignite easily but may burn with a smoke flame. Low volatile matters mean they are fire hardly but burn very cleanly.

3.5. Fixed content

The fixed carbon content implies to the carbon residue after volatile matter is released. High quality of briquette requires the higher amount of fixed carbon content. The volatile matters of briquettes produced using molasses at 5%, 10% and 15%w/w were 43.64, 45.06 and 40.16% respectively. The fixed carbon content influenced significantly with different molasses proportion used in briquette production ($p<0.001$). As compare to the volatile matters result, the briquettes having higher volatile content had lower fixed carbon content. The fixed carbon content adversely related to the volatile matters content of the briquettes.

3.6. Calorific value

Calorific value is the measurement of thermal efficiency released by the briquette during combustion. High calorific value means high quality briquette. The calorific values obtained from this study are in the range of 6,689.80-6,915.08 cal/g. The sequence of the calorific values was briquette produced using 5% w/w of molasses first, followed by 10% and finally 15%. The different molasses proportions affected significantly the calorific value of the briquettes ($p<0.01$). However, samples prepared using 10% and 15%w/w of molasses were no significant differences. High binder content resulted in decreased calorific value due to the calorific value of molasses, which is 2,450 cal/g [15], has lower than that of conventional charcoal. Furthermore, using high amount of molasses containing hygroscopic property also tends to reduce the calorific value. Similar results can be found in the other studies [3, 14]. Comparing with others, the calorific value of the briquette obtained from Palmyra palm waste having molasses binder was higher than that obtained from other charcoals [13, 16]. In addition, the calorific value of the briquette was higher in comparison with the Thai Community Product Standard (tcps 238/2547) indicating that the values obtained were acceptable especially for the application of the briquette as a fuel.

From above results infer that other effects like moisture content, ash content could also contribute to the lower in calorific value. Several studies discussed that high moisture and ash content in briquettes will lower calorific value [10, 16-17]. This is in agreement with this study which revealed that high moisture and ash content belonging to high molasses proportion resulted in low calorific value.

Briquettes produced in this study were tested cooking efficiency by boiling water. The results showed that water can be boiled with all conditions briquettes produced. Moreover, they could burn for a long time without spark, soot and were smokeless. All results in this study have shown the potential of converting Palmyra palm waste into an alternative fuel source.

4. Conclusions

Briquette produced from Palmyra palm waste with molasses as binder was developed at various binder proportions. Altering the molasses to charcoal proportions affected the quality and properties of the briquettes. Increasing the amount of molasses increased the moisture content and ash content but decreased calorific value and density. The briquette with 5% w/w molasses showed desirable characteristics in terms of density, moisture, ash content and calorific value. All briquettes in this study produced meet the criteria of the Thai Community Product Standard (tcps 238/2547). All results have revealed the potential of these briquettes developed from Palmyra palm waste and molasses binder into alternative fuel energy for household use.

Acknowledgement

This research work was supported by Budget Bureau. The authors would like to express their sincere appreciation to Mr. Sanit Iamcharoen for the assistance in the use of briquette extruder machine and Dr. Kittima Prukpousana for the suggestion of data analysis in SPSS statistical software. The authors would like to give special thanks to Faculty of Science and Technology, Kanchanaburi Rajabhat University, Thailand.

References

[1] Lubwama M and Yiga V A 2018 *Renew Energy* **118** 43-55
[2] Emerhi E A 2011 *Adv. Appl. Sci. Res.* **2(6)** 236-246
[3] Manyuchi M M, Mbohwa C and Muzenda E 2018 *S. Afr. j. chem.* **26** 70-73
[4] Ahmad K K, Sazali K and Kamarolzaman A A 2018 *Mater Today Proc.* **5** 21744-21752
[5] Mopoung S and Udeye V 2017 *Am. j. eng. appl. sci.* **10(2)** 353-365,
[6] Onchieku J M, Chikamai B N and Rao M S 2012 *Eur. j. sustain. dev.* **1(3)** 477-492
[7] Fikri E and Sartika C 2018 *Ecol Eng.* **19(2)** 81-88
[8] Promdee K, Chanvidhwatanakita J, Satitkunec S, Boonmee C, Kawichaid T, Jarernpraserte S and Vitidsanta T 2017 *Renew. sust. energ. rew.* **75** 1175-1196
[9] Tanpaiboonkul N and Budnumpetch T 2016 *SUSTJ.* **3(6)** 86-100
[10] Carnaje N P, Talagon R B, Peralta J P, Shah K and Ferreiro J P 2018 *PLOS ONE* **13(11)** e0207135
[11] Janchaiyapoom H 2018 *RTUNC the 3rd national conf. Thailand* 288-296
[12] Saidura R, Abdelaziza E A, Demirbas A, Hossain M S and Mekhilef 2011 *S Renew Sust Energ Rev* **15** 2262-2289
[13] Shukla S, and Vyas S 2015 *IOSR J. Environ. Sci. Toxicol. Food Technol.* **9** 37–44
[14] Tanui J K, Kioni P N, Kariuki P N and Ngugi J M 2018 *IJEEE.* **9** 314-350
[15] Henglein F H 1969 Chemical Technology (Oxford: Pergamo press) 656
[16] Tamilvanan A 2013 *Biofuels* **4(2)** 47-55
[17] Sotannde O A, Oluwadare A O, Ezekiel S and Anguruwa G T 2017 *AZOJETE.* 13(4) 489-501

Antioxidant activity and Total Phenolic compounds of Fresh and Blanching Banana blossom (*Musa ABB CV.Kluai "Namwa"*) in Thailand

Sarisa Thaweesang

Department of Food Science and Technology, Faculty of Technology, Udon Thani Rajabhat University, Udon Thani, Thailand.

E-Mail : Sarisa248@hotmail.com

Abstract. Fresh and blanching banana blossom were extracted by ethanol. Fresh banana florets crude extract had the highest antioxidant activity 42.74% and IC_{50} were 1275.50μg/ml respectively. For blanching banana florets crude extract, The antioxidant activity and IC_{50} value were 41.78 and 1306.67 μg/ml respectively. Antioxidant activity of fresh banana bract crude extract and blanching banana bract crude extract were only 28.43% and 13.33% respectively. Results also showed could not evaluation of IC_{50} value for banana bract crude extract. However, Banana blossom able to antioxidant activity. Banana florets crude extract showed effective antioxidant activity more than banana bract crude extract ($p<0.05$). Blanching banana florets and fresh banana florets contained the highest total phenolic compounds were 1380.58 μg GAE/g and 1091.30 μg GAE/g respectively. While banana bract had low phenolic compounds. The food processing for consumption did not affected to total phenolic compounds. It was therefore suggested that banana florets have potential as good source of natural antioxidant.

Keywords : banana blossom, banana bract , banana florets , Antioxidant activity, Phenolic compounds, minimally processing

1. Introduction

Bananas are popular consumed fruits in the world because of its simple taste, high nutritional value and potential health benefits [1]. Cultivated Banana (*Musa ABB CV.Kluai "Namwa"*) is an important fruit crop of Thailand, Banana blossom (widely known as flower or heart) is byproducts of banana cultivation and it is known to have health beneficial effect [2]. In Thailand, Banana blossom is usually considered as vegetable, also has potential to be regarded as a functional food due to its high nutrient content (excellent source of fiber, good source of protein, vitamin A,C and E, minerals like phosphorus, potassium bleeding, facilitates lactation, helps in overcoming diabetes, helpful in weight loss and good for gastrointestinal health) [1]. Therefore, the ingredient of Thai food contained the banana blossom in recipe or vegetable side dishes for example, Pad Thai. The method of consumption, for example fresh eating or cooking has effect to natural antioxidant [3]. Minimally processed foods such as Fresh fruits and vegetables are recommended daily intake for the health benefits. The advantages of minimally food processing are easy and quick preparation of meals, maintain quality as fresh or close to the fresh prepared meals or products and maintain products' nutritive values [4].

The objectives of the work were investigated the antioxidant activity and total phenolic compounds in the part of banana blossom (florets and bract) with the method of consumption or minimally food processing (fresh or blanching) by using DPPH assay analytical and Folin-Ciocalteu.

2. Materials and methods

2.1. Chemical and Reagents

All the chemicals and reagents used for experimention were of analytical grade.

2.2. Harvest of Banana blossoms and preparation

Banana blossoms were havested in June 2017. The sample of blossom of Cultivated Bananas (*Musa ABB CV.Kluai "Namwa"*) were optianed from cultivated local farmland in Udon thani province. The outer part of banana blossoms were discard (figure 1A). The banana blossoms were wash with water. The Sample were deviced in two part. The first part was fresh sample (figure 1B) and second part was blancing in boiling water for 5 minute (figure C), then florets and bract were separated. The fresh sample and blaching sample were ground by blender before expermentation

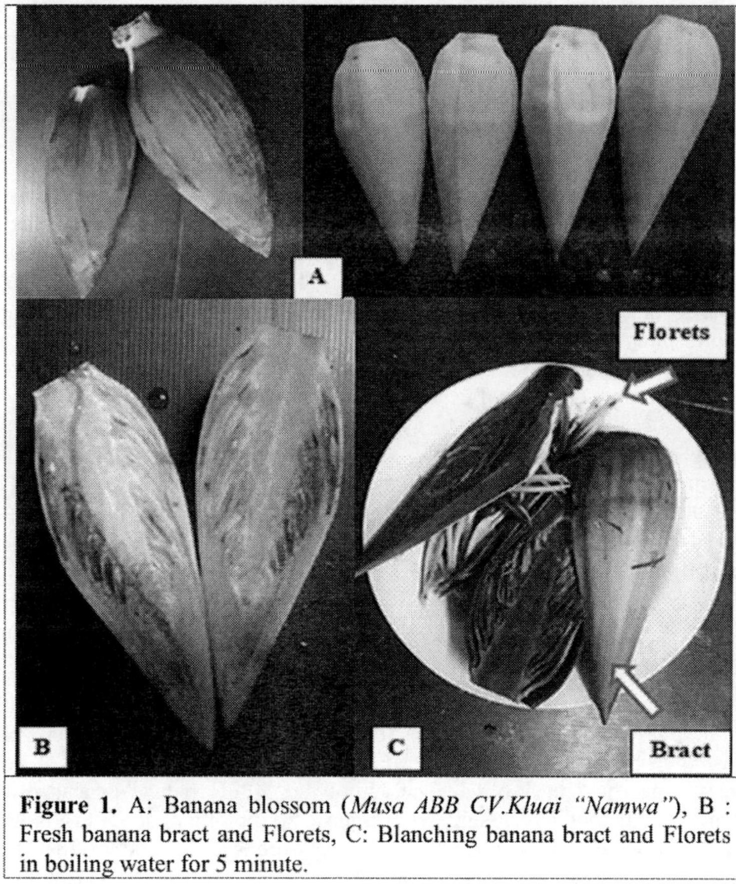

Figure 1. A: Banana blossom (*Musa ABB CV.Kluai "Namwa"*), B : Fresh banana bract and Florets, C: Blanching banana bract and Florets in boiling water for 5 minute.

Extraction was carried out by using ethanol (95%). 5 g of each part of banana blossom ; the fresh florets or blanching florets and the fresh bract or blanching were separately soaked in 100 ml of ethanol for 24 hours at room temperature to obtain successive extracts and filtered. The crude extracts obtained were evaluation of antioxidant activity and total phenolic compound.

2.3. Antioxidant Assays
The antioxidant activity of banana was evaluated using DPPH raidical scavenging assay was assessed according to method described by Roobha *et al.*[5] and Mahmood *et al.* [6] with some modifications. Briefly, Prepared 0.1µg/L DPPH solution (DPPH 0.0039 g in 100 ml absolute ethanol). DPPH solution of 2 ml was added in each of test tubes contaning the crude extracts sample of 2ml. After mixing the solution was incubated for 30 minute in the dark. Then, the absorpance was taken at 517nm using a spectophotometer (Cecil instrument, CE1011). Control contained ethanol instead of antioxidant solution and blank contained ethanol instead of DPPH solution. Acorbic acid solution was used as possitive control and establishing the standard calribarion curve at concentration in rang of 10 µg/ml – 30 µg/ml. Percent scavenging of the DPPH free raditcal was calculated using the follow equation :

$$\% \text{ Inhibition} = \{[A_C - A_S]/A_C\} \times 100$$

Where, A_C = absorbance of control
A_S = absorbance of sample solution

Percent of inhibition (IC_{50}) were plotted against respective concentration used. Linear regresstion analysis was used to calculate IC_{50} values of the extracts.

2.4. Total phenolic compounds
Banana blossoms were havested The total phenolic compouds of the extracts were determined by the Folin-Ciocalteau colorimetric method according to method described by Arya and Sinija [7] and Mohamed *et al.* [8] with some modifications. 0.4ml of the extract solution was mixed with 10%v/v the Folin-Ciocalteau reagent (2ml) and 7.5% Na_2CO_3 (1.6ml). After 30 minute of incubation at room temperature, the absorbace were measured at 765 nm (spectophotometer : Cecil instrument, CE1011). For each sample, three replicate assats were performed. Gallic acid was used for establishing the standard calribarion curve and the total phenalic compounds was calculated as gallic acid equivalents (GAE) and expressed as µg GAE /g of extract .

2.5. Statistic analysis
Analysis of variance (ANOVA) was carried out on the 2^2 Factorial in CRD. The independent variables were the part of banana blossom (bract and florets) and method of minimally food processing (fresh and blanching). The significance of differences between means was compared by Duncan's New Multiple Rang Test at $p<0.05$.

3. Result and discussion

3.1. Antioxidant Assays

Free radical scavenging is one of the know mechanism by which antioxidant inhibits lipid peroxidation .The DPPH radical scavenging activity has been widely used for screening antioxidants from fruit and vegetable juice or extract [5]. DPPH radical scavenging activities of fresh or blanching banana blossom were show in Table 1. The activity of DPPH radical depend on part of banana blossom and minimally food processing. The highest of inhibition of DPPH radical activity was observed that the fresh and blanching banana florets were 42.74% and 41.78% respectively, The results shown that the banana florets can effectively scavenge DPPH radicals ($p<0.05$). According to previous studies, China *et al.* [9] reported that the DPPH free radical scavenging activity of different cultivars of banana flower (*Musa paradicicus* L.) extracts in India depend on the cultivars. In addition, % inhibition of the extracts were shown in rang 10% to 65% similar to Divya *et al.* [10] who have found that the extract of banana flower (*Musa paradicicus* L.) exhibited radical scavenging activity of 80.07%. In the present study, the minimally food processing would have considered that had the effect of antioxidant properties of banana blossom. It was observed that DPPH free radical scavenging activities of the fresh or blanching were not significantly different for the banana florets extracts. Whereas the method of consumptions and % inhibition were different significant in the banana bract extracts. The results showed that % inhibition of blanching banana bract extract had decreased significantly when compared with fresh banana bract extract. A common perception is that thermally cooked foods have lower nutritional value of fresh food because of the loss of vitamin C and loss of particular physicochemical substance [3]. The previous studies have investigated the effect of various cooking processes on the level of antioxidant compound and antioxidant activities in fruits and vegetables. The results in this study were similar to Hwang *et al.* [3] who have found that boiling significantly reduced antioxidant activity in red pepper.

Table 1. Antioxidant assay in extract of fresh banana blossom and blanching blossom using DPPH assay

Part of banana blossom	Minimally food processing	% Inhibition[1]
Florets extract	Fresh	42.74±9.44[a]
	Blanching	41.78±8.03[a]
Bract extract	Fresh	28.43±7.92[b]
	Blanching	13.33±8.17[c]

[1] Values are expressed as the mean ± SD (N=3)
[a-c] Values followed by the different letter are statistically significant difference ($P<0.05$)

The activity of antioxidant in banana blossoms were shown in Table 2. IC_{50} value of fresh florets extract was 1275.50 µg/ml demonstrated that fresh banana florets required low concentration of substance that the crude extract to inhibit a 50% DPPH color. While, the banana bract extract could not calculated as IC_{50} value. This mean the fresh banana florets were stronger antioxidant activity than banana bract.

Table 2. Radical scavenging activity (IC_{50}) of fresh banana blossom and blanching blossom

Part of banana blossom	Minimally food processing	Antioxidant activity (IC_{50} value)[1] µg/ml
Florets extract	Fresh	1275.50
	Blanching	1306.67
Bract extract	Fresh	-
	Blanching	-

[1] IC_{50} = The concentration of substrate that cause 50% loss of the DPPH color

The antioxidant activity of banana blossom usually study on banana bract but does not the information of antioxidants activity of banana florets include method of processing. The antioxidant activity of banana flower were reported by Mahmood *et al.* [6] were investigated IC_{50} of the ethanol extract of banana bract (*Musa x Paradisiaca*) compared with aqueous extract and Arya and Sinija [7] reported that high antioxidant activity (82%) was shown by ethanolic extract of banana blossom of two cultivars in India. The result of this study were in line with those of Sheng *et al.* [11], who have studied the antioxidant properties of banana flower of two cultivars (*Musa spp.* 'Baxijiao' and 'Paradisiaca') in China and expressed antioxidant properties in EC_{50}. It appeared that Baxijiao cultivar revealed better antioxidant properties by present much lower EC_{50} value. On the other hand, Bibechana and Varalakshmi [12] have been reported that banana (*Musa paradisiaca*) flower was rich in phytochemicals and had antioxidant properties with IC_{50} value of <10 μg/ml. The previous studies indicate that antioxidant properties of banana blossom depend on cultivars and region of grown. On the basis of the result obtained, banana blossom are found to be a potential source of natural antioxidant, especially fresh florets of banana.

The naturally plant compounds normally correlate to their physiological function as antioxidant. However, it is quite difficult to compare the DPPH scavenging capacity data between different laboratories due to a few reasons such as the reaction time, the concentration of DPPH used and the antioxidants in the assay mixture. The presence of phenolic compounds in the extract could be associated with the antioxidant properties of fresh and blanching banana blossom (*Musa ABB CV.Kluai "Namwa"*)

3.2. Total phenolic compounds

Phenolic compounds exhibit considerable free-radical scavenging activity, which is determined by their reactivity as hydrogen-or- electron –donating agent [8]. The total phenolic compounds of fresh and blanching banana blossom were shown in Table 3. Blanching banana florets and fresh banana florets contained the highest total phenolic compounds were 1380.58 μg GAE/g and 1091.30 μg GAE/g respectively. Although, total phenolic compounds present in both florets and bract but the concentration of compound were different. Moreover, the food processing for consumption did not affected to total phenolic compounds.

Table 3. Total phenolic compounds of fresh banana blossom and blanching blossom

Part of banana blossom [s]	Minimally food processing	Total phenolic compounds (μg GAE/g)[1]
Florets extract	Fresh	1091.30± 156.98
	Blanching	1380.58±175.61
Bract extract	Fresh	742.42± 120.42
	Blanching	741.18 ±95.58

[1] Values are expressed as the mean ± SD (N=3)

[S] The main factor followed by the letter are statistically significant difference (P<0.05)

The main factors of this study were the part of banana blossom and the minimally food processing. Analysis of variance (ANOVA) was shown that no interaction between main factors. The minimally food processing including fresh and blanching did not impact on total phenolic compounds in banana blossom. On the other hand, total phenolic compounds depended on the part of banana blossom

(P<0.05). The total phenolic compounds of banana blossom were shown in Table 4. The banana florets contained the highest total phenolic compounds was 1235.94 µg GAE/g while total phenolic compounds of banana bract was 741.79 µg GAE/g.

Table 4. . Total phenolic compounds of part of banana blossom (main factors of experiment was significance)

Part of banana blossom	Total phenolic compounds (µg GAE/g)
Florets extract	1235.94± 217.48[a]
Bract extract	741.79± 97.34[b]

[1] Values are expressed as the mean ± SD (N=6)

[a-b] Values followed by the different letter are statistically significant difference (P<0.05)

The results of this study finding was in accordance with the previous findings obtained that phytochemicals contents in the banana blossom extract such as flavonoids and terpenoids observed only in the florets but not in the bract parts of the blossom [6]. Total phenolic compounds of the banana blossom were found to be within the part of blossom. In addition, pointed out that the cultivar and growing conditions effect of the variability in the phenolic compounds of banana blossom. According to Marikkar *et al.*[13] investigated the total phenolic content of banana flowers of six different Malaysian cultivars and found that the extracts of Malaysian banana flowers were potent sources of natural antioxidants.

4. Conclusion

This research work has investigated antioxidant activity and total phenolic compounds of banana blossom with considered the food processing for consumption. The analysis of banana blossom revealed their considerable antioxidant properties and it is well known that is a strong relationship between total phenolic compounds and antioxidant activity. From the results it can be concluded that cultivated banana (*Musa ABB CV.Kluai "Namwa"*) has high amount of antioxidant activity and total phenolic compounds on its florets both of fresh or blanching. These findings confirmed that the banana blossom may have use in food ingredient, pharmaceutical or utilized for functional foods with potential health-benefit effects.

References

[1] Sinhg S 2017 Banana blossom-an understated food with high functional benefits *International Journal of Current Research.* **9** 44516-44519

[2] Jamuna J B, Mahadevamma S, Nandini D C, Paramahan S V 2012 Banana (Musa sp. Var. elakki bale) flower and pseudostem : dietary fiber and associated antioxidant capacity *J. Agric.Food Chem.* **60** 427-432

[3] Hwang I G, Shin Y J, Lee S, Lee J, Yoo S M 2012 Effects of different cooking methods on the antioxdant properites of red peper (Capsicim annuum L.) *Prev Nutr Food Sci* **17** 286-292

[4] Vasudha B, Mohammed SW, Mohammad R S 2015 Minimally processed foods: Overview in *Minimally processed foods: technologies for safety, quality, and convenience*, (London:Springer) chapter 1 pp 1-16

[5] Roobha J J, Saravanakumar M, Aravinthan K M, Suganya devi P 2011 Antioxidant analysis of anthocyanin extracted from Musa acuminata bract *Journal of Phamacy Research* **4** 1488-1492
[6] Mahmood A , Ngah N, Omar M N 2011 Phytochemical constituent and antioxidant activity in Musa x Paradisiaca flower *European Journal of Scientific Research* **66** 311-318
[7] Arya K S and Sinija V R 2016 Proximate composition and antioxidant activity of banana blossom of two cultivars in India *International Journal of Agriculture and Food Science Technology* **7** 13-22
[8] Mohamed I S A, Amira A M, Ludger B 2011 Total phenol conent and antioxidant activity of standarized extract from leaves and cells cultures of three Callistemon species *American Journal of Plant Scienecs* **2** 837-850
[9] China R, Dutta S, Sen S, Chakrabarti R, Bhowmik D, Ghosh S, Dhar P 2011 In virto antioxidant activity of different cultivars of banana flower (*Musa paradicicus* L.) extracts available in India *Journal of Food Science* **76** C1292-C1299
[10] Divya R S, Venkatalakshmi P, Vadivel V, Brindha P 2016 In vitro studies on the biological activities of flowerd of banana (*Musa Paradisiaca* L.) *Der Phamacia Lettre* **8** 238-246
[11] Sheng Z W, Ma W H, Gao J H, Bi Y, Zhang W M, Dou H T, Jin Z Q, 2011 Antioxidant properties of banana flower of two cultivars in China using 2,2-diphenyl-1-picrylhydrazyl (DPPH,) reducing power, 2,2'-azinobis-(3-ehtylbenzthiazoline-6-sulphonate (ABTS) and inhibition of lipid peroxidation assays *Affrican Journal of Biotechnology* **10** 4470-4477
[12] Bibechana T and Varalakshmi K N 2014 Anti-cancer poetential of banana flower extract : an in vitro study *Bangaldesh J Phamacol* **9** 628-635
[13] Marikkar J M N, Tan S J, Salleh A, Azrina A, Shukri M A M 2016 Evaluation of banana (Musa sp.) flowers of selected varieties for their antioxidative and anti-hyperglycemic potentials *International Food Research Journal* **23** 1988-1995

Acknowledgments
The author is grateful to Udon Thani Rajabhat University (UDRU) for providing the facilities to carry out this research.

IOP Conf. Series: Materials Science and Engineering **639** (2019) 012048 doi:10.1088/1757-899X/639/1/012048

Detection of nosemosis in European honeybees (*Apis mellifera*) on honeybees farm at Kanchanaburi, Thailand

Samrit Maksong[1], Tanawat Yemor[2] and Surasuk Yanmanee[3]

[1]Department of General Science, Faculty of Science and Technology Kanchanaburi Rajabhat University, Thailand

[2]Department of Plant production technology, Faculty of agriculture and natural resources Rajamangala University of Technology Tawan-ok Chonburi, Thailand

[3]Department of Biology, Faculty of Science Naresuan University Phitsanulok, Thailand

E-mail: Samrit@kru.ac.th

Abstract. This study was aimed to the detection of *Nosema* in European honeybees at Kanchanaburi Province and Identify species of *Nosema* by Polymerase Chain Reaction technique. The ventriculus of bees was individually checked to verify the presence of *Nosema* spores under light microscope. The number of spores per bee were quantify on a haemocytometer for infectivity. It was studied for three periods of the year. The first period was studied between October 2015 to January 2016, the second period from February to May 2016 and the third period from June to September 2016. The results showed that the highest infection rate in June-September 2019 was 100 and the lowest infection rate was 40 in October 2015 and January 2019. The average number of *Nosema* spores infection level in an individual was infectivity. The highest infectivity was 2.47x106 spores/bee in February to May 2016 and the lowest infectivity was 0.04x106 spores/bee in October 2015 to January 2016. *Nosema ceranae* was found in this study by Polymerase Chain Reaction technique.

1. Introduction

The European honeybees (*Apis mellifera*) are a highly valued resource worldwide and importance to the ecosystem, not only as a honey or honeybees productivity but also as a pollinator of agricultural and horticultural crops and wild flora. A peculiar decline in honeybee colonies has gained worldwide. Much attention has been given to Colony Collapse Disorder (CCD). Several causes of these large-scale losses have been reported, including honey bee parasites, pesticides, environment, antibiotics, poor nutrition, and pathogens. *Nosema* are classified to class Microsporidia has a single cell spore-forming microorganisms. They are intracellular obligate parasites of eukaryotes including fishes, mammals, or especially in insects. The genus *Nosema* belongs to this group of fungi [1-4]. *Nosema apis* and *Nosema ceranae* are the two best known *Nosema* species because they were found in honeybees. *N. apis*, the historical microsporidian parasite of European honeybees, whilst *N. ceranae*, probably introduced into European honeybees from its Asian honeybees (*Apis cerana*) within the last few decades, and they invade the ventriculus epithelial cells of adult honeybees, and the best known of honey bee disease as nosemiasis or nosemosis. Honeybees become infected when they ingest *Nosema* spores, the fecal-oral and oral-oral routes being the main modes of transmission between honeybees. Once in the ventriculus, the parasites invade and develop within the cytoplasm of the mid gut epithelium cells. This disease negatively affects productivity and survival of honeybee colonies, and that may lead to colony depopulation and collapse [2-6].

In this study, the detection of *Nosema* on honeybee colonies needs data concerning its prevalence. To the best of our knowledge, little is reported about the prevalence of *Nosema* in Thailand apiaries and no reported in Kanchanaburi province was west of Thailand [7, 8]. Therefore, in this research we evaluated the prevalence of *Nosema* infection in Kanchanaburi province was west of Thailand apiaries. First, identify species of *Nosema* by polymerase chain reaction (PCR) assay based on 16S rRNA-gene-targeted species-specific primers to differentiate *N. ceranae* from *N. apis* and the second were light microscopy was carried out to screen the presence of Nosemosis in adult worker honeybees.

2. Materials and Method

2.1 DNA extraction, amplification and sequencing

Genomic DNA was extracted from a whole *Nosema* spores from the *A. mellifera* colonies at apiaries of Wanna Bee Farm using DNA Extraction Kit (Qiagen) according to the manufacturer's instructions. The 16S rRNA gene were amplified by polymerase chain reaction (PCR) using the following primers: Noce239L (5'–AGGGGCGAAACT TGACCTAT-3') and Noce950R, (5'–GGGCATAACKG ACCTGTTTA-3') [9]. PCR was carried out using 20 μl volumes containing 5 units of Taq DNA polymerase (biotechrabbit™), 2 μl of 10X PCR reaction buffer, 0.6 μl of 50 mM MgCl2 , 0.4 μl of 10 mM of each dNTP, 30 pmol of each primer, and 2 μl of the extracted DNA. The amplification profile comprised initial denaturation at 94°C for 3 min, 35 cycles at 94°C for 50 s, 55°C for 30 s, and 72°C for 1 min, and a final extension at 72°C for 10 min. The amplicons were then run on a 1% agarose gel, stained with ethidium bromide (0.5 lg/ml) and visualized with a UV illuminator (AB1500 Printgraph and AE 6905H Image Saver HR, ATTO, Tokyo, Japan). PCR products of 16S rRNA gene were purified with the PCR Purification Extraction Kit (Qiagen) and sent for sequencing using Sanger method (Macrogen, South Korea). The revealed sequences were verified by Bioedit software (version 7.0.5; Ibis Therapeutics, Carlsbad, USA).

2.2 Sequence and phylogenetic analysis

Sequences of 16S rRNA gene were searched against the GenBank nucleotide database (National Center for Biotechnology Information, http://www.ncbi.nlm.nih.gov) for gene homology Nosema sequences and their highest homologies obtained from GenBank, including N. apis (accession number. FJ789796). The sequences were aligned using the CLUSTALX multiple alignment program [10]. Gap sites were excluded from the following analysis. Genetic distances were estimated with the Kimura two-parameter method [11]. Construction of neighbor-joining trees [12] and the bootstrap test with 1,000 replications were conducted with the MEGA version 7 programs [13].

2.3 Detection of *Nosema* infection

The presence of nosemosis in honeybee colonies was investigated in three different seasons in Thailand. The first period was done on winter season, during October 2015 - January 2016 February, second period on dry season (February - May 2016) and the third period on raining season (June - September 2016). The honey bees used in this study were collected from the *Apis mellifera* colonies maintained at apiaries of Wanna Bee Farm, located at Kaeng Sian Subdistrict, Mueang District, Kanchanaburi Province, Thailand. Honeybee was randomly collected from 20 colonies, 50 bees from each colony in different season to verify the presence of *Nosema* spp. spores and the number of spores per individual bee. For this, the abdomens were individually separated, placed into a 1.5 ml Eppendorf tube and then homogenized thoroughly in 200 ml of deionized H_2O. Ten microliters of the homogenate were loaded onto a hemocytometer and the number of spores counted.

2.4 *Nosema* spore morphological identification using light microscopy
The mid gut of 50 bees was macerate in 1.5 ml Eppendorf containing 1,000 µl deionized water, the suspension was centrifuged at 6000g for ten minutes and the homogenate examined under the light microscope using × 400 magnification and photographed. Measurement is present in micrometers and data are expressed as the mean followed by the range in parentheses.

3. Results and discussion

3.1 *Nosema* species determination
The amplified product of 16S rRNA gene from two *Nosema* spores strains was successfully amplified by PCR "Fig.1" using Noce239L (5'–AGGGGCGAAACTTGACC TAT-3') and Noce950R (5'–GGGCATAACKGACCTGTT TA-3') described by Reference [9]. These primers generate PCR products approximately 710 and 685 bp for *N. ceranae* and *N. apis*, respectively. The results showed that the 710 bp DNA fragments were successfully obtained in one PCR reaction. While after PCR products were purified and sequenced, slight differences were found at five nucleotide positions (positions 240, 312, 350, 423, and 425; Fig. 6 shows only the 672-bp aligned sequences). One gap was present among these aligned sequences and the GenBank retrieved sequences (*N. ceranae* (DQ673615), *N. ceranae* France (DQ374655), *N. ceranae* Germany (DQ374656), *N. ceranae* (XR_002966746). These base differences may be due to recombination or mutation. When comparing the *Nosema* species to the *N. apis* 16S rRNA sequences obtained from GenBank (*N. apis* (FJ789796)) there were 32 additional positional differences and 21 gaps. These results indicate that *Nosema* strain tested in this study were *N. ceranae*. The phylogeny shows that a *N. ceranae* Noce950R strain cluster together with the GenBank *N. ceranae* strains and are clearly separated from *N. apis* 'Figs. 6 and 2".

3.2 Infection rate and infectivity
All 20 colonies of *A. mellifera* were monitor throughout the 3 different season period for *Nosema* infection level; there were significant differences in infection levels between periods. We observed the infections level were significantly lower in October 2015 - January 2016, a 40% of colonies was infection by *Nosema*. The infections level was increased in February-May 2016 with 80% of colonies were infected, while during June-September 2016 a 100% of colonies were infected with *Nosema* "Fig.3". However, a total number of *Nosema* spores counted from individual honeybee were not corresponding with the infection level. The highest number spores counted per bee was found in bees sampled during the warmer months were collected in February-May 2016 with 2.47 $x10^6 \pm 4.27$ spores per bee. The lower number spores counted per bee were showed in bees sampled in June-September 2016 with represent the average number of spores counted 0.27 $x10^6 \pm 0.43$ spores per bee, whereas it was the lowest in bees collected in February - May 2016 with 0.04 $x10^6 \pm 0.08$ spores per bee. There was a significant difference in number of spores counted in different periods sampled (Chi2 = 24.98, $P<0.05$), as expected "Fig.4".

3.3 Morphological identification of *Nosema* spp. spores
Spores of *Nosema* spp. were examined under the light microscope using 400x magnification. Light microscopy of the midgut content showed a large number of spores. *Nosema* spp. spores were oval to rod shaped with average length and width of 4-5.2 µm and 2-2.5 µm, N=50 "Fig.5" respectively.
In the last few years, infection of honeybees (*A. mellifera*) by *N. ceranae* has been reported in a number of European countries, including Sweden, Germany, Denmark, France, Finland, Switzerland, Greece, Holland, United Kingdom, Hungary, Italy, Serbia, Slovenia, Poland and Spain, soon being confirmed in worldwide including, Denmark, Finland, France, Germany, Greece, Italy, Serbia, Spain, Sweden, Switzerland, Republic of Croatia, Brazil, USA, Japan, Vietnam and Thailand [2, 7-8, 14]. Our result is the report on the *Nosema* disease in European honeybees at Kanchanaburi who west of Thailand and the result show that the highest infectivity was $2.47x10^6$ spores/bee in February to May 2016 and the lowest infectivity was $0.04x10^6$ spores/bee in October 2015 to January 2016 and

N. ceranae was found in this study by Polymerase Chain Reaction technique. Our findings, microscopy is still a valuable, relatively cheap and simple method to screen for the presence of *Nosema* infection in apiaries. Similar results were also reported by Reference [4, 15]

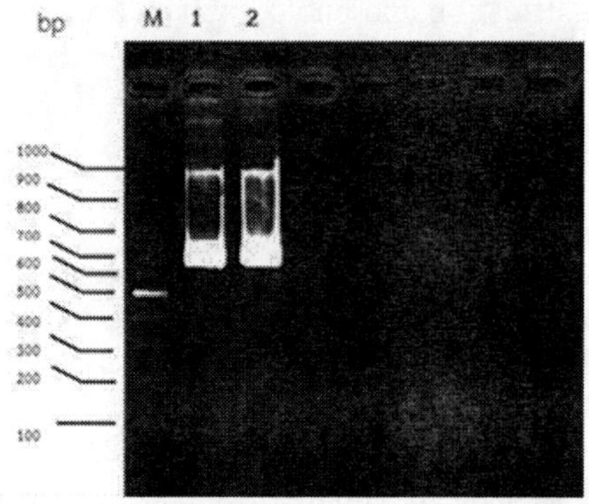

Figure 1. The amplified product of 16S rRNA of *N. ceranae* from *A. mellifera* runs on a 2% agarose gel. Lane 1 = Nosema 1, Lane 2= Nosema 2 respectively; M =100 bp ladder

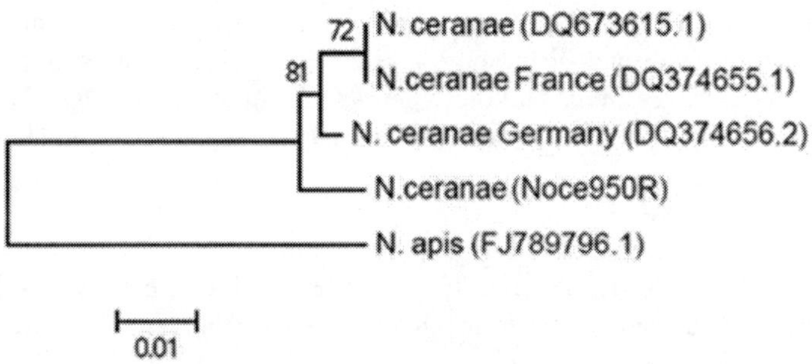

Figure 2. Phylogeny of *N. ceranae* based on 16s rRNA. The tree was generated by neighbor-joining analysis. Numbers on the nodes indicate probabilities based on 1,000 bootstrap replicates. A probability of more than 50% is shown. Branch lengths are proportional to genetic distance (scale bar)

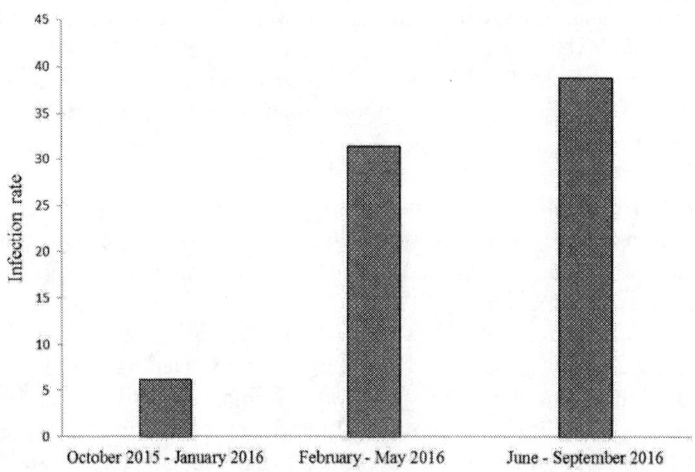

Figure 3. The infection rate of European honeybees on honeybees farm at Kanchanaburi Province

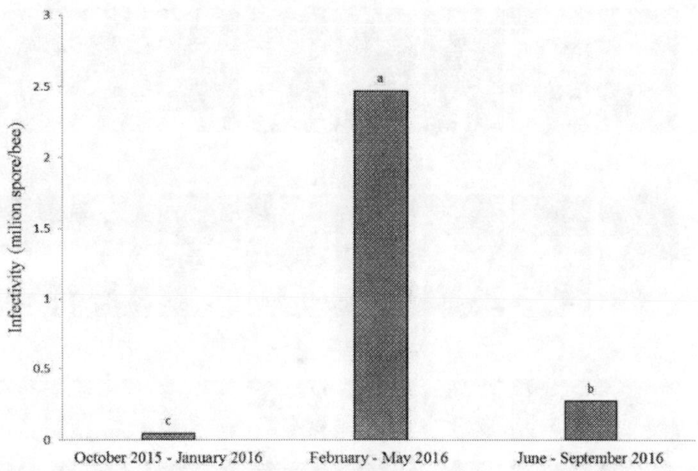

Figure 4. The infectivity of European honeybees on honeybees farm at Kanchanaburi Province (Chi2 = 24.98, $P<0.05$)

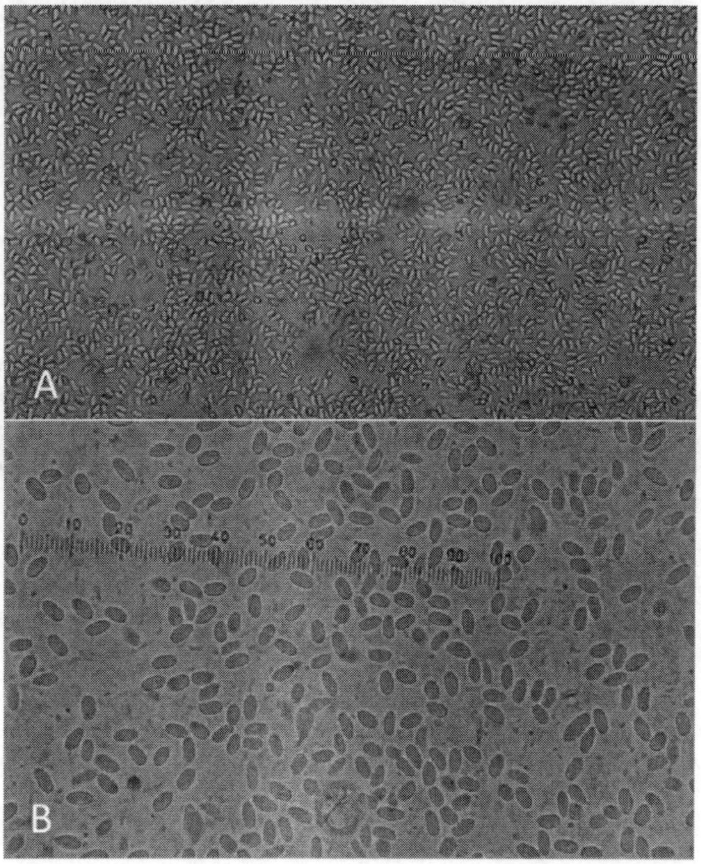

Figure 5. Spores of *N. ceranae* under light microscope (A=400x), (B=1000x)

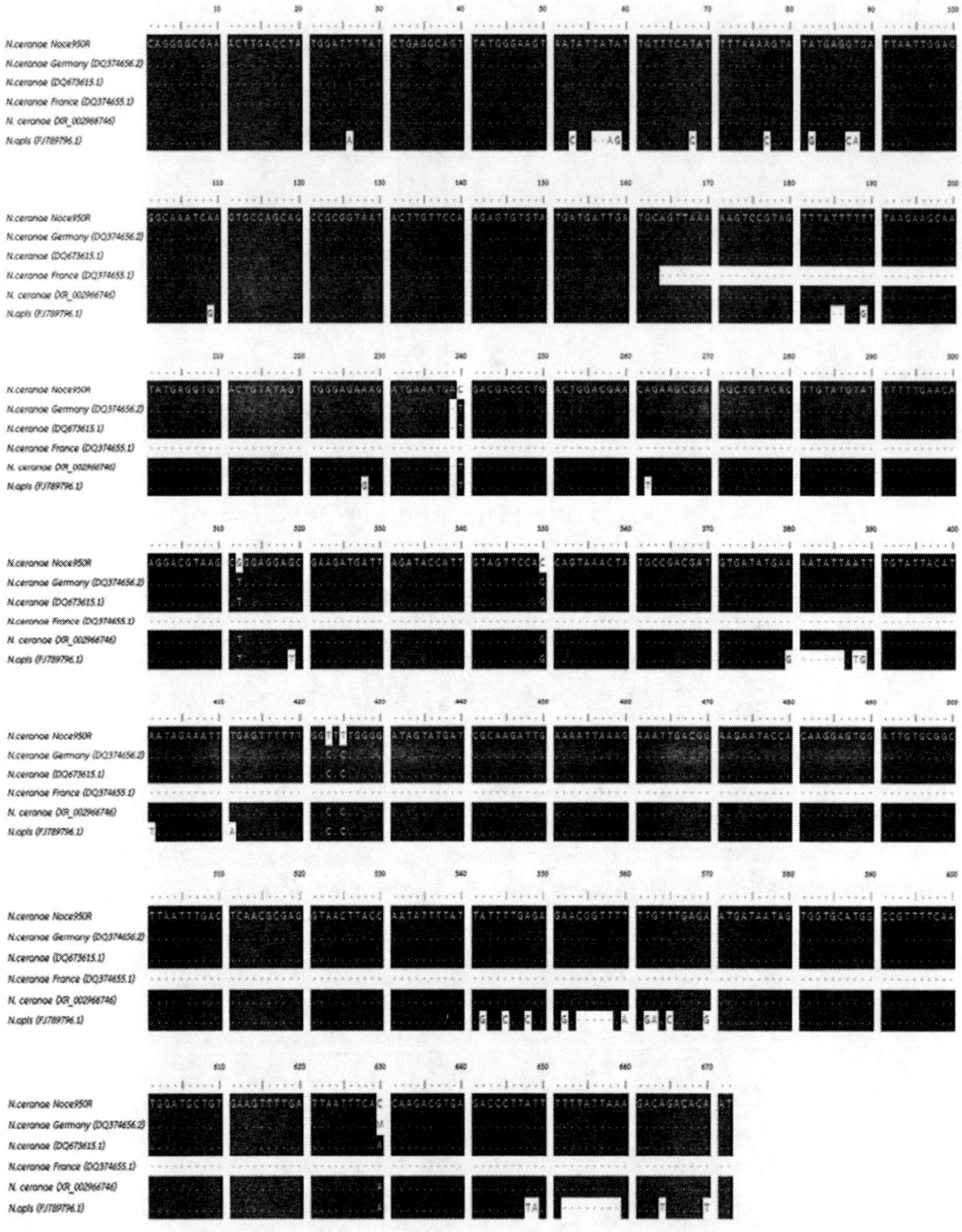

Figure 6. Comparisons of *Nosema* 16S rRNA sequences. The *Nosema* 16S rRNA sequences were aligned using ClustalX. Numbers on the top refer to nucleotide position. The nucleotide differences are denoted and similarities are shaded and (–) indicates gap. Abbreviations: *N. ceranae* Noce950R = *N. ceranae* isolated from *A. mellifera*; *N. ceranae* (DQ673615, *N. ceranae* France (DQ374655), *N. ceranae* Germany (DQ374656), *N. ceranae* (XR_002966746) = *N. ceranae* retrieved from GenBank accession number. DQ673615, DQ375655-6, and XR_002966746, respectively; *N. apis* (FJ789796) = *N. apis* retrieved from GenBank accession number. FJ789796

4. Conclusions

From our studies, light microscopy is a simple, cheap and valuable method to check the spread of Nosemosis in honeybee farm. However, morphological identification of *Nosema* spp. spores under light microscope is difficult to distinguish the species, because *N. ceranae* and *N apis* are very similar. To identification to species level, molecular techniques such as PCR are needed to reliable. The average number of *Nosema* spores infection level in individual was infectivity. The highest infectivity in February to May 2016 and the lowest infectivity in October 2015 to January 2016.

5. References

[1] Keeling P 2009 Five questions about microsporidia, *PLoS Pathog.* **5** 1-3

[2] Higes H, Martín-Hernández R and Meana A 2010 Nosema ceranae in Europe: an emergent type C nosemosis *Apidologie* **41** 375–392

[3] Fries I 2010 Nosema ceranae in European honey bees (Apis mellifera) *J. Invertebr. Pathol.* **103** 573–579

[4] Papini R, Mancianti F, Canovai R, Cosci F, Rocchigiani G, Benelli G and Canale A 2017 Prevalence of the microsporidian *Nosema ceranae* in honeybee (*Apis mellifera*) apiaries in Central Italy *Saudi. J. Biol. Sci.* **24** 979–982

[5] Paxton R J 2010 Does infection by *Nosema ceranae* cause ''Colony Collapse Disorder'' in honey bees (*Apis mellifera*) *J. Apic. Res.* **49** 80–84

[6] Paris L, El Alaoui H, Delbac F D R and Diogon M 2018 Effects of the gut parasite *Nosema ceranae* on honey bee physiology and behavior *Curr. Opin. Insect. Sci.* **26** 149-154

[7] Chaimanee V, Warrit N and Chantawannakul P 2010 Infections of *Nosema ceranae* in four different honeybee species *J. Invertebr. Pathol.* **105** 207–210

[8] Suwannapong G, Maksong S, Phainchajoen M, Benbow M E and Mayack C 2018 Survival and health improvement of *Nosema* infected *Apis florea* (Hymenoptera: Apidae) bees after treatment with propolis extract *J. Asia-Pac. Entomol.* **21** 437-444

[9] Suwannapong G, Yemor T, Boonpakdee C and Benbow M E 2011 *Nosema ceranae*, a new parasite in Thai honeybees *J. Invertebr. Pathol.* **106** 236-241

[10] Jeanmougin F, Thompson J D, Gouy M, Higgins D C and Gibson T J 1998 Multiple sequence alignment with Clustal X *Trends Biochem. Sci.* **23** 403-405

[11] Kimura M 1980 A simple method for estimating evolutionary rates of base substitution through comparative studies of nucleotide sequences *J. Mol. Evol.* **16** 111–120

[12] Saitou N and Nei M 1987 The neighbor-joining method: A new method for reconstructing phylogenetic trees *Mol. Biol. Evol.* **4** 406–425

[13] Kumar S, Stecher G and Tamura K 2016 MEGA7: molecular evolutionary genetics analysis version 7.0 for bigger datasets *Mol. Biol. Evol.* **33** 1870–1874

[14] Szalanski A, Whitaker J, Tripodi A D and Cappy P 2013 Prevalence of *Nosema* from managed honey bee colonies in South Dakota and New York *J. Agr. Urban. Entomol.* **29** 99-104

[15] Michalczykl M, Sokół R, Szczerba-Turek A and Bancerz-Kisiel A 2011 A comparison of the effectiveness of the microscopic method and the multiplex PCR method in identifying and discriminating the species of *Nosema* spp. Spores in worker bees (*Apis mellifera*) from winter hive debris *Pol. J. Vet. Sci.* **14** 385–391

6. Acknowledgement

The authors would like to thank Mr. Manus Prachakal for his help. This work was supported by the Budget bureau of Thailand no. 227678 and special thanks also go to Faculty of Science and Technology, Kanchanaburi Rajabhat University.

Effect of passion fruit juice and pectin on characteristics of Purple Yard Long Bean jam

T Natnicha[1], L Kanyarat[2] and S Kraiyot[2]

[1] Faculty of Science and Technology, Kanchanaburi Rajabhat University Nong Bua, Muang, Kanchanaburi ,Thailand 71190.

[2] Faculty of Agricultural Technology Burapha University, Sakaeo Campus Suwannasorn Road, Watthana Nakhon, Sa Kaeo, Thailand 27160.

E-mail : forest029@gmail.com, natnicha@kru.ac.th

Abstract. The present study aims at studying the amounts of passion fruit juice (21.1, 23.3 and 25%) and the amounts of pectin (0.5, 1 and 1.5) optimum to produce Purple Yard Long Bean jam (*Vigna unguiculata* subsp. *sesquipedalis*) cv. Sirindhorn No.1. The results showed that jam with 21.1% of passion fruit juice and 0.5% of pectin had the highest pH value and the lowest Rupture Strength. The water activity value and total soluble solids of 3 formulas jam were not statistically different. Jam with 25% of passion fruit juice and 1.5% of pectin had increasing of L* value and decreasing of a* value. Sensory evaluation showed that the jam with 21.1% of passion fruit juice and 0.5% of pectin gained the highest hedonic score in terms of odor, taste, texture, spreadability and overall preference. Purple Yard Long Bean jam with 21.1% of PJF and 0.5% of pectin had with higher fiber content than in commercial products.

1. Introduction

Sirindhorn No.1 purple yard long bean was a long bean hybrid from the long-stemmed species, green pods and mixed with red long pods. There was a characteristic of the species is the reddish purple pod and the seeds have red and white stripes. Sirindhorn No.1 purple yard long bean is source of antioxidant phenolic compounds, specifically flavonoids.[1] And source of high fiber because dietary fiber is a component of the cell wall in plants which are complex carbohydrates containing cellulose, pectin, hemicellulose, lignin and inulin. Consumption of dietary fiber provides other health benefits and decrease risk of diabetes, obesity, coronary heart disease, lowering of blood glucose and prevention of colon cancer.[2,3]

The Passion fruit (*Passiflora edulis var. flavicarpa*) containing vitamin C, provitamin A, phenolic compounds, flavonoids such as quercetin catechins, carotenoids can contribute to beneficial human. These compounds present antioxidant capacity.[4] Fruits and vegetables are source of high antioxidant can protect against several human diseases such as cancer hypertension, Alzheimer's cardiovascular diseases and prevention of the evolution of some diseases.[4,5]

Jam is an intermediate moisture food prepared by fruit and vegetables with sugar, acid, pectin other ingredients.[6] Fruits and vegetables contain acidity and pectin content that can improve quality of the texture of jam. However, the amount of pectin and acidity suitable for the gel formation by pectin in the range of 0.5-1% and 2.8-3.2 of pH value.[7] The jam should contain 60-65% of total soluble-solids content and at least 45% fruit.[3]

Therefore, the objective of the present study was to develop Sirindhorn No.1 purple yard long bean jam incorporated with passion fruit juice and study the amount of passion fruit juice and the amount of pectin on jam characteristics.

2. Materials and Methods

2.1 Materials

Sirindhorn No.1 purple yard long bean (Vigna unguiculata subsp. sesquipedalis) in the preparation of jams, were obtained from Ratchaburi Province. Sirindhorn No.1 purple yard long bean were ground using a blender and stored at -20 °C until processing. The passion fruit in the preparation of jams, were purchased from the local market. The passion fruit were washed and opened to remove the pulp with seeds and then it was squeezed until the total juice was removed. The passion fruit juice (PFJ) stored at -20 °C until processing.

2.2 Jam processing

Purple Yard Long Bean jam were prepared using different amounts of passion fruit juice (21.1, 23.3 and 25%) and the amounts of pectin (0.5, 1 and 1.5%) at 45% Purple Yard Long Bean and 55% sugar. After that, they were heated until about 64-65 % brix. The jams were produced in duplicate.

2.3 Chemical and physicochemical analysis

2.3.1 pH and Total soluble solids

The pH was measured using a S220 pH meter (Mettler Toledo, Switzerland). The water activity was measured by a water activity serie 4TE. (AquaLab, United States of America). Total soluble solids determined by Hand refractometer ATC-20E. (ATAGO, Japan).

2.3.2 Proximate analysis

The moisture content, fat, ash, fiber, protein and carbohydrate contents of the purple yard long bean jam were estimated by using standard method of [8] .

2.3.3 Color

Color was measured in terms of L*, a*, b* values using a Hunterlab Colorflex Spectrophotometer (CIE Lab Hunter Associates Inc, USA).

2.3.4 Textural analysis

The textural properties of Purple Yard Long Bean jam samples were measured using Texture Analyzer, TA.XT Express Enhanced (Charpa Techcenter Co.,Ltd.). The textural was measured in terms Rupture Strength and Brittleness. A cylindrical probe (cylinder probe p/20) was used to Compression, Pre - test speed : 1 mm/sec, Test speed : 0.50 mm/sec, Post- test speed : 10 mm/sec, Target mode : Distance 8 mm.

2.3.5 Anthocyanins analysis

Analysis of anthocyanins was measured by acidified ethanol (1% solution of HCl in ethanol 95%) was added to 10 g homogenized jam. The extract was kept room temperature, and filtered the whatman filter paper number 1. The extracts were estimated using the pH-differential method [8] using hydrochloric acid / potassium chloride buffer (0.025M, pH 1.0) and sodium acetate buffer (0.4 M, pH 4.5) and the absorbance was measured at 510 nm and 700 nm on a UV–vis spectrophotometer (specord 250 plus, Jena, Germany). The concentration of anthocyanins in the extract was calculated and expressed as cyanidin-3-O-glucoside equivalent per g of jam according to Eq. (1)

$$TA\ (mg/L) = \frac{A \times MW \times DF \times 10^3}{Ma \times L} \quad (1)$$

$A = (A_{510} - A_{700})$ pH 1 - $(A_{510} - A_{700})$ pH 4.5; Mw = molecular weight (449.2 g/mol); DF= dilution factor ; Ma = extinction coefficient 26,900 L/(cm*mol) L= path length (1 cm).

2.4 Sensory evaluation

The sensory evaluation of Purple Yard Long Bean jam were evaluated color, taste, odor, spreadability, and overall acceptability following the nine-point hedonic scale test (9= like extremely, 8 = like very much, 7 = like moderately, 6= like slightly, 5= neither like nor dislike, 4 = dislike slightly, 3 = dislike moderately, 2 = dislike very much, 1 = dislike extremely) by an untrained panel 30 subjects. Served at room temperature within 30 ml cups with three-digit randomized codes and bread pieces and spoons were provided to the panelists.

2.5 Statistical analysis

The study of Chemical and physicochemical analysis in completely randomized design (CRD) and the sensory analysis in randomized complete block design (RCBD). Statistical analysis was carried out using SPSS software (SPSS, Chicago, Ill., U.S.A.). Treatments were compared using analysis of variance (ANOVA) followed by Duncan's New Multiple Range Test (DMRT) was used to determine significant differences among results, and statistical significance was accepted at 95 %, with probability ($P \leq 0.05$).

3. Results and discussion

3.1 Chemical and physicochemical analysis

The results of studying the production of Purple Yard Long Bean jam by varying the amount of passion fruit juice at 21.1, 23.3 and 25% and the amount of pectin at 0.5, 1 and 1.5%. The results as shown in Table 1. It was found that increasing the amount of passion fruit juice the pH trended to decrease. Purple Yard Long Bean jam formula 3 (25:1.5) has the lowest pH ($P \leq 0.05$). The pH value affects the gel formation of the jam the appropriate pH value for the gel formation of the jam was 3.2 however the gel increases of hardness as the pH decreases. Acidity was an important, influence pectin gelation, texture of jams.[9] Yard Long Bean jam had water activity (a_w) values range between 0.58±0.00-0.64±0.02. Which was in the lower range than the microbial growth, the yeast begins to grow when the food is a_w in the range of 0.70-0.80. Mold can growth as a_w value was high than 0.70 and the bacteria can growth when the a_w value was high than 0.80. Yard Long Bean jam had total soluble solids in the range 64.40±0.70-65.50±0.70 % brix. Jam and all formulas are not statistically different. [10] Reported that the total soluble solids of the product should be in the range of 60-65 % brix and the product should contain at least 45% fruit.

Table 1. pH value, water activity (a_w) and Total soluble solids of Purple Yard Long Bean jam

Formulas jam (% PFJ:Pectin)	pH	a_w	Total soluble solids (% brix)
1 (21.1: 0.5)	3.22 ± 0.01^a	0.61 ± 0.01^a	65.50 ± 0.70^a
2 (23.3: 1)	3.21 ± 0.00^a	0.64 ± 0.02^a	65.50 ± 0.70^a
3 (25: 1.5)	3.18 ± 0.00^b	0.58 ± 0.00^a	64.40 ± 0.70^a

a,b,c…in each the column with different letters were significantly different at ($P \leq 0.05$).

Table 2 showed that the increase the amount of passion fruit juice in Purple Yard Long Bean jam increased the L* value and the a* value was decreased. [9] Similarly reported that the added to fibers increasing of pomegranate peel powders the L* value increased from 38.97 to 43.49 and a* value decreased from 13.13 to 10.62. In addition its to affect on L* and a* value were increasing proportions of fruits. Jam with higher amount of fruit were a* value decreased which could be due to destruction of the anthocyanins during processing [11].

Table 2. Color of Purple Yard Long Bean jam

Formulas jam (% PFJ:Pectin)	L*	a*	b*
1 (21.1: 0.5)	$21.45{\pm}0.62^{c}$	$19.80{\pm}0.49^{ab}$	$19.49{\pm}0.28^{a}$
2 (23.3: 1)	$25.75{\pm}0.86^{b}$	$20.32{\pm}0.59^{a}$	$19.56{\pm}0.73^{a}$
3 (25: 1.5)	$29.39{\pm}0.57^{a}$	$18.35{\pm}0.48^{b}$	$18.14{\pm}0.71^{a}$

a,b,c…in each the column with different letters were significantly different at ($P \leq 0.05$).

Rupture Strength and Brittleness value of Purple Yard Long Bean jam as shown in Table 3 the results showed that rupture strength was increased with increased pectin and PFJ concentration from 21.1- 35 and 0.5-1.5% respectively. [9] Similarly reported that an increase in pectin concentration from 0.2% to 1.2% the firmness of jams fruits increased. [12] Similarly reported that the hardness of mango jam increased with pectin concentration. Pectin and sugar were responsible for gel formation in fruit jam due to the number of junction zones with pectin concentration result in number of the elasticity active polymeric chains of gel pectin structure stronger. [9,12] Brittleness of the Purple Yard Long Bean jam, all formulas are not statistically different.

Table 3. Rupture Strength and Brittleness value of Purple Yard Long Bean jam

Formulas jam (% PFJ:Pectin)	Rupture Strength (g)	Brittleness (mm)
1 (21.1: 0.5)	$407.63{\pm}40.23^{c}$	$16.81{\pm}0.38^{a}$
2 (23.3: 1)	$684.07{\pm}4.57^{b}$	$16.99{\pm}0.11^{a}$
3 (25: 1.5)	$992.91{\pm}53.40^{a}$	$17.96{\pm}0.80^{a}$

a,b,c…in each the column with different letters were significantly different at ($P \leq 0.05$).

3.2 Sensory evaluation

The sensory evaluation of Purple Yard Long Bean jam were evaluated color, taste, odor, spreadability and overall acceptability following the nine-point hedonic scale test are reported in Figure 1. The scores of sensory attributes in color of the Purple Yard Long Bean jam, all formulas are not statistically different. Jam formulas 1 with 21.1% of PJF and 0.5% of pectin gained the highest hedonic score in terms of odor, taste, texture, spreadability and overall preference ($P \leq 0.05$). These results also indicates that Purple Yard Long Bean jam formulated 1 it could be concluded that acceptable and produce good quality. Therefore, 21.1% of PJF and 0.5% of pectin were chosen and used in next steps for proximate and anthocyanins analysis.

3.3 Proximate and anthocyanins analysis

Figure 2. shows the values of protein, moisture, fat, ash, fiber, carbohydrate and anthocyanins content of Purple Yard Long Bean jam the results shown that Purple Yard Long Bean jam with 21.1% of PJF and 0.5% of pectin had protein, moisture, fat, fiber, ash, carbohydrate content at 1.38±0.28, 29.34±0.03, 0.08 ± 0.02, 25.18±1.01, 0.47±0.00, 43.54 ±1.06 respectively. [3] Similarly reported that jams produced from tomato pomace with increased amount of dietary fiber, proximate composition of low calorie. Purple Yard Long Bean jam had total anthocyanins (TA) content 3.48±0.08 mg/L.[13] reported that the anthocyanins were the most abundant flavonoid compounds these are highly unstable pigments and processing can greatly influence their content in the final product. Purple Yard Long Bean jam had with higher fiber content and lower value aw than in commercial products.

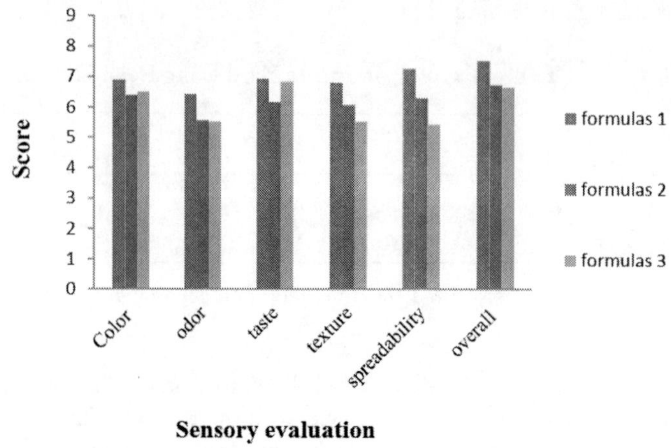

Figure 1. Sensory evaluation of Purple Yard Long Bean jam.

formulas 1 (PFJ 21.1 : Pictin 0.5) ; formulas 2 (PFJ 23.3 : Pictin 1); formulas 3 (PFJ 25 : Pictin 1.5)

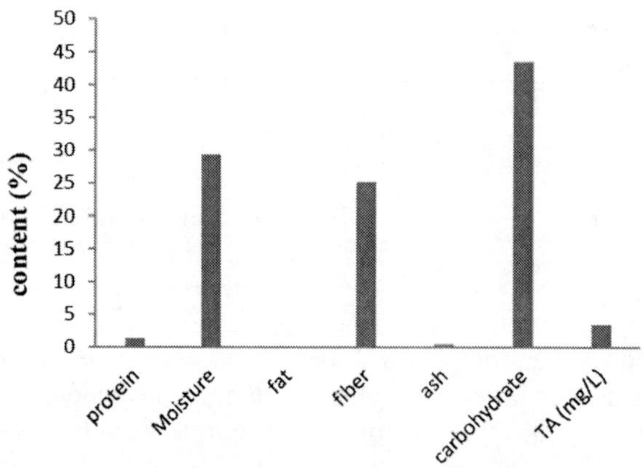

Figure 2. Proximate and anthocyanins content of Purple Yard Long Bean jam

4. Conclusions

In this study of the amount of passion fruit juice and the amount of pectin optimum to produce Purple Yard Long Bean jam showed that the passion fruit juice (PFJ) and pectin had effect on pH value, L*, a* Rupture Strength and score of odor. The formula that most panellists was that better accepted was 21.1% of PJF and 0.5% of pectin. Purple Yard Long Bean jam had with higher fiber content than in commercial products and aw value lower with to a longer shelf life.

5. References

[1] Thaweeseang N and Lueangprasert K 2018 Substitution of Wheat Flour with Purple Yard Long Bean Flour (*Vigna unguiculata* subsp. *sesquipedalis*) cv. Sirindhorn No.1 in cookies. Thai Science and Technology J., vol 26 (7), Supplement Issue pp. 1222-1233.

[2] Mudgil D and Barak S 2013 Composition properties and health benefits of indigestible carbohydrate polymers as dietary fiber: A review. Int. J. Biol. Macromol. pp. 61 : 1-6.

[3] Belovic M, Torbica A, Pajic I and Mastilovic J 2017 Development of low calorie jams with increased content of natural dietary fibre made from tomato pomace, Food Chemistry. 237, pp. 1226–1233.

[4] Reisa L C R, Faccob E M P, Floresa S H, Riosa A O 2018 Stability of functional compounds and antioxidant activity of fresh and pasteurized orange passion fruit (*Passiflora caerulea*) during cold storage. Food Research International.106. pp. 481–486.

[5] Pertuzatti P B, Sganzerla M, Jacques A C, Barcia M T and Zambiazi R C 2015 Carotenoids, tocopherols and ascorbic acid content in yellow passion fruit (*Passiflora edulis*) grown under different cultivation systems LWT-Food Science and Technology. 64 pp. 259-263.

[6] Basu S, Shivhare U S and Singh T V 2013 Effect of substitution of stevioside and sucralose on rheological, spectral, color and microstructural characteristics of mango jam, Journal of Food Engineering. 114, pp. 465–476.

[7] Shinwari K J and Rao P S 2018 Stability of bioactive compounds in fruit jam and jelly during processing and storage : A review, Trends in Food Science & Technology. 75, pp. 181–193.

[8] AOAC, Official methods of analysis of AOAC international 2005 18th Ed. International.

[9] Abid M, Yaich H, Hidouri H, Attia H, Ayadi M A 2018 Effect of substituted gelling agents from pomegranate peel on colour, textural and sensory properties of pomegranate jam, Food Chemistry. 239 pp. 1047–1054.

[10] CODEX 2009 Codex standard 296: Standard for jams, jellies and marmalades, 2009, Website: <http://www.codexalimentarius.net>. (Accessed on January 16, 2017).

[11] Kopjar M, Pilizota V, Tiban N N, Subaric D, Babic J, Ackar D and Sajdl M 2009 Strawberry jams: Influence of different pectins on colour and textural properties, Czech J. of Food Sciences 2, pp. 20–28.

[12] Basu S and Shivhare U S 2010 Rheological, textural, micro-structural and sensory properties of mango jam, J. of Food Engineering. 100, pp. 357–365.

[13] Kovacevic D B, Putnik P, Dragovic-Uzelac V, Vahcic N, Babojelic M S and Levaj B 2015 Influences of organically and conventionally grown strawberry cultivars on anthocyanins content and color in purees and low-sugar jams. Food Chemistry. 181 pp. 94–100.

6. Acknowledgments

This research was financially supported by the Research Grant of Burapha University through National Research Council of Thailand (Grant no. 107/2561). And the author would like to thank the Faculty of Agricultural Production, Maejo University, Chiang Mai for assistance the purple yard long bean (cv. Sirindhorn No.1) seeds.

IOP Conf. Series: Materials Science and Engineering **639** (2019) 012050 doi:10.1088/1757-899X/639/1/012050

Adsorption of Ammonia Nitrogen in Aqueous Solution Using Zeolite A

P Assawasaengrat and R Rueangdechawiwat

Department of Chemical Engineering Faculty of Engineering, King Mongkut's Institute of Technology Ladkrabang Bangkok, Thailand 10520

pornsawan.as@kmitl.ac.th

Abstract. The capacity of alternative adsorbents including rice straw, delignified rice straw and zeolite A for ammonia nitrogen removal from aqueous solution was investigated. The adsorption experiments were carried out in batch system with initial ammonia nitrogen concentration of 0.5 mg-N/l, 1000 ml ammonia nitrogen solution for 1 g of adsorbent. Experimental data indicated that rice straw and delignified rice straw cannot adsorb ammonia nitrogen. However, zeolite A provided rapidly adsorption capacity for ammonia nitrogen removal. The amount of ammonia nitrogen adsorbed by zeolite A was 0.256 mg/g at 1 minute after the experiment was initially conducted and slightly stable at 0.275 mg/g after 10 minutes. The adsorption experiments were carried out at 0.25, 0.5, 1.0, 2.0, 4.0, and 8.0 mg-N/l of ammonia nitrogen concentration. The results indicated that zeolite A had the adsorption capacity of 0.102, 0.294, 0.551, 1.153, 2.303, and 3.897 mg/g respectively at equilibrium time. The Langmuir and Freundlich adsorption models were applied to describe the equilibrium isotherms. Based on R^2, Freundlich model was more promising to represent the adsorption data than Langmuir model.

1. Introduction

Ammonia is widely used in the chemical industries as a bleaching agent in the production of fertilizers, plastics and explosives. The discharge of ammonia from wastewater treatment and industrial processes has been an important environmental issue and has been gained major worldwide concerns. Besides, the production of commercial fertilizers and other industrial applications, natural sources of ammonia include the decomposition of organic waste matters, human and animal wastes, and food scraps from aquatic lives. Ammonia does not only cause pollution of lakes and seas, but its excess level from the recommended limit dangerously harms aquatic lives and other animals. Therefore, it is necessary to remove ammonia from wastewater before discharge to environment.

Various methods have been employed to eliminate ammonia from wastewater including physical methods such as reverse osmosis and distillation, chemical methods such as chemical precipitation. However, these methods have some limitations and disadvantages. Reverse osmosis method requires large amount of energy input to provide hydraulic pressure driving water through the membrane.

IOP Conf. Series: Materials Science and Engineering **639** (2019) 012050 doi:10.1088/1757-899X/639/1/012050

Moreover, semi-permeable membrane can be easily damaged since suspended solids in wastewater are high. Distillation method requires large amounts of energy and water during the removal process. Chemical precipitation method requires working with corrosive chemicals, increasing operator safety concerns. Adsorption is one of the treatment process which is a highly-efficient technique and has been gained considerable attention for treating ammonia from aqueous solution due to low initial cost, flexibility and simplicity of design. In recent years, most of attention has been considered and focused on the use of alternative adsorbents to replace the conventional adsorbents. Various unwanted plant wastes such as rice husk, water hyacinth and waste tea.

Rice straw is an agricultural waste, low cost, abundance, reliability, and availability. The surface of rice straw normally contains a lot of pores. Thus, rice straw has the potential to be used as an adsorbent. Zeolite A has been reported as an adsorbent used to remove cations from aqueous solution. Zeolite A can be synthesized by silica extracted from local rice straw using a simple extraction method. In this work, rice straw, delignified rice straw, and zeolite A was considered as an adsorbent for ammonia removal process and the adsorption isotherm was studied.

2. Experiment methods

2.1. Adsorbents Preparation and Characterization

The rice straw was boiled in distilled water at 80 °C for 3 hours and was extracted in acetone at 50 °C 3 hours. Then it was placed in oven to dry 24 hours at 100 °C. The sample was milled to pass through a 8 mesh screen. The rice straw was immersed in 1.4% acidified $NaClO_2$ (250 ml), with pH adjusted to 4 by CH_3COOH, at 70 °C for 5 h to dissolve lignin. The delignified rice straw washed and dried. Silica extracted from rice straw was used as silica source for Zeolite A synthesis. 1 g of NaOH dissolved in 100 ml of distilled water was added to 4 g of silica (from rice straw). Then sodium silicate solution was heated at 80 °C until clear solution. 6 g of sodium aluminate and 3.4 g of NaOH dissolved in 100 ml of distilled water was stirred until clear solution. The sodium silicate solution was dropped into the mixture of sodium aluminate and stirred for 1 hour. The mixture was refluxed for 1 hour at 250 °C. Centrifugation at 2000 rpm was used to separate the precipitate and the liquid phase was removed by filtration. The precipitate was dried at 100 °C for 24 hour. Analysis by X-ray diffractometer (XRD) revealed major crystalline phases in Zeolite A.

2.2 Ammonia Nitrogen Adsorption

1 g of adsorbent was added in to 50 ml of ammonia solution (0.25, 0.5, 1.0, 2.0, 4.0, and 8.0 mg-N/l). The sample was stirred with speed 200 rpm at room temperature by various contact time. After stirring, the adsorbent was removed and ammonia nitrogen concentration was determined by spectrophotometer according to Phenolnypochlorite Method [1]. The adsorption capacity was defined as the amount of ammonia nitrogen adsorbed onto unit mass of adsorbent (mg/g).

3. Results and Discussion

3.1 Adsorbents Characterization

The overall crystalline phases of zeolite A were determined by XRD measurement. The XRD pattern in Figure1 indicated that the presence of peak with 2θ= 7.14, 10.12, 12.40, 16.06, 21.62, 23.94, 27.06, 29.90, and 34.14 respectively matching with the data base of zeolite A [2]. Therefore, it could be confirmed that the sample characterized by XRD was zeolite A.

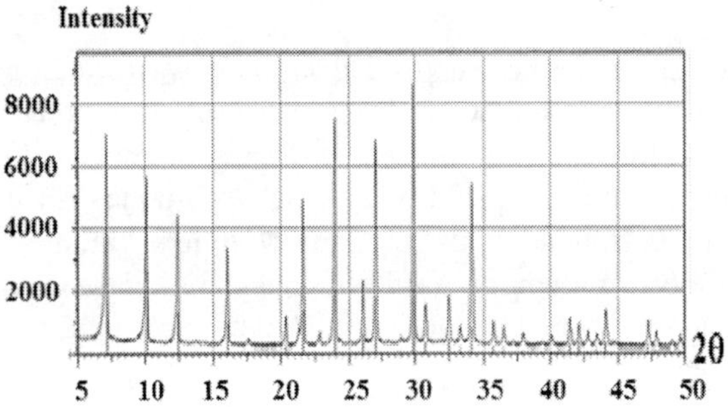

Figure 1. The XRD pattern of zeolite A

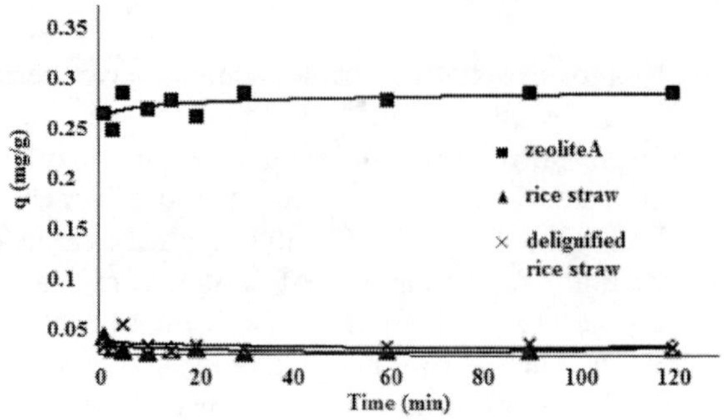

Figure 2. Ammonia adsorption capacity of rice straw, delignified rice straw and zeolite A

3.2 Ammonia Nitrogen Adsorption

Ammonia nitrogen removal of different adsorbents is presented in Figure 2. Preliminary experiments were carried out to initially investigate the efficiency of rice straw, delignified rice straw, and zeolite A adsorbents for ammonia removal with initial ammonia nitrogen concentration of 0.5 mg-N/l, 1000 ml ammonia nitrogen solution for 1 g of adsorbent. The result showed that ammonia nitrogen concentration remained constant after adsorption process using rice straw and delignified rice straw adsorbents. However, the amount of ammonia nitrogen adsorbed by zeolite A was 0.256 mg/g at 1 minute after the experiment was initially conducted and slightly stable at 0.275 mg/g after 10 minutes. The high efficiency of zeolite A adsorbent is due to the high surface area of adsorption sites and the capacity to enhance the cation exchange with NH_4^+ from ammonia solution [3, 4].

3.3 Effect of Initial Concentrations

The adsorption experiments onto zeolite A adsorbent were carried out at 0.25, 0.5, 1.0, 2.0, 4.0, and 8.0 mg-N/l of ammonia nitrogen concentration considering removal capacity at 1, 3, 6, 10, 15, 20, 30, 60, 90, and 120 minutes, using 1 g/l of zeolite A adsorbent. The rate of ammonia nitrogen removal was very rapid during the initial 1 minute and remained nearly constant thereafter for every initial concentration due to the number of vacant sites on the adsorbent were much available to adsorb the adsorbate molecules in the solution. As shown in Figure 3, the results indicated that zeolite A had the adsorption capacity of 0.102, 0.294, 0.551, 1.153, 2.303, and 3.897 mg/g respectively at equilibrium time. Differences of initial ammonia concentration clearly showed that a gradual increase in the ammonia nitrogen concentration leaded to high adsorption capacity using zeolite A adsorbent since the more increased of ammonia nitrogen molecule in the solution, the higher driving force of ammonia nitrogen molecules adsorbed onto available sites of the zeolite A particles [5].

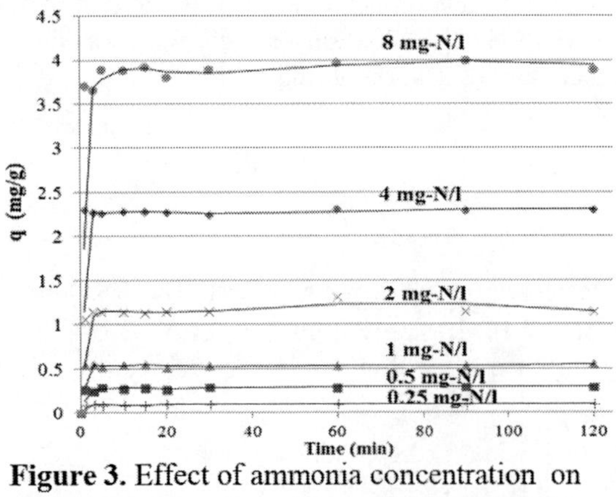

Figure 3. Effect of ammonia concentration on adsorption capacity using zeolite A adsorbent

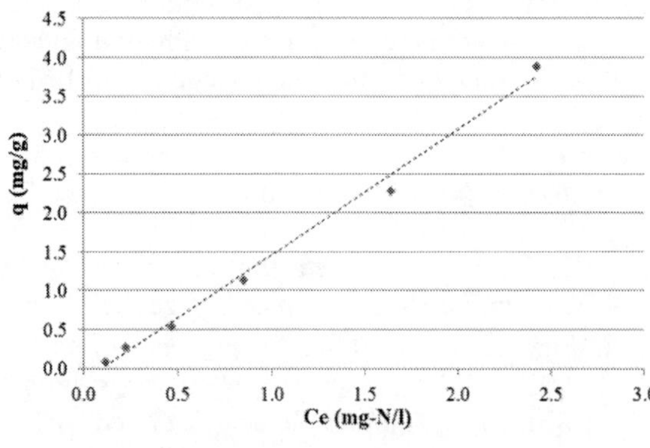

Figure 4. Adsorption isotherm of ammonia nitrogen using zeolite A adsorbent

3.4 Adsorption Isotherms

The amount of ammonia nitrogen adsorbed onto unit mass of zeolite A adsorbent at equilibrium time (q_e) and the ammonia nitrogen concentration at equilibrium time (C_e) were calculated and correlated for the adsorption isotherm as shown in Figure 4. The correlation of adsorption capacity with equilibrium ammonia nitrogen concentrations after adsorption process conducted at different initial concentrations indicated that the ammonia nitrogen removal capacity of zeolite A increased with the increased of ammonia nitrogen concentration. The maximum capacity of ammonia nitrogen adsorbed onto unit mass of zeolite A adsorbent was about 3.897 mg/g.

The adsorption isotherms can be represented by several numerous theoretical models where Langmuir and Freundlich models have been commonly used to fit experimental data. The equilibrium experimental data for the adsorption of ammonia nitrogen from aqueous onto zeolite A adsorbent was compared using Langmuir and Freundlich models as shown in Figures 5. and Figures 6. respectively.

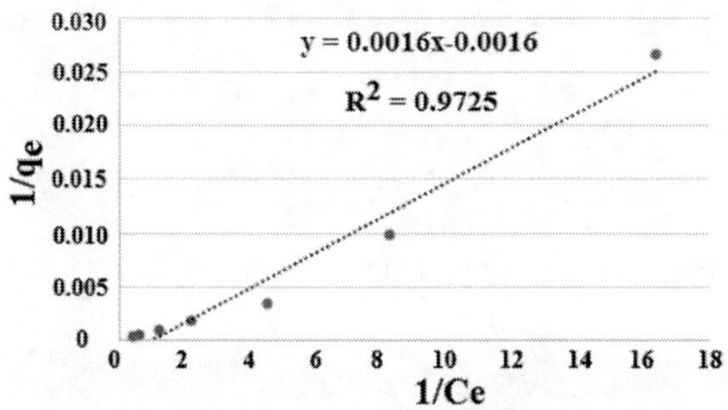

Figure 5. Langmuir plots for adsorption of ammonia nitrogen onto zeolite A adsorbent

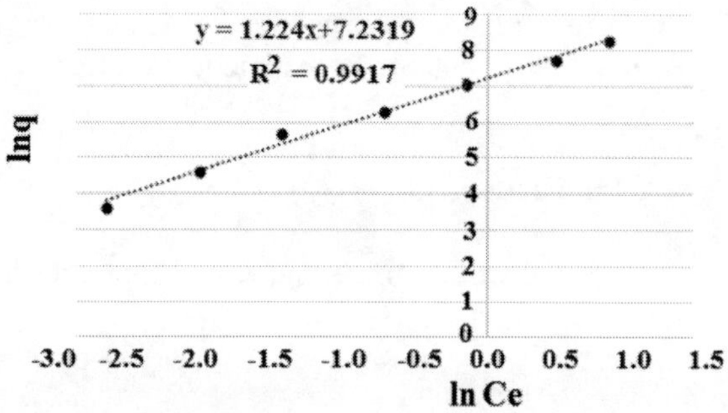

Figure 6. Freundlich plots for adsorption of ammonia nitrogenonto zeolite A adsorbent

The model parameters are tabulated in Table 1. The Regression coefficients (R^2) were 0.9723 and 0.9917 for Langmuir and Freundlich model, respectively. Based on R^2, Freundlich model was more promising to represent the ammonia nitrogen adsorption data than Langmuir model. It is evident from these data that the adsorption sites occurring between zeolite A and ammonia nitrogen molecules are multilayer adsorption. Charge on the surface of zeolite A is a determining factor of ammonia nitrogen removal efficiency. Ammonium ion (NH_4^+) which has a positive charge will attract negative charge on the surface of zeolite A adsorbent [6].

Table 1. Equation and R^2 value of Langmuir isotherm and Freundlich isotherm

Isotherm	Linear equation	R^2
Langmuir	$q_e = \frac{0.625C_e}{1 + 0.16C_e}$	0.9723
Freundlich	$q_e = 1.978C_e^{1.223}$	0.9917

4. Conclusion

The presented study was carried out to investigate the adsorption capacity of rice straw, delignified rice straw, and zeolite A as alternative adsorbents for ammonia nitrogen removal. The experimental results clearly indicated that zeolite A was an effective adsorbent for the removal of ammonia nitrogen from aqueous solution. However, rice straw and delignified rice straw were not able to provide the adsorption capacity of ammonia nitrogen. The influence of ammonia nitrogen concentration had significant effect on the removal capacity of zeolite A. The rate of ammonia nitrogen removal was rapid after adsorption processes conducted due to the high surface area of adsorption sites and the capacity to enhance the cation exchange with NH_4^+ from ammonia solution. For Langmuir and Freundlich isotherm, the experimental results showed that the Freundlich model gives the best correlation coefficient value based on the obtained determination coefficients (R^2) of Freundlich model gives the highest value. Therefore, it employed to describe the adsorption sites between zeolite A and ammonia nitrogen are multilayer adsorption. Zeolite A provides the application potentials for effective ammonia nitrogen removal and may display the advantages of adsorption capacity for other hazardous pollutants from aqueous solution.

Acknowledgment

The Faculty of Engineering, King Mongkut's Institute of Technology Ladkrabang is gratefully acknowledged.

References

[1] Weatherburn M W 1967 Phenol-hypochlorite reaction for determination of ammonia *Anal. Chem.*, vol 39, pp 971–974
[2] Treacy M J and Higgins J B, Collection of Simulated XRD Powder Patterns for Zeolites Fourth Revised Edition, p213
[3] Fulya A and Ayse K 2011 The Effect of Modification and Initial Concentration on Ammonia Removal from Leachate by Zeolite *International Journal of Environment, Chemical, Ecological and Geophysical Engineering,* vol 5, pp336-339
[4] Indrek K, Robert H and Eric M, 2001 Adsorption of Ammonia on Coal Fly Ash *International Ash Utilization Symposium*, p 147
[5] Chen Y, Nie J and Pol J 2016 Removal of Ammonia Nitrogen from Wastewater Using Modified Activated Sludge *Environ. Stud.*, pp 410-42
[6] Burgess R M, Perron M M and Cantwell M G 2004 Use of Zeolite for Removing Ammonia and Ammonia-Caused Toxicity in Marine Toxicity Identification Evaluations *Arch. Environ. Contan. Toxical.,* vol 47, pp440-447

Shelf Life Evaluation of Instant Pasta in Two Different Packaging.

S Murathathunyaluk and P Jindawanich

Chemical Engineering Department, Faculty of Engineering, King Mongkut's Institute of Technology Ladkrabang, Bangkok, Thailand 10520

siripan.mu@kmitl.ac.th

Abstract. The objective of this study is to evaluate shelf life of instant pasta which had been packed in two different packaging, type I is an OPP/MCPP sachet and type II is an OPP/CPP sachet, under accelerated condition (37°C, 85%RH) for 12 weeks. There were 2 index which were used to evaluate the shelf life, one was sensory tests and another one was moisture content in pasta. Assessed by sensory tests, the shelf life of instantly cooked pasta packed in type I and type II sachet, were revealed more than 12 weeks and equal 12 weeks, respectively. Likewise, the shelf life predicted on the basis of moisture sorption isotherm and water permeability of packaging material, were estimated at 119 and 46 days in OPP/MCPP and OPP/CPP sachets, respectively. According to these results, the quality changing of instant pasta measured during storage suggested that OPP/MCPP sachet had better protectively effect than the OPP/CPP sachet.

1. Introduction

Food packaging plays a vital role in maintaining the food's properties. If food is not packed in packaging, it may change both in physical and chemical properties, occurring the contamination in food. Moreover, Importantly, It may lead to taste changing and consumer will not be able to accept. Keeping food in a proper package will prolong the shelf life. For this reason, research and development of packaging are required to maintain food quality and extend the shelf life of food [1]. The shelf life of a food is defined by the time for which the product, stored under determined temperature conditions, presents alterations considered, up to a certain point, acceptable by the manufacturer, consumer and current food legislation. Many products show prolonged shelf lives, making their experimental determination difficult [2]. Although shelf life tests under the accelerated environmental condition represent an alternative, the accuracy of the shelf life evaluation also depends on the expertise of the sensory panelist. Using the critical moisture content, the shelf life of a dry food can be predicted. A model for shelf life prediction that is based on moisture content, permeability of the packaging material and storage temperature will be useful for identifying the quality storage life of food [3].

One of the company operating as a consumer goods would like to change the packaging of instant pasta from plastic cup made by polyprepylene to be packing instant pasta in plastic bag first before packing another layer into the cup made from paper. Two types of plastic bags, which have the different qualities of oxygen and moisture permeation, are brought to study about the packaging of instant pasta. The objectives of this study are to develop a prediction model for the shelf life of instant pasta based on moisture content and permeability of packaging material. Moreover, the results of these are brought to affirm by sensory test which can help to select the appropriate packaging.

2. Materials and methods

2.1. sample preparation

Instant pasta, purchased from local market, was packaged in two different packaging materials i.e. type I OPP/MCPP sachet, made from orient polypropylene plastic film laminated with metalized cast polypropylene plastic film, and type II OPP/CPP sachet, made from orient polypropylene plastic film laminated with cast polypropylene plastic film. Samples of 22 g were taken in the sachet (8 cm*11 cm size) which were closed by heat sealing taking care that minimum possible air space remained in sachet. The sealing was carefully inspected to avoid any possibility of leakage. The sealed sachets were stored under accelerated condition (incubator that maintain temperature and humidity levels at 37°C, 85%RH) and inside a home refrigerator maintained at 4-10°C. The experiments were carried out for 12 weeks storage period at an interval of 1 week taking samples from each of the packaging and storage conditions. These samples were analyzed for multiple comparison sensory tests. Raw material characterization had been done by measuring moisture content (% d.b.), measured with moisture analyzer (Mettler Toledo HB43), and water activity, determined using a water activity meter (Aqua Lab Series 3).

2.2. Statistical analysis

The moisture content of instant pasta in sachet was analyzed by using MINITAB-18 statistical software. The factors, included in the ANOVA, were packaging material (OPP/MCPP and OPP/CPP), storage period and their interaction.

2.3. Sensory analysis

The method used for sensory testing is the multiple comparison test, which tasting instantly cooked pasta to find the difference between instant pasta that stored in accelerated condition and in low temperature condition .Three sensory specialists evaluated the instantly cooked pasta for overall, color, aroma and flavor acceptability on a 6-point hedonic scale. (1 = extreme difference, 2 = very much difference, 3 = distinctly difference, 4 = moderately difference, 5 = slightly difference, 6 = equal to stardard). However, the regulation of acceptable minimum score by evaluated from average overall score was more than 2.

2.4. Moisture isotherm model

The GAB equation [4] was used to describe the dry basis moisture content (X) as a function of water activity (a_w):

$$X = \frac{X_0 C K a_w}{(1-Ka_w)(1-Ka_w+CKa_w)} \quad (1)$$

where C and K are constants and X_0 is described in the literature as the monolayer moisture content on dry basis (kg/kg)

2.5. Measurement of permeability of packaging material

Five grams of dehydrated silica gel, packed in three OPP/MCPP sachets and three OPP/CPP sachets, were kept in the environment at 85%RH and 37°C. The weight of each package samples, determined by weighing the whole individual package using a laboratory level weighing balance having 0.0001 g accuracy, was measured at an interval of 24 h for 7 days and the respective weight gain was computed. The water vapor permeability, k (kg/m^2 day Pa) of both the sachets was determined by using Eq.(2).

$$k = \frac{(Q/t)/A}{p} \quad (2)$$

where Q/t (kg/day) is slope of graph between weight gain of silica gel and time. A (m^2) is a surface area of the sachet that water vapor can penetrate. p (Pa) is saturation vapor pressure of water at 37°C

2.6. Assessment of shelf-life of instant pasta

Shelf life [5] of instant pasta (θ) was calculated using Eq.(3)

$$\theta = \frac{W_s}{pkA}\int_{X_i}^{X_c} \frac{1}{RH-a_w} dX \qquad (3)$$

where w_s is initial weight of instant pasta. The moisture content (X) was considered as critical moisture content and corresponding water activity was calculated by fitting the GAB model (Eq.(1)).

3. Result

3.1 Sensory analysis

Average sensory scores of instant pasta packed in type I OPP/MCPP sachet and type II OPP/CPP sachet under accelerated conditions; overall, color, aroma and flavor acceptability, are given in Figure 1. Instant pasta packed in type I sachet was shown better score; overall, color, aroma and flavor than instant pasta packed in type II sachet. During 12 weeks, the overall score of instant pasta packed in type I sachet was more than 2 points which was within the acceptable range of consumers, and the overall score of pasta packed in type II sachet was 2 points, which was the limit of the range that consumers can accept. Therefore, the shelf life of instant pasta at the accelerated condition packed in type I sachet showed over 12 weeks and packed in type II sachet showed 12 weeks.

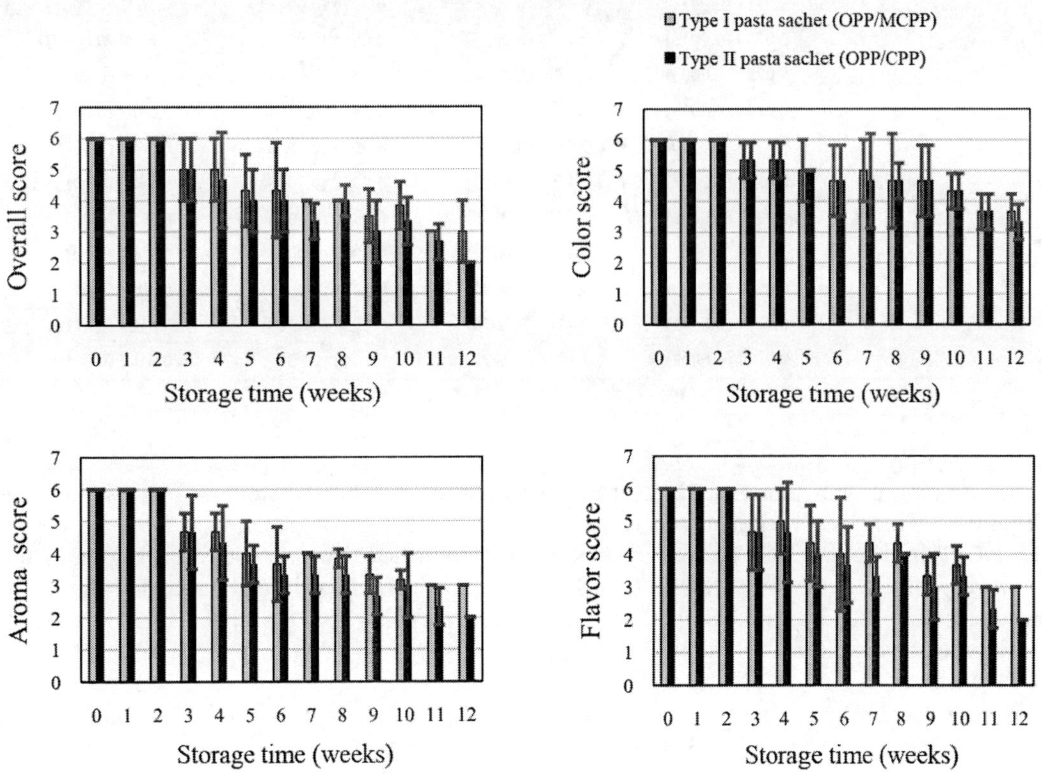

Figure 1. Mean sensory score of instant pasta under different packaging condition

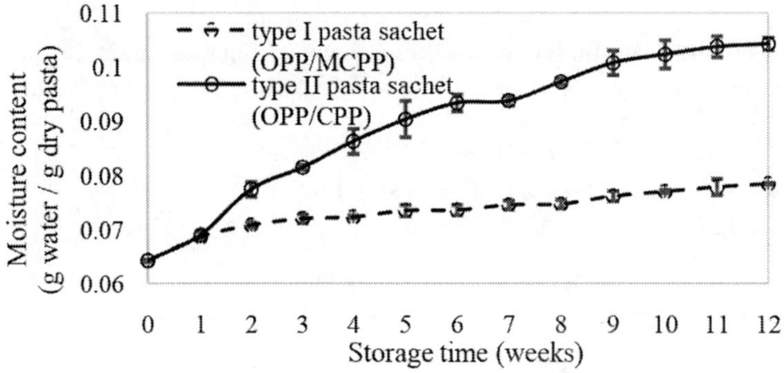

Figure 2. Variation in moisture content of instant pasta packed in OPP/MCPP and OPP/CPP sachet under accelerated storage conditions

Table 1. Moisture content and water activity of instant pasta packed in OPP/MCPP and OPP/CPP sachet under accelerated storage conditions

Storage period (weeks)	Type I pasta sachet (OPP/MCPP)		Type II pasta sachet (OPP/CPP)	
	Moisture content (g water / g dry pasta)	Water activity	Moisture content (g water / g dry pasta)	Water activity
0	0.0642 (0.0002)	0.4053 (0.0081)	0.0642 (0.0002)	0.4053 (0.0081)
1	0.0706 (0.0002)	0.4357 (0.0110)	0.0690 (0.0006)	0.4503 (0.0050)
2	0.0717 (0.0015)	0.4497 (0.0096)	0.0774 (0.0014)	0.5023 (0.0067)
3	0.0722 (0.0016)	0.4523 (0.0055)	0.0815 (0.0005)	0.5200 (0.0010)
4	0.0723 (0.0003)	0.4530 (0.0096)	0.0863 (0.0024)	0.5380 (0.0027)
5	0.0735 (0.0010)	0.4560 (0.0017)	0.0904 (0.0034)	0.5483 (0.0031)
6	0.0736 (0.0010)	0.4617 (0.0038)	0.0934 (0.0016)	0.5600 (0.0080)
7	0.0745 (0.0008)	0.4663 (0.0050)	0.0938 (0.0009)	0.5617 (0.0015)
8	0.0747 (0.0008)	0.4703 (0.0148)	0.0973 (0.0006)	0.5723 (0.0038)
9	0.0762 (0.0009)	0.4723 (0.0021)	0.1008 (0.0023)	0.5993 (0.0057)
10	0.0770 (0.0003)	0.4763 (0.0080)	0.1024 (0.0026)	0.6097 (0.0035)
11	0.0778 (0.0015)	0.4817 (0.0076)	0.1038 (0.0020)	0.6123 (0.0110)
12	0.0784 (0.0002)	0.4847 (0.0012)	0.1043 (0.0011)	0.6160 (0.0010)

3.2 Shelf-life of instant pasta

The variation in moisture content of instant pasta during accelerated storage is shown in Table 1 and Figure 2. There was a gradual increase in moisture content of samples in both OPP/MCPP and OPP/CPP packages. ANOVA, for moisture content, indicated that the type of packaging material, and storage period, significantly affected ($p \leq 0.05$) the moisture gain of instant pasta. The moisture gain by instant pasta packed in the OPP/CPP was more than the OPP/MCPP sachet. Moisture content after 3 months of storage, was 10.43% and 7.84% (d.b.), respectively for samples packed in OPP/CPP and OPP/MCPP.

An isotherm graph of instant pasta is presented in Figure 3. The constant in GAB model; C, K and X_0, was calculated by fitting the GAB model (second order polynomial regression) between moisture content (X) and water activity (a_w). The GAB model (Eq.(1)) showed the regression coefficient (R^2) = 0.9998 which can be considered as a good fit with experimental data. The C, K and X_0 are 3.6006, 0.8051 and 0.0677, respectively.

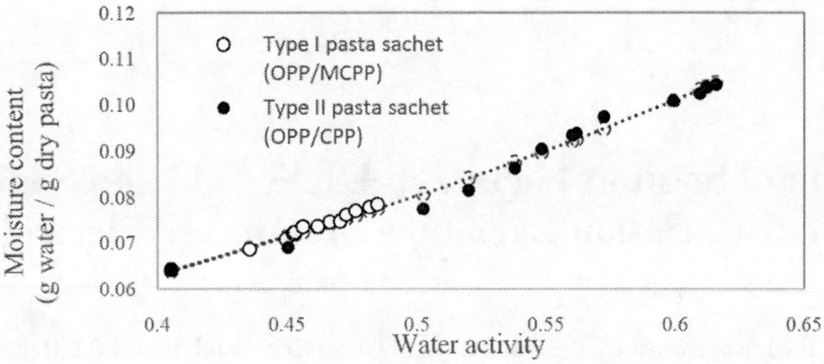

Figure 3. GAB model between water activity and moisture content of instant pasta.

The cumulative moisture gain of silica gel with time in OPP/MCPP and OPP/CPP sachets at 37°C and 85%RH was used to fit straight line. By inserting the slope of the straight line (Q/t), the surface area of the sachet (A) = 0.0176 m^2 and the saturation vapor pressure of water (p) at 37°C = 6275.47 Pa, into Eq.(2), the water vapor permeability (k) of OPP/MCPP and OPP/CPP are 1.34×10^{-7} and 3.44×10^{-7} kg/(m^2 day Pa), respectively.
Water activity and moisture content are the most valuable characteristics for assessing the stability of dried foods. Increasing of the moisture content during storage causes the product to become damaged. Requirements for storing dry food, water activity of instant pasta is set to not excess 0.55. In this research, the initial moisture content of instant pasta was 6.24%. The critical moisture content, corresponding with water activity of 0.55, calculated by the GAB model, is 0.090. The shelf-life of instant pasta in OPP/MCPP and OPP/CPP sachets, estimated by Eq.(3), were calculated as approximate 119 and 46 days, respectively.

4. Conclusions

Comparing shelf life of instant pasta packed in type I and type II pasta sachet, type I pasta sachet was made from orient polypropylene plastic film laminated with metalized cast polypropylene plastic film (OPP/MCPP) and type II sachet was made from orient polypropylene plastic film laminated with cast polypropylene plastic film (OPP/CPP), at accelerated condition. The shelf life, predicted from sensory tests and model for the shelf life based on moisture content and permeability of packaging material, revealed in the same way. As a result of, OPP/MCPP sachet had better protectively effect than OPP/CPP sachet.

5. References

[1] Maddah H A 2016 Polypropylene as a Promising Plastic: A Review American Journal of Polymer Science, 6(1): pp 1-11
[2] Breda C A, Argandona E J S, Correia C A C 2012 Shelf life of powdered Campomanesia adamantium pulp in controlled environments Food Chemistry 135 pp 2960–2964
[3] Jaja S and Das H 2005 Accelerated Storage, Shelf Life and Color of Mango Power Journal of Food Processing and Preservation 29 pp 45–62
[4] Prothon F and Ahrne L M 2004 *Application of the Guggenheim, Anderson and De Boer model to correlate water activity and moisture content during osmotic dehydration of apples* Journal of Food Engineering 61 pp 467–470
[5] Dak M, Sagar V R and Jha S K 2014 *Shelf-life and kinetics of quality change of dried pomegranate arils in flexible packaging* Food Packaging and Shelf Life 2 pp 1-6

IOP Conf. Series: Materials Science and Engineering **639** (2019) 012052 doi:10.1088/1757-899X/639/1/012052

Conversion of Sodium Lactate to Lactic acid and Sodium Hydroxide with Cation Exchange Membrane Electrolytic Cell

R Sirisangsawang, P Samaikaew, B Chotiviriyavanich and P Kitchaiya

Department of Chemical Engineering, King Mongkut's Institute of Technology Ladkrabang, Thailand

prakob.ki@kmitl.ac.th

Abstract. This research conversion of sodium lactate to lactic acid and sodium hydroxide with electrolysis technique combined with cation exchange membrane (CEM). The electrolytic cell has effective membrane area of 144 cm^2. Experiments were carried out by feeding sodium lactate solution which was prepared with alkaline degradation of invert sugar. This solution was feed through the anolyte. The process was studied in terms of the initial concentration of sodium lactate (i.e. 0.75, 1.20 and 1.50 M) and the applied voltage (i.e. 4, 6 and 8 volt) to assess the efficiency of separation. The results suggested that the transport of sodium ion through the membrane was proportional to the applied voltage but it was not depend on sodium lactate concentration which was used as feed. The current efficiency increased as the applied voltage increased. Conversely, the energy consumption with respect to sodium hydroxide production increased with both factors. From this experiment, it was found that the highest lactic acid recovery was 55.5% at 4 volt and initial lactate concentration of 0.75 M.

1. Introduction

Polylactic acid is a bioplastic which produced from polymerization of lactic acid ($CH_3CHOHCOOH$). Lactic acid is organic acid which composed of carboxyl group and hydroxyl group. Lactic acid can be produced from renewable resource) such as corn, cassava. Lactic acid can be produced from fermentation or chemical synthesis. In present, the most lactic acid was produced with fermentation because it got high yield of lactic acid more than 90% depended on raw material, the type of microorganism or the method of fermentation. [1]. However, the fermentation is difficult to control microorganism and used long fermentation time. Lactic acid from chemical synthesis got lower yield than fermentation, but it was simple and not take long time of reaction.

Alkaline degradation, a chemical synthesis, it is reaction between monosaccharide, di-saccharide or polysaccharide with alkaline. The most used alkali was sodium hydroxide ($NaOH$), potassium hydroxide (KOH), calcium hydroxide ($Ca(OH)_2$) [2, 3]. The main product was lactate salts, which could be convert to lactic acid, the other products were the salt of saccharinic acid (>C6) and higher molecular weight compound [2].

Many techniques were proposed to convert lactate salt to lactic acid, such as adsorption [4], reactive distillation [5] or electrodialysis [6,7]. The adsorption process requires regenerated of ion-exchange resin and use more chemicals. Reactive distillation required high energy consumption and necessary large amount of chemicals. Electrodialysis is an attractive because this process didn't generate waste, and no cost for waste disposal.

This research study on conversion of sodium lactate, from alkaline degradation of invert sugar and sodium hydroxide, to lactic acid with electrodialysis and cation exchange membrane. Moreover, alkali which didn't react occurred can recycled to alkaline degradation. It will decreased the production cost.

2. Material and Methods

2.1. Material and Electrodialysis Equipment

2.1.1.Material : Sodium lactate was synthesized in our laboratory. Sodium hydroxide (NaOH) and sulfuric acid (H_2SO_4) were analytical grade.

2.1.2 Electrodialysis Equipment : The single stage in-house electrodialysis unit consisted of cation exchange membrane (Selemion CSO, AGC Engineering Ltd., Japan). The effective membrane area was 144 cm^2 and located between anolyte and catholyte compartment. The two electrodes made from stainless steel 304. The schematic diagram was shown in Figure.1.

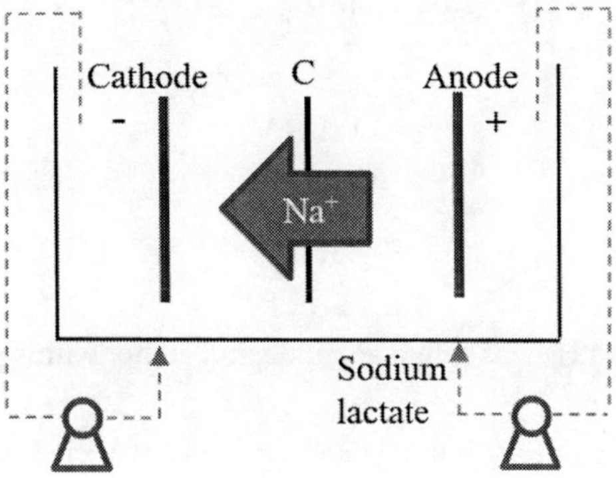

Figure 1. Schematic diagram for electrolysis. C is cation exchange membrane

2.2. Method

Sodium lactate was added with CO_2 until pH about 8-9. The solution was filtered and the filtrated was use for electrodialysis.

2.2.1 Factors effect on electrodialysis :

Sodium lactate and 0.001N sodium hydroxide (NaOH) at flow rate 0.3 ml/s were circulated though anode and cathode side, respectively. The samples from anode side and cathode side were taken periodically for analyze the concentration of lactic acid and NaOH, respectively. The effect of initial concentration of sodium lactate varied at 0.75, 1.2 and 1.5 mol/L. The voltage was supplied and varied at 4, 6 and 8 V. The current during run were recorded for calculate current efficiency and energy consumption as shown in eq.(1) and (2), respectively.

$$CE\ (\%) = \frac{96{,}485 N_{NaOH}}{Q} x100 \quad (1)$$

$$W = \int_0^t \frac{VI}{N_{NaOH}} dt \quad (2)$$

Where CE = current efficiency (%)

W = energy consumption (kWh/kg NaOH)

N_{NaOH} = mol of NaOH in cathode side (mol)

Q = electrical (coulomb)

V = voltage (volt)

A = current (amp)

t = time (hr.)

2.2.2 Analytical Procedure :

Samples were taken for analyze the lactic acid concentration with High performance liquid chromatography (HPLC). The column was Plastisil ODS column and the detector was UV at 210 nm. The 5 mmol sulfuric acid was mobile phase at flow rate of 0.8 ml/min.

The samples were analyze the concentration of NaOH with titration with 0.01 mol/L of hydrochloric acid.

3. Result and Discussions

3.1. Effect of initial concentration of sodium lactate

The sodium lactate solution was feed to anode side at constant voltage. The sodium ion (Na^+) will transport though cation exchange membrane to cathode side and form NaOH in cathode side. While lactate ion ($CH_3CHOHCOO^-$) moved to anode and form lactic acid in anode side. The reaction at anode and cathode showed in eq. (3) and (4), respectively.

At Anode:

$$H_2O \longrightarrow 2H^+ + \frac{1}{2} O_2 + 2 e^- \qquad E^0 = +1.229 \text{ V} \qquad (3)$$

At Cathode:

$$2H_2O + 2e^- \longrightarrow 2OH^- + H_2 \qquad E^0 = -0.828 \text{ V} \qquad (4)$$

From the experiment, the O_2 and H_2 was found at anode and cathode, this phenomenon was explained in eq.(3) and eq.(4)

The effect of initial concentration of sodium lactate was study at varied time as shown in Figure 2. The concentration of NaOH increased when time increased. It showed that Na^+ can pass though cation exchange membrane. When the initial concentration of NaOH varied from 0.75, 1.2 and 1.5 mol/L, it got NaOH as 0.7, 0.9 and 1 mol/L, respectively.

Considering the initial concentration of sodium lactate showed that concentration had no significant impact on the NaOH which was produced at cathode side. This results was similar to Simon et.al [8] found that the transport of Na^+ through the membrane didn't vary when using different brine concentration.

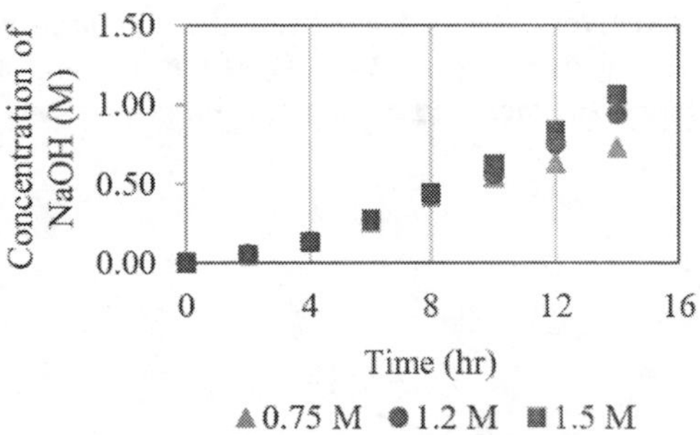

Figure 2. The concentration of NaOH at cathode side at constant 8 volts.

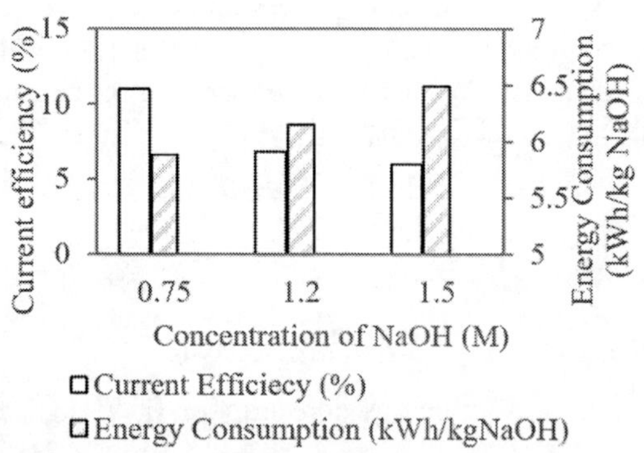

Figure 3. The current efficiency and enegry consumption at 0.75, 1.2 and 1.5 M of sodium lactate.

The depletion of Na^+ occurred at anode side and increase at cathode side. The increasing of Na^+ at cathode side was used to calculate current efficiency. The current efficiency and energy consumption as a function of using varied concentration of sodium lactate as shown in Figure 3. The current efficiency decreased, when concentration of sodium lactate increased.

When high concentration of sodium lactate was used, there are more Na^+ in anode side but the transfer of Na^+ wasn't depend on concentration of sodium lactate, see Figure 2. Furthermore, it used long time in transfer, so the current efficiency decreased and energy consumption increased. Moreover, the current lose due to the resistance of electrode, membrane and solution in both sides.

The percentage of recovery lactic acid decrease with different concentration of sodium lactate as shown in Figure 4. The highest recovery lactic acid was 48% at initial concentation of 0.75 mol/L and 8 volts.

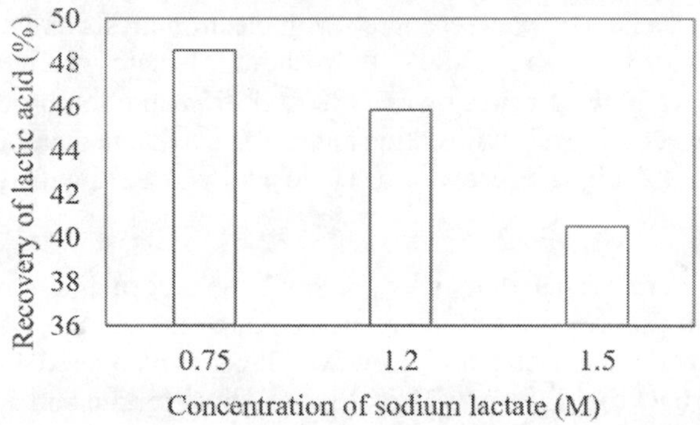

Figure 4. The recovery of lactic acid from sodium lactate at constant voltage at 8 Volts.

3.2. Effect of voltage

The rate of production NaOH in cathode side increased according to voltage as shown in Figure. 5. From the result, the voltage will improve Na^+ transfer. The current efficiency and energy consumption increased as voltage was increased as shown in Figure.6. At low voltage, the rate of H^+ or OH^- which was produced was low. On the other hand, the rate of H^+ or OH^- will increase at high voltage. However, energy consumption has recession trend because the more gas production (formed bubble) at high voltage, it reduced the area of electrode and the effective membrane area [9]. Moreover, the bubble will decrease conductivity of electrolyte [10].

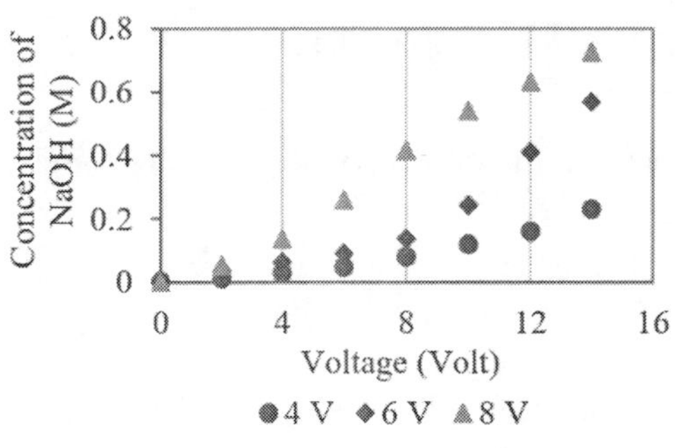

Figure 5. The concentration of NaOH at cathode side at constant 4, 6 and 8 volts.

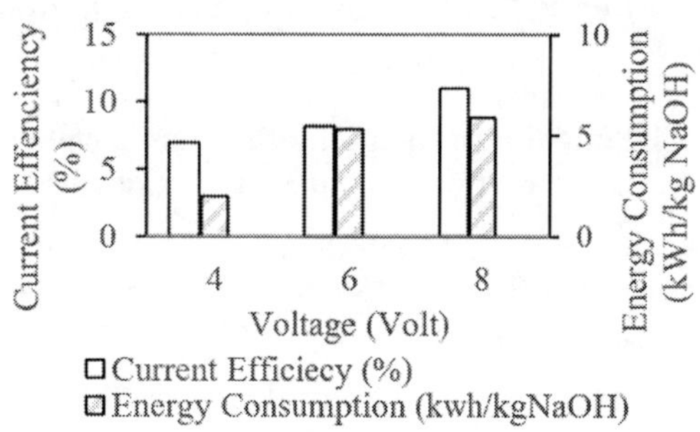

Figure 6. The current efficiency and enegry consumption at 0.75 M of sodium lactate at 4, 6 and 8 volts

When the voltage was incresed, the lactic acid was decrease as shown in Figure 7. The highest recovery of lactic acid was 55.5% at 4 V. At high voltage lactate ion might be oxidized to pyruvate ion and proton as shown in eq.(5), so lactic acid decreased. Moreover the produced sodium hydroxide increase at higher voltage, sodium hydroxide was produced 0.23, 0.57 and 0.73 M from 0.75 M of sodium lactate at 4, 6 and 8 V, respectively.

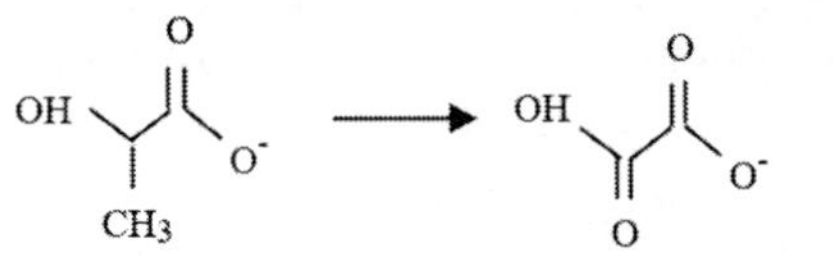

lactate → pyruvate $+ 2H^+ + 2e^-$ (5)

In previous research, the lactic acid was recovered with electrodialysis and anion exchange membrane, the recovery of lactic acid was about 70-80% from sodium lactate which was prepared with titrated lactic acid with NaOH, while the recovery was 78% from fermentation broth [7]. Moreover, the other research used in-house electrodialysis with cation and anion exchange membrane to recovery lactic acid from fermentation broth, the highest recovery of lactic acid was 92% with 100% the purity at optimal condition [11].

Although, the results from this research may be got lower the percentage of recovery of lactic acid than previous research due to this research used cation exchange membrane which may more suitable to recovered sodium hydroxide than lactic acid. However, it could recovered 49% lactic acid and got high quatity of NaOH about 0.73 mol/L from 0.75 mol/L of sodium lactate and applied voltage at 4 volts.

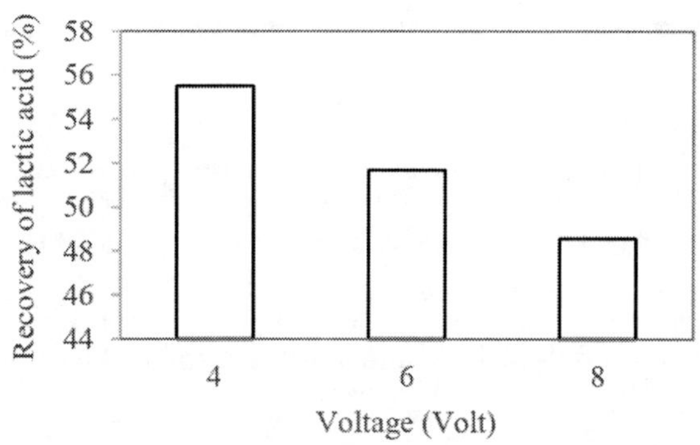

Figure 7. The recovery of lactic acid from sodium lactate at constant voltage at 4, 6 and 8 Volts.

4. Conclusions

The initial concentration of sodium lactate didn't influence on the rate of transfer Na^+, conversely the voltage has effected on the rate of transfer of Na^+.

When initial concentration of sodium lactate was increased, the energy consumption increased but the current efficiency decreased. While the voltage increased, the energy consumption and the current efficiency increased.

This process is feasibility alternative for conversion lactic acid from sodium lactate which was synthesis with chemical reaction. The lactic acid can be recovered 50%, furthermore it can produced NaOH which can recycle for alkaline degradation, and decrease cost of production.

References

[1] Abdel-Rahman M A Tashiro Y and Sonomoto K 2013 *Biotechnol. Adv* **31** 877
[2] Yang B Y and Montgomery R 1996 *Carbohydr Res* **280** 27

[3] Shaffer P A and Friedemann T E 1930 *J. Biol. Chem.* **86** 345
[4] Kaufman E N Cooper S N and Daison B H 1994 *Appl. Biochem. Biotech.* **45** 545
[5] Seo Y Hong W H and Hong T H 1999 *Korean J. Chem. Eng.* **16(5)** 556
[6] Madzingaidzo L, Danner H and Braun R 2002 *J. of Biotech.* **96** 223
[7] Saxena A Gohil G S and Shahi V K 2007 *Ind. Eng. Chem. Res.* **46** 1270
[8] Simon A Fujioka T Price W E and Nghiem L D 2014 *Separ Purif Tech* **127** 70
[9] Xiong Y Jialing L Hong S 1992 *J. Appl. Electrochem.* **22** 486
[10] Chikhi M Rakib M Viers P Laborie S Hita A and Durand G 2002 *Desalination* **149** 375
[11] Boonkong W Sangvanich P Petsom A and Thongchul N 2009 *Chem Eng Technol* **32(10)** 1542

IOP Conf. Series: Materials Science and Engineering **639** (2019) 012053 doi:10.1088/1757-899X/639/1/012053

Effect of Distillation Methods on Citronella Oil (Cymbopogon nardus) Content

S Phovisay[1], X Briatia[1], V Chanthakoun[2] and S Savathvong[3]

[1]Division of Food Science and Technology, Faculty of Agriculture and Forest resource, Souphanouvong University, Lao PDR
[2]Division of Animal Science, Faculty of Agriculture and Forest resource, Souphanouvong University, Lao PDR
[3]Division of Forest Resource, Faculty of Agriculture and Forest resource, Souphanouvong University, Lao PDR

Corresponding author. E-mail: vongpasith@yahoo.com

Abstract. Citronella grass (Cymbopogon nardus) is a permanently growing grass belonging to Poaceae grass family. Which plant is native to the tropical region, especially the south-eastern Asia regions. The Citronella oil is used wildly in industrial product as insect repellent, perfumery, cosmetic, diffusion aromatherapy and health benefit. The aim of this study was investigated the efficiency of distillation methods on citronella oil content by laboratory scale, modified alcohol boiler and extractor machine. In addition, possibility of Citronella oil was reviewed to further product development. The three distillations processing were compared on essential oil content that found the distillation method significantly influenced essential oil content ($p<0.05$). The extractor machine showed highest yield 6.85 ± 0.07 ml/Kg followed by laboratory scale 4.55 ± 0.31 ml/Kg and modified alcohol boiler 2.35 ± 0.16 ml/Kg. The possibility of product development was found citronella oil using on mosquito repellent is recommended at a concentrations of 5-10% and massage oil should not higher than 1.0%. The strong concentration can cause skin sensitivity. Therefore, the optimized concentration should be observed in product development, and value investing. \

1. The first section in your paper

The The citronella oil was obtained from citronella grass (*Cymbopogon nardus*) extraction. This plant is native growing in tropical and semitropical areas of Asia. Citronella essential oil is required for aromatherapy, medicinal treatment, food industry and insect repellent agent [1, 2]. The main compositions contain citronellol, citronellal and geraniol which had function of antifungal, antimicrobial, Anti-inflammatory, volatile, and well-being [3, 4, 5]. On previous research citronella oil has ability against 7 bacterial strains (*Escherichia coli, Salmonella typhi, Pseudomonas aeruginosa, Gram-positive Streptococcus pneumonia, Bacillus cerues, Staphylococcus aureus, Bacillus subtilis*) and 3 fungi strains (*Aspergillus flavus, Candida albicans, Geotrichum candidum*) on food applications. In addition, Sensory evaluation was performed non-significantly on adding essential oils [6].

Citronella oil is an option prevention for mosquito repellent which lead of malaria fever and danger fever [7]. Dengue Hemorrhagic Fever (DHF) have been importance illness in Laos. Local Daily Vientiane Times reported that total of dengue fever had 5.584 cases in 2016, 4.550 cases 2017 and 1.089 half year 2018 [8]. Thus, citronella oil is attractive for product development in local area. Hence, this study aims to compare distillation methods on citronella oil (Cymbopogon nardus) content, moreover, its utilization and benefits of was also reviewed briefly.

2. Materials and Methods

2.1. Materials
The citronella grass (*Cymbopogon nardus*) was collected from the local area of Luangprabang and used as a raw material. The essential oil was distilled by hydro-stream by 3 methods differently (Laboratory scale, Extractor machine and Lao traditional alcohol boiler). The extractor machine (Dongbang, CO. KR, Korea) was supported by Lao-Korea Science and Technology Center (LKSTC). Traditional alcohol boiler was modified for essential oil distillation. Heat mantle was used for heat source of Laboratory scale distillation.

2.2. Essential oil distillation process
The leaves of citronella grass were cut in 20-25 cm for each method. The capacity of raw material was contained about 75-80% for boiler volume and the essential oil was distilled by hydro-stream in each process. Traditional alcohol boiler was used for citronella oil distillation which compared with laboratory distillation and extractor machine. The experimental condition was conducted in different distillation machine (Table 1). The yields of operation were compared on essential content to evaluate efficiency distillation methods. Essential oil contents were performed in triplicate and reported as mean + standard deviation.

Table1. Conditions of distillation process

No.	Distillation methods	Citronella grass (Kg)	Temperature (°C)	Time (minutes)
1	Extractor machine	10.00	100	120
2	Laboratory scale	0.25	100	120
3	Alcohol boiler	7.00	100	120

Remark: Capacity input depend on boiler volume (Extractor machine 9-11 Kg, Laboratory scale 0.2-0.3 and Alcohol boiler 6-7 Kg)

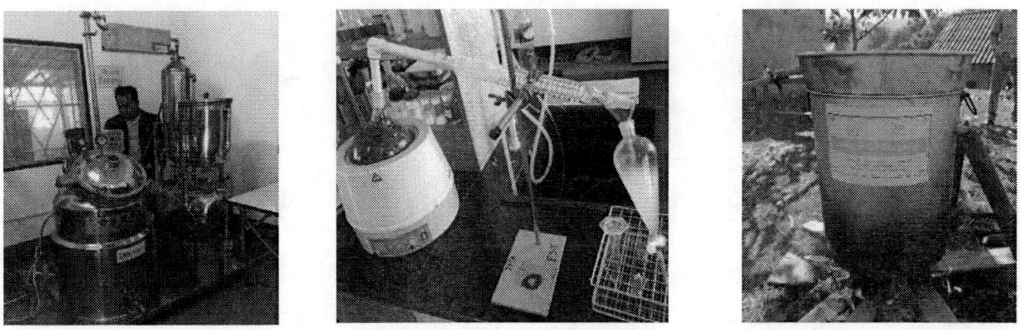

a. Extractor machine b. Laboratory distillation c. Alcohol boiler distillation

Figure 1. Distillation equipment (a. Extractor machine, b. Laboratory distillation c. Alcohol boiler distillation)

2.3. Citronella oil utilization and possibility of product development in local market
The utilization and benefits of citronella oil were reviewed briefly for product development. Presently, citronella oil is one of a nature therapies for relievers and well-being. This review aims to study on citronella (Cymbopogon nardus) utilization in commercial and applications for product development in local market. In addition, citronella grass cultivation and chemical composition were also concluded in briefly reviews.

2.4. Statistical analysis

The effect of distillation methods on oil content was analyzed for significance ($p<0.05$) by using one-way analysis of variance (ANOVA) and Duncan's post hoc test. The statistical analysis was displayed using statistical software SPSS Ver.23 (Chicago, Illinois, USA). The citronella oil utilization was reviewed for possibility of product development in local area.

3. Results and Discussion

3.1 Effect of distillation methods on citronella oil content

Citronella oil content was distilled by Hydro-stream processes (extractor machine, Laboratory scale, alcohol boiler). The operation conditions were conducted at same temperature and time (100 °C for 2 h) in three replications. The result found that essential oil of citronella grass was significantly affected by the distillation method of uses ($p < 0.05$). The highest yields of essential oil content was distilled by extractor machine (6.85 ± 0.07 ml/Kg), following by laboratory distillation (4.55 ± 0.31 ml/Kg) and higher than that alcohol boiler distillation (2.35 ± 0.16 ml/Kg), respectively (Table 2). The lower yield of essential oil content distilled by alcohol boiler machine maybe due to heat convection system was not equilibrium inside the boiler which is leads to evaporate efficiency of volatile oil. While, the extractor machine and laboratory distillation provided steady heating and condenser cooling system which contrast to distillate by alcohol boiler machine.

Previous research reported that yield of citronella essential oil content distilled by laboratory distillation standard was 5.0-18.0 ml/Kg [9]. Incomparison, our experiment showed yield of citronella essential oil content distilled by laboratory distillation and alcohol boiler machine were lower. However, the yield of essential oil content distilled by extractor machine was range similarly to the reported of Oyen (1999) [9]. On the other hand, citronella oil recovery was also depended on various factors such as raw material state, harvest season, age of leave and soil fertilizer [10, 11]. Likewise, the steady of temperature in process was affected on essential oil content in operation. Yield and quality of essential oil was increased by pressure [12]. In addition, alcohol boiler distillation provided lower yield of citronella oil content of this experiment, which was due to equilibrium and leak of stream in distilled tank.

Table2. Effect of distillation methods on citronella oil content

No.	Distillation methods	Raw material (Kg)	Yield of oil content (ml/Kg)
1	Extractor machine	10.00	$6.85^{a} \pm 0.07$
2	Laboratory scale	0.25	$4.55^{b} \pm 0.31$
3	Alcohol boiler	7.00	$2.35^{c} \pm 0.16$

Note: The different superscript letters are significantly different at column ($p < 0.05$).

3.2 Citronella oil utilization and possibility of product development in local market

Plantation: citronella grass is native to the tropical region, especially the south-eastern Asia regions. The cultivation is popularly during May to June. Plantation period should be long term because of cultivation citronella grass remains harvesting for 5 to 7 years [11]. The suitable citronella cultivation requests well drained, fertile soil, pH~6-7, 150-700 m of sea level, temperature 20-38°C of climate and steady rainfall. In case of industry plantation land should limit and avoids of termite. Plant propagation is separated from old plant or called "Slips". Distance of plantation is about 60x60 cm to 90x90 cm, the distance depends on soil fertilizer, weed control, irrigation and area [4, 11].

Harvesting: all parts of citronella grass carries essential commonly, but essential oil quality and yield are contained by leaves. Thus, citronella leaves should be harvested in plantation. The citronella leaves can harvest after farm installing 4-6 months, and every 3-4 months after first harvesting in new plantation. The frequency of harvests is belonged to farmer-attention and climate. Moreover, number of cutting is lead to inferior essential oil [2, 13].

Citronella oil distillation: Hydro-distillation is traditional process to achieve essential oil from citronella leaves, this method is low investment cost and non-advanced technology. Stream distillation is used to extract citronella oil basically, fresh leaves was harvested form plantation and semi-dry making 12 to 24 hours on sunlight for moisture decreasing or extraction with fresh leaves, but oil content is lower more than semi-dry leaves. Distilled operation was taken for 4 hours by distiller machine then oil and water were separated by layer [2, 11]. Yield of citronella oil is dependent on various factors which are ranged from 5.0 – 18.0 ml/Kg. The application of supercritical extraction, ohmic heating, pressure and ultrasound were provided higher oil content [14, 15, 16, 17].

Utilization of Citronella Oil: citronella is used wildly product such as fragrances in perfumes, aromatherapy, traditional medication for anti-spasmodic, insect repellent, soaps, cosmetic and food product [14, 18]. In Laos citronella grass is native plant, it is used for herb and aromatherapy in Sapa. In commercially, citronella is natural repellents on the market, used at concentrations of 5-10%. Massage oil is should not higher than 1.0%. The higher concentration can effect on skin sensitivity. Although citronella oil is non-toxicity with human but pregnant and children should be avoided from citronella oil using [19, 20].

4. Conclusion

Methods of distillation process influenced all citronella oil content. Extractor machine showed best distilled efficiency of oil content. Traditional alcohol boiler shown lowest on oil content compared to extractor machine and laboratory distillation, however, this is the beginning of application with Lao traditional equipment, which lead to improve apparatus perfectly in future.

Citronella oil is used on different product as insect repellent, aromatherapy, medical, cosmetic and food product. However, citronella oil is still had various benefit and offered potential essential options which could be further applied to other applications.

References

[1] Burdock GA 2002 *Fanarali's Handbook of Flavor Ingredients* (Boca Raton FL CRC Press)

[2] Brian P Baker, Jennifer A Grant and Raksha Malakar-Kuenen 2018 Citronella and Citronella Oil Profile *New York State IPM Program.*

[3] Camacho SC, Crandang AP, Camacho LD, Gevaña DT, Carandang MG, Lorida RE, and Bandian MN 2015 Economic potential of small-scale citronella (Cymbopogon winterianus) production in the Philippines. *Philippine Journal of Crop Science. 1;40*(**3**):73-81.

[4] Cassel E, Vargas RM 2006 Experiments and modeling of the Cymbopogon winterianus essential oil extraction by steam distillation *Journal of the Mexican Chemical Society 50*(**3**): 126-129.

[5] Donelian A, Carlson LH, Lopes TJ, Machado RA 2009 Comparison of extraction of patchouli (Pogostemon cablin) essential oil with supercritical CO2 and by steam distillation *The Journal of Supercritical Fluids 48*(**1**): 15-20.

[6] El-Kholany E A 2017 Utilization of Essential Oils from Citronella and Geranium as Natural Preservative in Mayonnaise *International Journal of Microbiology and Biotechnology 1*(**1**): 49.

[7] Hasfalina CM, Hishamuddin J and Zurina ZA 2012 Optimization and kinetics of essential oil extraction from citronella grass by ohmic heated hydro distillation *International Journal of Chemical Engineering and Applications 3*(**3**): 173.

[8] Kongkaew C, Sakunrag I, Chaiyakunapruk N, and Tawatsin A 2011 Effectiveness of citronella preparations in preventing mosquito bites: systematic review of controlled laboratory experimental studies *Tropical Medicine & International Health 16*(**7**): 802-810.

[9] Kowalski R, Kowalska G, Jamroz J, Nawrocka A and Metyk D 2015 Effect of the ultrasound-

assisted preliminary maceration on the efficiency of the essential oil distillation from selected herbal raw materials *Ultrasonics sonochemistry* **24**: 214-220.

[10] Lao Department of Communicable Disease Control, Ministry of Health 2018 online: Xinhuanews 2018-06-18 12:06:17, *1.089 cases of dengue fever reported so far this year in Laos, 2 killed.* http://www.xinhuanet.com/english/2018-06/18/c_137262283.htm

[11] Maia, M. F., & Moore, S. J. (2011). Plant-based insect repellents: a review of their efficacy, development and testing. *Malaria journal, 10*(1), S11.

[12] Maia MF and Moore SJ 1985 Response of java citronella (Cymbopogon winterianus Jowitt.) to harvesting intervals with different nitrogen levels *V International Symposium on Medicinal* Aromatic and Spice Plants **188**.

[13] Oyen LP and Dung NX 1999 Plant resources of South-east Asia no. 19: Essential-oil plant

[14] Silva CF, Moura FC, Mendes MF and Pessoa FL 2011 Extraction of citronella (Cymbopogon nardus) essential oil using supercritical CO2: Experimental data and mathematical modeling *Brazilian Journal of Chemical Engineering 28*(**2**): 343-350.

[15] Simic A, Rančic A, Sokovic MD, Ristic M, Grujic-Jovanovic S, Vukojevic J and Marin PD 2008 Essential oil composition of Cymbopogon winterianus and Carum carvi and their antimicrobial activities *Pharmaceutical Biology 46*(**6**): 437-441.

[16] Singh A, Kumar A 2017 Cultivation of Citronella (Cymbopogon winterianus) and evaluation of its essential oil yield and chemical composition in Kannauj region *International Journal of Biotechnology and Biochemistry 13*(**2**): 139-146.

[17] Singh AK and Naqvi AA and Ram G K Effect of hay storage on oil yield and quality in three cymbopogon species (C. winterianus, C. martinii and C. flexuosus) during different harvesting seasons *Journal of Essential Oil Research 6*(**3**): 289-294.

[18] Tewari, P. A. and R. Mohan 2014 Cultivation of citronella and its oil extraction *Innoriginal: International Journal Of Sciences 3*(**1**):1-2

[19] Torres FC, Lucas AM, Ribeiro VL, Martins JR, Poser GV, Guala MS, Elder HV and Cassel E 2012 Influence of essential oil fractionation by vacuum distillation on acaricidal activity against the cattle tick *Brazilian Archives of Biology and Technology 55*(**4**): 613-621.

[20] Wany A, Jha S, Nigam VK and Pandey DM 2013 Chemical analysis and therapeutic uses of citronella oil from Cymbopogon winterianus A short review *International Journal of Advanced Research* (**6**): 504-521.